New Frontiers in Ultrasensitive Bioanalysis

CHEMICAL ANALYSIS

A SERIES OF MONOGRAPHS ON ANALYTICAL CHEMISTRY
AND APPLICATIONS

Editor
J.D. WINEFORDNER

VOLUME 172

A complete list of the titles in this series appears at the end of this volume.

New Frontiers in Ultrasensitive Bioanalysis

Advanced Analytical Chemistry Applications in Nanobiotechnology, Single Molecule Detection, and Single Cell Analysis

Edited by

XIAO-HONG NANCY XU

Old Dominion University

Wiley-Interscience
A John Wiley & Sons, Inc., Publication

Copyright © 2007 by John Wiley & Sons, Inc. All rights reserved.

Published by John Wiley & Sons, Inc., Hoboken, New Jersey
Published simultaneously in Canada

No part of this publication may be reproduced, stored in a retrieval system, or transmitted in any form or by any means, electronic, mechanical, photocopying, recording, scanning, or otherwise, except as permitted under Section 107 or 108 of the 1976 United States Copyright Act, without either the prior written permission of the Publisher, or authorization through payment of the appropriate per-copy fee to the Copyright Clearance Center, Inc., 222 Rosewood Drive, Danvers, MA 01923, (978) 750-8400, fax (978) 750-4470, or on the web at www.copyright.com. Requests to the Publisher for permission should be addressed to the Permissions Department, John Wiley & Sons, Inc., 111 River Street, Hoboken, NJ 07030, (201) 748-6011, fax (201) 748-6008, or online at http://www.wiley.com/go/permission.

Limit of Liability/Disclaimer of Warranty: While the publisher and author have used their best efforts in preparing this book, they make no representations or warranties with respect to the accuracy or completeness of the contents of this book and specifically disclaim any implied warranties of merchantability or fitness for a particular purpose. No warranty may be created or extended by sales representatives or written sales materials. The advice and strategies contained herein may not be suitable for your situation. You should consult with a professional where appropriate. Neither the publisher nor author shall be liable for any loss of profit or any other commercial damages, including but not limited to special, incidental, consequential, or other damages.

For general information on our other products and services or for technical support, please contact our Customer Care Department within the United States at (800) 762-2974, outside the United States at (317) 572-3993 or fax (317) 572-4002.

Wiley also publishes its books in a variety of electronic formats. Some content that appears in print may not be available in electronic format. For more information about Wiley products, visit our web site at www.wiley.com

Library of Congress Cataloging-in-Publication Data:
New frontiers in ultrasensitive bioanalysis : Advanced analytical chemistry applications in nanobiotechnology, single molecule detection, and single cell analysis / [edited by] Xiao-Hong Nancy Xu.
    p. ; cm. – (Chemical analysis ; v. 172)
    Includes bibliographical references and index.
    ISBN: 978-0-471-74660-7 (cloth)
    1. Biomolecules–Analysis. 2. Nanotechnology. 3. Microchemistry.
I. Xu, Xiao-Hong Nancy. II. Series.
    [DNLM: 1. Chemistry, Analytical–methods. 2. Cytological Techniques.
    3. Molecular Probe Techniques. 4. Nanotechnology. QY 90 N532 2007]
    QPJ19.7.N49 2007
    572′.33 – dc22   2006036935

Printed in the United States of America

10 9 8 7 6 5 4 3 2 1

CONTENTS

PREFACE	vii
CONTRIBUTORS	xi

CHAPTER 1. IS ONE ENOUGH?   1
Andrew C. Beveridge, James H. Jett, and Richard A. Keller

CHAPTER 2. DISSECTING CELLULAR ACTIVITY FROM SINGLE GENES TO SINGLE mRNAs   29
Xavier Darzacq, Robert H. Singer, and Yaron Shav-Tal

CHAPTER 3. PROBING MEMBRANE TRANSPORT OF SINGLE LIVE CELLS USING SINGLE-MOLECULE DETECTION AND SINGLE NANOPARTICLE ASSAY   41
Xiao-Hong Nancy Xu, Yujun Song, and Prakash Nallathamby

CHAPTER 4. NANOPARTICLE PROBES FOR ULTRASENSITIVE BIOLOGICAL DETECTION AND IMAGING   71
Amit Agrawal, Tushar Sathe, and Shuming Nie

CHAPTER 5. TAILORING NANOPARTICLES FOR THE RECOGNITION OF BIOMACROMOLECULE SURFACES   91
Mrinmoy De, Rochelle R. Arvizo, Ayush Verma, and Vincent M. Rotello

CHAPTER 6. NANOSCALE CHEMICAL ANALYSIS OF INDIVIDUAL SUBCELLULAR COMPARTMENTS   119
Gina S. Fiorini and Daniel T. Chiu

CHAPTER 7. ULTRA SENSITIVE TIME-RESOLVED NEAR-IR FLUORESCENCE FOR MULTIPLEXED BIOANALYSIS   141
Li Zhu and Steven A. Soper

CHAPTER 8. ULTRASENSITIVE MICROARRAY DETECTION OF DNA USING ENZYMATICALLY AMPLIFIED SPR IMAGING   169
Hye Jin Lee, Alastair W. Wark, and Robert M. Corn

CHAPTER 9. ULTRASENSITIVE ANALYSIS OF METAL IONS AND SMALL MOLECULES IN LIVING CELLS   195
Richard B. Thompson

CHAPTER 10. ELECTROCHEMISTRY INSIDE AND OUTSIDE SINGLE NERVE CELLS   215
Daniel J. Eves and Andrew G. Ewing

CHAPTER 11. ELECTROCHEMILUMINESCENCE DETECTION IN BIOANALYSIS   235
Xiao-Hong Nancy Xu and Yanbing Zu

CHAPTER 12. SINGLE-CELL MEASUREMENTS WITH MASS SPECTROMETRY   269
Eric B. Monroe, John C. Jurchen, Stanislav S. Rubakhin, and Jonathan V. Sweedler

CHAPTER 13. OUTLOOKS OF ULTRASENSITIVE DETECTION IN BIOANALYSIS   295
Xiao-Hong Nancy Xu

INDEX   301

PREFACE

This book aims to provide an overview of current exciting research topics in ultrasensitive bioanalysis; to provide an in-depth understanding of objectives, motivations, and future directions of the new frontiers of ultrasensitive bioanalytical research; and to introduce new ideas, new technologies, and new applications of ultrasensitive bioanalysis to a wide spectrum of research communities, including biological, biomedical, clinical, chemical, environmental, and materials science. The book strives to provide sufficient fundamentals and research background for readers to learn and apply these tools. The book is also structured in a way that can serve as a textbook or a primary reference book for an advanced analytical chemistry course or a special topic course for graduate and senior undergraduate students. I discovered the difficulty of finding effective teaching materials and suitable reference books for my students and collaborators while I taught the course and engaged joint research projects with colleagues in the department of biological sciences and engineering.

Ultrasensitive detection plays a vital role in advancing analytical chemistry and has been a primary driving force for the development of new analytical instrumentation and methodology. As analysis of biological samples (e.g., genomics, proteomics) and living systems (e.g., cellular and subcellular function) becomes more demanding, new platforms of ultrasensitive analysis using multiplexing, single nanoparticle sensing, nano-fluidics, and single-molecule detection have emerged and have become indispensable tools to advance biological, biomedical, and biomaterials sciences and engineering. Furthermore, the emerging research field of nanoscience and nanotechnology provides new possibilities for development of new analytical tools and instrumentation for bioanalysis. For example, the unique properties of nanoparticles offer enormous opportunities to develop new probes for real-time *in vivo* and *in vitro* imaging and sensing of individual biomolecules with sub-100-nm spatial resolution and millisecond to nanosecond time resolution. Such powerful capabilities will lead to the development of new analytical techniques to improve disease diagnosis and treatment, as well as to advance our fundamental understanding of important phenomena such as membrane transport, gene expression, enzyme activities, and intracellular and intercellular signaling. It is undeniable that analytical chemistry has evolved to be a vital player in all cutting-edge scientific research fields.

In this book, a diverse group of analytical chemists working in the forefront of ultrasensitive bioanalysis share their insights, visions, and latest results. The book is structured to offer appreciable background on the fundamentals and to provide brief overviews of current status of ultrasensitive bioanalysis. More importantly, this book focuses on showcasing the exciting opportunities of the new frontiers of ultrasensitive bioanalysis. Thus, this book is much more for the future than for the summary of the past and present.

This book includes 13 chapters and starts, in Chapter 1, by addressing the most fundamental question of single-molecule detection (SMD): How many measurements of a single molecule are required to obtain representative molecular properties of ensemble distributions? It describes the advantages of SMD and its applications to virtual sorting and DNA fingerprinting of single genomes. Chapter 2 focuses on presenting a representative application of SMD to directly observing gene expression in living cells, illustrating potential applications of SMD for better understanding of living cellular mechanism and function. Chapter 3 offers an overview of SMD using fluorescence microscopy and spectroscopy, and it gives an example of studying membrane transport of living cells using single fluorophor molecules and single-nanoparticle optics to illustrate the distinguished advantages of SMD and non-photobleaching noble metal nanoparticles for probing living cellular mechanism and function. This chapter aims to introduce the basic detection schemes and experimental configurations of SMD and single-nanoparticle optics for bioanalysis and single-living-cell imaging.

Chapter 4 is devoted to the recent development of multifunctional nanoparticle probes, especially quantum dots (QDs), for ultrasensitive analysis of biomarkers in living cells and *in vivo* imaging, providing the latest advances and future possibility of this exciting research arena. To develop nanoparticle probes for living-cell imaging, smart drug delivery, and *in vivo* diagnosis, the primary challenge is to rationally design the functional nanoparticle surfaces that can specifically recognize biomolecules of interest and be biocompatible to living systems. Thus, Chapter 5 offers examples of the latest advances in designing surface functionality of nanoparticles that can specifically recognize protein and DNA surfaces. This chapter presents an overview of this exciting research topic and provides a wealth of experimental details and new insights into the future research direction.

Miniaturization and multiplexing have become important technology platforms in ultrasensitive analysis of biomolecules and profiling contents of individual living cells. Chapters 6 through 8 present several interesting research projects, reporting the recent development of this particularly exciting research field. Chapter 6 describes the new platform of using the state-of-the-art research tools, such as a laser-based system and droplet-based microfluidics, to selectively extract organelles from living cells for subsequent analysis. Chapter 7 presents the development of near-IR fluorescent dyes, time-resolved fluorescence spectroscopy, and multiplexing approach for DNA sequencing, offering more sensitive and informative genomic measurements. In contrast, Chapter 8 describes a new approach of directly determining genomic DNA samples with no need of labeling or PCR amplification. In this approach, surface plasmon resonance imaging (SPRI) is used as a label-free detection means, while nucleic acid microarrays serve as a multiplexing platform, which is coupled with surface enzymatic method to amplify SPRI imaging signal.

Like proteins and DNA, metal ions and small molecules play a key role in cellular function. However, quantitative and qualitative analysis of small molecules, especially metal ions, in living cells in real time remains extremely challenging. Chapter 9 gives a glimpse of research activities and emerging approaches in this extremely important research area and offers new prospects of the future research possibility and direction.

Chapter 10 describes the development and application of electrochemical methods to monitor the release of neurotransmitters in real time for better understanding communication among individual nerve cells, demonstrating the unique and powerful detection capability of electrochemistry. Chapter 11 introduces electrochemiluminescence (ECL) detection for ultrasensitive bioanalysis, describes its latest development and applications, and presents the potential challenges and exciting future research opportunities of ECL detection in ultrasensitive bioanalysis.

Mass spectrometry offers an unparalleled detection capability for identifying unknown analytes (e.g., tumor markers) and has demonstrated the possibility of profiling the chemical compositions of individual living cells. However, mass spectrometry currently provides very limited detection sensitivity. Chapter 12 illustrates the latest research development of using state-of-the-art mass spectrometry for single-cell analysis, highlights several representative examples, and presents the unique research challenges and opportunities in this exciting emerging new research field. Finally, the book is concluded with Chapter 13: it summarizes current cutting-edge ultrasensitive detection means in bioanalysis, along with their unique features and potential applications. This final chapter also offers a brief outlook on ultrasensitive bioanalysis in terms of emerging technologies and methodologies, along with new challenges and opportunities.

I hope that this book will serve as a valuable tool for beginning and well-established scientists, especially graduate students and the researchers who consider learning about and using ultrasensitive detection tools. I have selected one example of particular detection techniques and research approaches to illustrate the concepts, ideas, experimental design, and potential applications of research field, rather than providing a comprehensive review of an entire research field. It is hoped that through these selected examples, the readers will feel the vibrant and excitement of entire research field and will pursue so many other distinguished research works that are unable to be included in this book. Thus, I hope that I will have the forgiveness and understanding of my colleagues whose research work has not been presented here.

I am very grateful to all contributors for their enthusiastic support and for taking precious time from their demanding busy schedule to prepare the chapters. I would like to acknowledge each and every one of them for their generous effort and inputs. I wish to extend my gratitude to James Winefordner at University of Florida for his persistent invitations over years, as well as to Heather Bergman at Wiley-Interscience for her valuable assistance.

It has been a very pleasant and valuable experience for me to work on this book. I gained the first-hand appreciation and better understanding of the effort, time, energy, and persistence that are essential for an editor of a book. With an already saturated daily schedule of young academics, I had to devote all my weekends and holidays this past year to working on this book in the hope that it would reach readers in time to meet the demands of this very rapidly expanding research arena. I will be extremely delightful if readers, especially students and new investigators, find that the book stimulates their fascination in ultrasenstive bioanalysis and helps them to steer toward new research directions that ultimately lead to new scientific discoveries.

Despite my best effort, I am mindful that I could have done more in editing this book. Therefore, I am eager to hear the constructive comments and suggestions of my colleagues and readers. I am determined to continue to engage in vigorous research and teaching activities in this thriving research field, and I plan to prepare follow-up volumes in the coming years.

It is fitting to take this opportunity to acknowledge several funding agencies for their generous support of my research program: NIH (R21 RR15057; R01 GM0764401), NSF (BES 0507036; DMR 0420304), DoD-MURI (AFOSR #F49620-02-1-0320), DOE (DE-FG02-03ER63646), and Old Dominion University. This financial support allows me to actively participate in this exciting cutting-edge research field, acquire the first-hand research experience, widely interact with colleagues in the field, closely follow the advance of research field, and gain in-depth understanding of the research direction, which are essential to the successful construction of this book.

Finally, I would like to thank my family, especially my parents, for their tireless support of this endeavor. Without their patience, love, and guidance, this mission would have been impossible.

XIAO-HONG NANCY XU

Norfolk, Virginia
January 2007

CONTRIBUTORS

Amit Agrawal, Departments of Biomedical Engineering and Chemistry, Emory University and Georgia Tech, Atlanta, Georgia

Rochelle R. Arvizo, Department of Chemistry, University of Massachusetts, Amherst, Massachusetts

Andrew C. Beveridge, Bioscience Division, Los Alamos National Laboratory, Los Alamos, New Mexico

Daniel T. Chiu, Department of Chemistry, University of Washington, Seattle, Washington

Robert M. Corn, Department of Chemistry, University of California—Irvine, Irvine, California

Mrinmoy De, Department of Chemistry, University of Massachusetts, Amherst, Massachusetts

Xavier Darzacq, Biologie Cellulaire de la Transcription, Ecole Normale Supérieure, Paris, France

Daniel J. Eves, Department of Chemistry, Pennsylvania State University, University Park, Pennsylvania

Andrew G. Ewing, Department of Chemistry and Department of Neural and Behavioral Sciences, Pennsylvania State University, University Park, Pennsylvania

Gina S. Fiorini, Department of Chemistry, University of Washington, Seattle, Washington

James H. Jett, Bioscience Division, Los Alamos National Laboratory, Los Alamos, New Mexico

John C. Jurchen, Department of Natural Sciences, Concordia University, Seward, Nebraska

Richard A. Keller, Bioscience Divison, Los Alamos National Laboratory, Los Alamos, New Mexico

Hye Jin Lee, Department of Chemistry, University of California—Irvine, Irvine, California

Eric B. Monroe, Department of Chemistry, University of Illinois, Urbana, Illinois

Prakash Nallathamby, Department of Chemistry and Biochemistry, Old Dominion University, Norfolk, Virginia

Shuming Nie, Departments of Biomedical Engineering and Chemistry, Emory University and Georgia Tech, Atlanta, Georgia

Vincent M. Rotello, Department of Chemistry, University of Massachusetts, Amherst, Massachusetts

Stanislav S. Rubakhin, Department of Chemistry, University of Illinois, Urbana, Illinois

Tushar Sathe, Departments of Biomedical Engineering and Chemistry, Emory University and Georgia Tech, Atlanta, Georgia

Yaron Shav-Tal, The Mina and Everard Goodmen, Faculty of Life Sciences, Bar-Ilan University, Ramat-Gan, Israel

Robert H. Singer, Departments of Anatomy and Structural Biology, Albert Einstein College of Medicine, Bronx, New York

Yujun Song, Department of Chemistry and Biochemistry, Old Dominion University, Norfolk, Virginia

Steven A. Soper, Department of Chemistry, Louisiana State University, Baton Rouge, Louisiana

Jonathan V. Sweedler, Department of Chemistry, University of Illinois, Urbana, Illinois

Richard B. Thompson, Department of Biochemistry and Molecular Biology, School of Medicine, University of Maryland, Baltimore, Maryland

Ayush Verma, Department of Chemistry, University of Massachusetts, Amherst, Massachusetts

Alastair W. Wark, Department of Chemistry, University of California—Irvine, Irvine, California

Xiao-Hong Nancy Xu, Department of Chemistry and Biochemistry, Old Dominion University, Norfolk, Virginia

Li Zhu, Department of Chemistry, Louisiana State University, Baton Rouge, Louisiana

Yanbing Zu, Department of Chemistry, The University of Hong Kong, Hong Kong, China

CHAPTER

1

IS ONE ENOUGH?

ANDREW C. BEVERIDGE, JAMES H. JETT, AND RICHARD A. KELLER

1.1. INTRODUCTION

A concern often expressed about single-molecule measurements can be stated as, "Single-molecule measurements are 'cool' but what good are they if you have to measure a million single molecules to estimate the properties of the system?" This is a reasonable concern. Thus, a major focus of this chapter is modeling the effect of sample size (typically a few molecules) on the accuracy of analytical measurements and the extrapolation of properties measured at the single-molecule level to properties derived from ensemble measurements. We include a summary of DNA fingerprinting of single viral particles and bacterial cells that demonstrates that only a few measurements are required. The contrast between single-molecule and bulk measurements is that single-molecule measurements yield physical properties of *individual* molecules whereas individual molecular properties are difficult or impossible to obtain from bulk measurements; often the heterogeneous nature of single measurements is masked in bulk systems. We finish with (a) a description of "virtual sorting," where the data stream is sorted instead of the sample stream, and (b) the implementation of virtual sorting to statistical chemistry to either eliminate or reduce the necessity to purify a sample.

1.1.1. Significance of a Single-Molecule Measurement

There is a fundamental postulate of statistical mechanics that states, "The (long) time average of a mechanical property M in the thermodynamic system of interest is equal to the ensemble average of M, in the limit as N goes to infinity" (Hill, 1960). (Here and throughout this chapter, N refers to the number of molecules.) If this true, and indeed it must be, why do single-molecule experiments? It is clear that the average fluorescence intensity of a single molecule recorded over many cycles, as it undergoes changes between a fluorescent species and a nonfluorescent species, is equivalent to the average intensity from multiple molecules over one cycle.

We address two different but related aspects of this postulate: (1) How many times should the same single molecule be analyzed for the measured value(s) to be a good representation of the ensemble average? (2) Can single-molecule properties be extracted from bulk data?

The first question relates to studying the properties of single molecules free from averaging effects associated with bulk measurements. A good example is the quantification of the

---

New Frontiers in Ultrasensitive Bioanalysis. Edited by Xiao-Hong Nancy Xu
Copyright © 2007 John Wiley & Sons, Inc.

thermodynamics and dynamics of protein folding where subtle differences in the response curves indicate the presence of multiple folding pathways, barriers in the exit channel, and deviations from a simple two-state model, which are hidden in ensemble measuements (Schenter et al., 1999; Yang and Xie, 2002a,b; Yang and Cao, 2001). A related question is, How many identical single molecules are needed to estimate a sample's composition? For example, DNA fingerprinting for pathogen detection and identification by gel electrophoresis typically takes several days. Most of the time is spent culturing the sample to acquire sufficient material for analysis. The ability to size accurately *single* DNA fragments reduces sample preparation time considerably (Ferris et al., 2005; Larson et al., 2000). A similar problem exists with nonculturable bacteria where it is difficult to grow enough sample for standard analyses (Leadbetter, 2003; Pennisi, 2004).

The second question—How can you extract single-molecule properties from data collected from probe volumes that contain more than one emitting molecule?—is more difficult. The magnitude of the problem is apparent by looking at Figure 1.1, which shows time-dependent fluorescence intensity fluctuations as a function of the number of molecules in the probe volume. Even for two molecules, it is impossible to deconvolute the signal into the contribution from each molecule.

It is often stated that the difference between single-molecule and ensemble measurements is just a synchronization problem; but it is more than a synchronization problem. Due to the stochastic nature of fundamental processes, following a synchronized start, the molecules

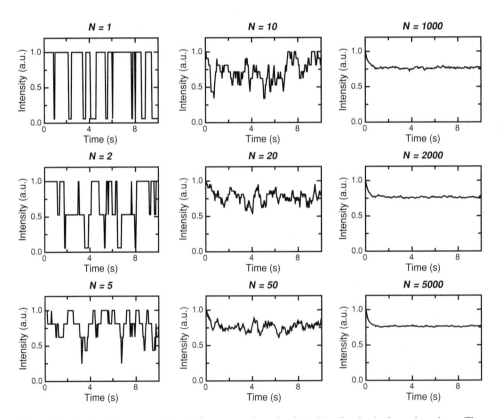

Figure 1.1. Simulated data stream for the fluorescence intensity from N molecules in the probe volume. The parameters for the simulation are listed in Table 1.1. Note that the intensity in each plot is normalized and that all the molecules start in the "on" state. The number of molecules present in the probe volume varies from 1 to 5000.

Table 1.1. Values for the Simulations Computed Here

Species	$k(s^{-1})$	τ (s)	Intensity
A	1.111	0.9	1.00
B	3.333	0.3	0.06

k is the rate constant, τ is the inverse of k (that is, the lifetime of the state), and the intensity represents the fluorescence intensity of each state. The average cycle time, τ_{cyc}, is 1.2 s and the mean fluorescence intensity is 0.765.

get "out of step" in approximately one cycle and distributions of the measured parameters of the sample are difficult to discern. Fortunately, properties of the system can be measured without a synchronous start.

1.2. MODEL SYSTEM

1.2.1. Equilibrium and Kinetics

We choose a simple system of a nondiffusing single molecule oscillating between a strongly fluorescent state and a weakly fluorescent state to characterize the particularities associated with the study of single molecules in small probe volumes. We focus on the problems and opportunities that result from using probe volumes less than or equal to a femtoliter. Equations (1.1a) and (1.1b) describe the equilibrium behavior of two conformations of the same molecule, A and B:

$$A \underset{k_2}{\overset{k_1}{\rightleftarrows}} B \tag{1.1a}$$

$$K_{eq} = \frac{[B]}{[A]} = \frac{k_1}{k_2}, \tag{1.1b}$$

where [A] and [B] are the respective concentrations of each species, k_1 and k_2 are the rate constants, and K_{eq} is the equilibrium constant. For a small displacement from equilibrium, the differential rate law is

$$\frac{d\Delta f_A(t)}{dt} = -(k_1 + k_2)\Delta f_A(t) = -\frac{d\Delta f_B(t)}{dt}, \tag{1.2}$$

where $\Delta f_A(t)$ and $\Delta f_B(t)$ denote the deviation of the concentration from equilibrium for species A and B at time t. For the closed, two-state system modeled here, the sum of [A] plus [B] is constant; thus $\Delta f_A(t)$ equals $-\Delta f_B(t)$. The solution to Eq. (1.2) is

$$\Delta f_A(t) = \Delta f_A^0 \exp[-(k_1 + k_2)t], \tag{1.3}$$

where Δf_A^0 is the initial deviation from equilibrium (Cantor and Schimmel, 1980). The probability of finding the system in A or B is

$$P_A = \frac{k_2}{k_1 + k_2}, \quad P_B = \frac{k_1}{k_1 + k_2}, \tag{1.4}$$

where P_A and P_B are the probabilities for A and B, respectively. The mean fluorescence intensity, $\langle I_f \rangle$, is given by

$$\langle I_f \rangle = \alpha \phi_A [A] + \beta \phi_B [B] = \alpha \phi_A N \frac{k_2}{k_1 + k_2} + \beta \phi_B N \frac{k_1}{k_1 + k_2}, \quad (1.5)$$

where α and β depend upon the absorption, emission, and detection factors for A and B, and ϕ_A and ϕ_B are the fluorescence quantum yields for A and B.

Single-molecule measurements of the system give k_1, k_2, I_A, and I_B; therefore the parameters α and β can usually be determined by calibration of the apparatus allowing one to obtain values for ϕ_A and ϕ_B.

1.2.2. Generation of Synthetic Data

The parameters used in the following Poisson simulation of a data stream (Demas, 1983; Matthews and Walker, 1964), chosen to represent a typical photophysical process, are listed in Table 1.1. The conclusions drawn from data synthesized from these particular values may change significantly for a different choice of parameters. Here a cycle is defined as the total of the time period that a molecule spends in each state.

$$\tau_{\text{cyc}} = \tau_1 + \tau_2, \quad (1.6)$$

where τ_1 and τ_2 are the inverse of k_1 and k_2, and τ_{cyc} is the cycle time. The probability for the forward reaction in Eq. (1.1), $P(t)$, that A changes its conformation at time t within dt is

$$P(t) \approx k_1 \, dt. \quad (1.7)$$

The approximation is valid for small dt. For the simulation, a molecule changes states if a uniform random number is less than $k_1 dt$. The reaction time is determined by the number of trials needed for a newly generated random number to meet the specified condition above; the number of trials is then multiplied by dt. In the case for simulations of multiple molecules, data from individual molecules were added together. In Figure 1.1, we choose to start all of the molecules in A in order to simulate a perturbation-jump experiment. Alternatively, when a random start is desired, a uniform random number is compared to one of the probabilities in Eq. (1.4) to determine in which state each molecule starts.

1.2.3. Data Analysis

1.2.3.1. Single Molecule in the Probe Volume

Figure 1.1, $N = 1$, is a simulation of the time-dependent fluorescence intensity for a single molecule, fixed in the probe volume, that oscillates between a fluorescent and a weakly fluorescent conformation; thermodynamic driving forces cause the molecule to change conformations. The time that the molecule spends in each conformation is stochastic. Confining the molecule to the probe volume is useful because it extends the range of kinetic parameters that can be studied and improves the accuracy of the parameters extracted from fits of the data to curves derived from the reaction mechanism and eliminates complications due to diffusion (Baldini et al., 2005; Jia et al., 1999; Rhoades et al., 2003; Talaga et al., 2000). In practice, the sampling time is limited to less than 100,000 photon absorptions by photobleaching (Soper et al., 1993) and other considerations (Vazquez et al., 1998; Zander et al., 2002).

We describe three methods for data analysis when only one molecule is in the probe volume: (1) averaging, (2) autocorrelation, and (3) distribution analyses.

1. The time averages of the "on and off" periods are measured for a chosen number of cycles; the averages are then inverted to give the rate constants, k_1 and k_2. The rate constants then can be summed to give the relaxation time, $(k_1 + k_2)^{-1}$, of the system. Here, the assumption is made that differences between states are discernible such that their respective time dependence can be accurately determined (Verberk and Orrit, 2003).

2. The fluorescence autocorrelation function for a simple first-order kinetic process, undergoing fluctuations around the equilibrium concentration, exhibits an exponential decay with a decay rate that is the sum of the rate constants $(k_1 + k_2)$. This type of analysis is called fluorescence correlation spectroscopy (FCS). Autocorrelation requires an analysis time long enough for both states to be sampled such that their average is independent of the time—that is, constant within statistical variation. Note that the time to reach a stationary system is independent of the number of molecules in the system. The fluorescence intensity autocorrelation function is defined as

$$C(t) = \frac{\langle \delta I(t) \delta I(t+\tau) \rangle}{\langle I \rangle^2}, \qquad (1.8)$$

where δI are the zero-mean intensity fluctuations, $\delta I(t) = I(t) - \langle I \rangle$, at time t and $t + \tau$, respectively (Aragon and Pecora, 1976, Elson, 1985, Krichevsky and Bonnet, 2002, Zander et al., 2002).

3. Distribution analyses include both binning and cumulative distribution functions. In the binning method, "on and off" time periods are binned separately and plotted individually in histogram format; rate constants are determined from each histogram. For the simulations here, a minimum of 10 cycles is needed to define a rudimentary histogram. Alternatively, the data can be fit to the cumulative distribution function for the probability density function that describes the data. The cumulative distribution method has an advantage over binning in that as few as two to three points are needed. However, for cases more complicated than considered here, the cumulative distribution function may be difficult to compute analytically.

1.2.3.2. Multiple Molecules in the Probe Volume

In the study of biological systems within a fixed probe volume (cells, lipid vesicles, cell membranes, etc.), it is often of interest to determine the number of a particular species in the probe volume. The ability to measure the number of molecules and determine individual molecular properties leads to new insights regarding basic intracellular functions. Here, two limits of multiple molecules in the probe volume are examined: a bulk system, $N > 10$, where discrete contributions from individual molecules are not apparent, and a system where discrete contributions from individual molecules are obvious, $2 \leq N \leq 10$, herein referred to as a "prebulk" system.

Figure 1.1 illustrates the change from single-molecule system to prebulk system to bulk system as N is increased from 1 to 5000. This corresponds to a concentration range from 1 nM to 10 μM for a probe volume of 1 fL. Although each molecule is in either A or B, the system is not constrained to a number normalized intensity of either 1 or 0.06 because the molecules get out of phase. For example, in the case where the number of molecules is five, all of the molecules must be in B for a normalized intensity of 0.06; this is improbable, approximately one out of a thousand, for the simulation parameters used here. For five molecules, six normalized intensities are possible with frequencies determined by probability theory. The corresponding plateaus are clearly visible for N equals 2 and 5, and occasionally 10 in Figure 1.1. Discrete steps are not evident for N greater than 10.

Bulk Systems and Synchronous Starts. Consider the simulations in Figure 1.1 for the bulk case. Here, at $t = 0$ all of the molecules are in A and relax to the equilibrium distribution in less than the average of one cycle. As stated previously, the simulations are analogous to rapid perturbation experiments—for example, optical pumping, temperature jump, pH jump, or the addition of an appropriate chemical. For this type of experiment, the time dependence of the signal relaxation gives the sum of the rate constants, $k_1 + k_2$, whereas the long-time limit of the signal gives the average fluorescence intensity as given in Eq. (1.5). In the special case where one of the states is "dark," the equilibrium constant is determined from the ratio of the intensity at time zero to the intensity at times greater than the chemical relaxation time. Depending on the type of experiment, bulk measurements give the equilibrium fluorescence intensity and the sum of the rate constants; individual values of I_A, I_B, k_1, and k_2 are typically difficult to measure.

Prebulk System. For the prebulk system we refer to two distinct states of the system: *all on* and *all off*. (There are a total of $N + 1$ distinct states; the number of configurations is $(A + B)^N$.) Here *all on* refers to the case where *every* molecule in the probe volume is A concurrently; correspondingly, *all off* indicates that every molecule is B at the same time. Assuming that all the fluorescing molecules in the probe volume are identical, probability theory predicts that the individual rate constants can be determined by the average of the period when *all the molecules* are either *all on* or *all off*; the average period multiplied by the number of molecules gives the inverse of the corresponding rate constant (Riley et al., 1997). This measurement depends on explicit knowledge of the number of molecules in the probe volume. Figure 1.2 gives an example for 2 to 10 molecules. Conversely, if the rate constants are known and assuming identical molecules, the average of either the *all on* or *all off* periods can be used to determine the number of molecules in the probe volume. If neither the rate constant nor the number of molecules is known, the ratio of the average lifetime of the *all off* state to the *all on* state gives the equilibrium constant. For all types of measurements in a prebulk system the system state must be known—that is, whether the system is in the *all on* or *all off* state. There are two possible experimental procedures that

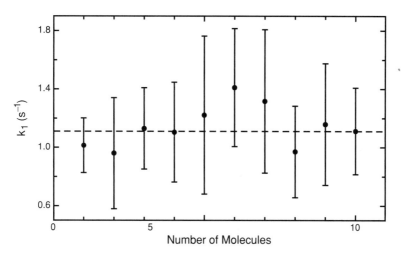

Figure 1.2. Determination of k_1 in a prebulk system from 1 to 10 identical molecules in the probe volume. The average of *all on* period is computed for a data stream where the average number of *all on* periods is 10 in the data stream; see Eq. (1.9). The average *all on* period is then multiplied by the number of molecules and inverted to give k_1. The error bars represent the standard deviation from 10 independent numeric simulations. The dashed line indicates the true value for k_1.

can be used with this method, and both methods can determine the system state:

1. If there is a distribution of N in the sample, then one can look for a probe volume where $N = 1$; then use the data to determine the rate constants. Once the rate constants are known, N can be determined for other probe volumes when N is less than or equal to 10.
2. After sufficient data are collected, the light intensity can be increased to facilitate photobleaching. For noninteracting molecules, photobleaching occurs in discrete steps; the steps are counted to determine N (Park et al., 2005). The rate constants can be calculated once N is known.

Table 1.2 gives the probabilities of finding the system in either *all on* or *all off* states as a function N for the parameters in Table 1.1. The time trajectory for either a simulation or an

Table 1.2. The Probabilities of Finding the System with Either *All the Molecules On* or *Off* as Function of the Number of Molecules for the Simulation Parameters of k_1 and k_2 Equal to 1.111 and 3.333s^{-1}, Respectively (see Table 1.1)[a]

Molecules	Probability (All On)	Probability (All Off)	Simulation Time (s) (All On)	Simulation Time (s) (All Off)
2	0.56	6.3×10^{-2}	8.0	14.0
4	0.32	3.9×10^{-3}	7.2	192.0
6	0.18	2.4×10^{-4}	8.5	2.1×10^3
8	0.10	1.5×10^{-5}	11.3	2.5×10^4
10	0.06	9.5×10^{-7}	16.0	3.2×10^5

[a]The simulation time is the time needed for either a simulation or real experiment, assuming one makes 10 measurements of either the "*all on* or *all off*" periods. Figure 1.2 is the determination of k_1 using the simulation times in this table.

experiment can be computed from the following formulae:

$$T_A = \left(\frac{m}{Nk_1}\right)(P_A)^{-N}, \tag{1.9a}$$

$$T_B = \left(\frac{m}{Nk_2}\right)(P_B)^{-N}, \tag{1.9b}$$

where T_A and T_B are the times needed to measure a specific number, m, of *all on* and *all off* periods. For example, we anticipate 10 separate periods of *all on* (*all off*) for $m = 10$. For the *all on* case where $N = 3$ and $m = 10$, along with the parameters in Table 1.1, T_A is 7.2 s; 100 independent simulations with the time set to 7.2 s give $m = 10.1 \pm 1.9$, which agrees with the initial premise of 10 independent measurements in the data stream. The interesting part of this method is that times are measured rather than intensities. Finally, autocorrelation analysis can also be applied to this system to determine the sum of the rate constants. Neither method requires a synchronous start.

1.2.4. Comparison of Data Analysis Techniques

Both averaging and distribution methods give the mean value of the histogram and the variance; thus there is little difference between the two. Because averaging analysis and distribution analysis are closely related, only the differences between the averaging method and autocorrelation analysis are considered; the differences between distribution and autocorrelation analyses are assumed to the be same as the former case.

The major difference between distribution analysis and averaging is that distribution analysis gives a visual representation of the values that may help distinguish between two distinct states, whereas averaging gives a single point. For example, the distinction between two closely separated peaks in a distribution analysis is evident, whereas the average would only give one point, the mean of the two peaks. In the limit of a normal distribution, averaging and distribution analysis give identical answers.

1.2.4.1. Single-Molecule Case

There is a fundamental difference between an autocorrelation analysis and measuring the distributions of on and off periods for the case where both states are fluorescent. The time dependence of the autocorrelation of the fluorescence intensity fluctuations provides the sum of the forward and reverse reaction rate constants; in the special case, where only one state is fluorescent and the other state is dark, the zero-time offset gives the equilibrium constant (Zander et al., 2002). However, the averaging method gives rate constants for each state even if both states are fluorescent. In addition, the averaging method is computationally simpler to implement than autocorrelation analysis. For a series of Poisson events, as in the reactions modeled here, probability theory predicts that maximum likelihood for the series is the average value of the Poisson process (Matthews and Walker, 1964). Thus, a reasonable estimation of the real value results from measuring only a few cycles; however, autocorrelation analysis depends on the system being in a stationary state; therefore, measurement of only a limited number of cycles may not meet this requirement.

A graphical comparison of the two data analyses methods is shown in Figure 1.3 for simulated data sets of the same single molecule. Figure 1.3 clearly shows that as the

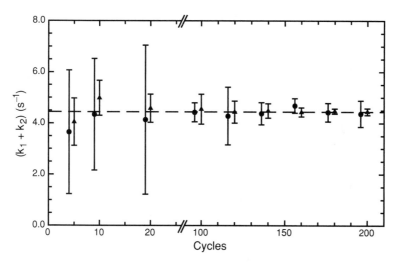

Figure 1.3. Comparison of the sum of the rate constants for a single molecule calculated either by autocorrelation analysis or averaging the "on and off" times for a given number of cycles as indicated on the x axis. For clarity, the points for the autocorrelation analysis are on one side of the points for the averaging method; the points for the averaging method give the true number of cycles for the other method. The circles (•) are the autocorrelation fit values and the triangles (▲) are the values determined by averaging for the rate constant sum. The error bars are the standard deviations for five independent simulations. The dashed horizontal line at 4.444 s^{-1} is from the parameters used in the simulation of the data stream in Table 1.1 and represents the true value.

number of cycles increases, both methods give similar values for the sum of the rate constants $(k_1 + k_2)$. For example, at 150 cycles, the values are 4.58 ± 0.53 and 4.46 ± 0.19 for autocorrelation and averaging, respectively. However, for a limited number of cycles, generally less than 20, the accuracy of the averaging method is usually greater than that for autocorrelation analysis. For example, in Figure 1.3, the averaging method for the five cycles gives the sum of the rate constants within 10% of the true value, whereas autocorrelation analysis gives a value within 18%; in addition, the standard deviations of the simulations are greater for autocorrelation analysis.

Another possible advantage of the averaging method is the analysis of long-time data sets. A "moving" average, a nonweighted average over successive subsets of the data, which is commonly used in assembly line quality control, identifies any slow processes, such as conformation changes, occurring on a time scale long with respect to changes between states. Because the averaging method gives a reasonable result with between 5 and 15 measurements, a moving average can be applied to a subset within this range to determine where any changes are occurring. Once any changes are identified, the data can be reanalyzed by adjusting the subset size to give favorable results. Autocorrelation analysis of a system with slow conformation changes gives deviations from simple exponential decay; however, it may be difficult to determine when the deviations occurred. More complicated autocorrelation functions are needed to fit the deviations (Schenter et al., 1999; Yang and Xie, 2002b). To our knowledge, the use of a moving average method—a simple, powerful technique—to single molecule studies has not been reported previously.

Noise Considerations. The noise considerations for each type of analysis are quite different. The averaging method measures time widths whereas autocorrelation analysis measures

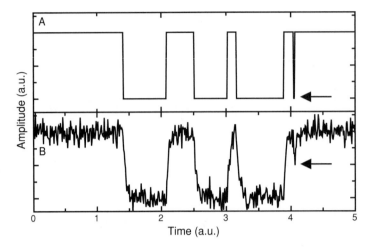

Figure 1.4. (A) Simulated data for an on and off process. (B) Identical data as in part A except noise and an RC constant of 0.04 time units are added; the amount of noise added gives a signal-to-noise ratio of approximately 5. Because the averaging method requires distinguishable states, the signal-to-noise ratio is important; a threshold amplitude value can be chosen to determine the molecule's state. If the noise level approaches the threshold value, discrimination between states is difficult. For example, in part B, if the threshold value for being in part A is set to greater than 0.5, then time spent in part B for a cycle, as indicated by the arrow, is overlooked. Alternatively, setting the threshold value to a smaller value can lead to noise signals interpreted as changes in state.

intensity fluctuations. Assuming a reasonable signal-to-noise ratio (i.e., the ability to distinguish states), the uncertainty for measurement of the widths of the on and off pulses depends largely on the RC constant of the detector and measurement system (see Figure 1.4). A large RC constant has the effect of noise reduction and data smoothing, but fast transitions between states are integrated and are potentially not resolvable. Because the autocorrelation function analyzes fluorescence intensity fluctuations as the molecule cycles between A and B, it is subject to shot noise associated with photon counting, white noise, $1/f$ noise, and other technical noise; however, most noise is uncorrelated, and thus there is a reduction of noise proportional to the square root of the number of time bins.

The method of choice for data analysis depends largely on the data itself:

1. If the signal-to-noise ratio is low (i.e., it is difficult to distinguish individual states), autocorrelation analysis is the preferred method.
2. If individual states are distinguishable and the amount of data is limited, then the averaging method generally yields more accurate results.
3. If individual states are distinguishable and the amount of data is not limited, then both the averaging method and autocorrelation analysis give equivalent results.

1.2.4.2. Prebulk Case

Both the averaging method and autocorrelation analysis assume that the chemical system is known in the probe volume. In addition to the former requirement, the averaging of the periods when all the molecules are either *all on* or *all off* involves knowledge of the molecular brightness if either individual rate constants or N is to be ascertained from the system. The best type of analysis for the prebulk system, for two fluorescent states, is a combination of both methods. For this case, autocorrelation analysis gives $k_1 + k_2$, whereas the averaging method gives K_{eq}. By combining the results of both types of analysis, individual rate constants and N are determined. The noise considerations are analogous to those for the $N = 1$ case.

1.2.4.3. Bulk Case

In the bulk case, data analysis is limited to either the time dependence of the fluorescence intensity or autocorrelation analysis. Generally, for under 100 molecules, autocorrelation analysis is preferred because of its noise reduction feature. Because the relative intensity fluctuations become smaller as the number of molecules in the probe volume increases, autocorrelation analysis of bulk systems is not appropriate due to signal-to-noise considerations. Thus, time analysis of the relaxation following a perturbation jump experiment for bulk systems for large numbers of molecules may be the only method to determine the sum of the rate constants. In this case, technical noise is the main source of noise.

1.3. PROBE VOLUMES

In many interesting biological cases—for example, cells and lipid vesicles—choosing the probe volume is not possible. The determination of N is often the object of the experiment. In the previous section we discuss the treatment of data for multiple molecules in the probe volume. Here we examine the importance of the choice of a probe volume for single-molecule measurements and methods for attaining probe volumes smaller than can be obtained with diffraction optics.

The requirement that the average occupancy of the probe volume be less than one limits the range of reactions that can be studied by single-molecule studies (Laurence and Weiss, 2003). Consider the reaction

$$A \underset{k_a}{\overset{k_d}{\rightleftarrows}} A + B \tag{1.10a}$$

$$K_D = \frac{k_d}{k_a} = \frac{[A][B]}{[AB]}, \tag{1.10b}$$

where K_D is the equilibrium constant, k_a is the association rate constant, and k_d is the dissociation rate constant.

The magnitude of K_D is inversely related to the strength of the complex; K_D is often called the affinity to reflect this relationship. A small affinity value indicates a tightly bound complex. Here, affinity values of AB where the complex is half-dissociated are examined:

$$[B] = [AB] \tag{1.11a}$$

$$[B] = C_0/2 = K_D, \tag{1.11b}$$

where C_0 is the concentration corresponding to one analyte molecule in the probe volume. At equilibrium, k_d equals $(k_a \times C_0/2)$ and the recombination rate is $(k_a \times C_0/2)$. Assuming

Table 1.3. Representative Parameters for the Different Probe Volumes[a]

Method	Volume (fL)	C_0 (nM)	Dissociation Time (s)	Affinity (nM)
Focused flow	1.0×10^3	1.7×10^{-3}	1.2×10^4	8.3×10^{-4}
Droplets	1.0	1.7	12.1	0.8
Two-photon	1.0	1.7	12.1	0.8
Nanochannels	0.3	5.5	3.6	2.8
Confocal	0.2	8.3	2.4	4.2
TIR	0.1	16.6	1.2	8.3
TIR/FCS	5.0×10^{-2}	33.2	0.6	16.6
Two-plate	5.0×10^{-3}	3.3×10^2	0.1	1.7×10^2
Waveguide	1.0×10^{-6}	1.7×10^6	1.2×10^{-5}	8.3×10^5
Human cell	6.6×10^4	2.5×10^{-5}	1.6×10^6	1.3×10^{-5}

[a] Column 1 lists the method used to attain the probe volume listed in column 2. Column 3 lists the concentration, C_0, where there is, on average, one molecule in the probe volume. The dissociation time is the time needed for the complex to split—that is, the inverse of k_d. The affinity, K_D, is defined as the concentration where the complex is half-dissociated, [B] = [AB], and the affinity is $C_0/2$.

that the forward reaction is diffusion controlled, k_a is approximately 10^8 M$^{-1} \cdot$ s^{-1} for a typical protein in water (Cantor and Schimmel, 1980).

A particular probe volume is appropriate for a limited range of affinities. For example, a volume of 0.2 fL, typically used in single-molecule confocal spectroscopy, requires a concentration less than 8.3 nM for an average occupancy of less than 1. This corresponds to an affinity of approximately 4 nM. It would be difficult to study molecules with larger affinities in this probe volume because the lower C_0 would result in unacceptable dissociation times resulting in fewer observable events. (See Table 1.3.) Two ways around this problem are smaller probe volumes and multiple occupancy of the probe volume.

Various ways of attaining probe volumes smaller than 1 fL are described below and are listed in Table 1.3; the volumes cover the range from picoliters to zeptoliters. This table illustrates several important points. Using the correct probe volume is critical when studying probe/target binding. While the range of affinities for a particular probe volume is limited, affinities from 10^{-3}–10^6 nM can be studied by choosing the appropriate probe volume.

It is mistakingly perceived that the diffraction of light limits the smallest probe volume to approximately a femtoliter (1 μm × 1μm × 1μm). Indeed, the most commonly used probe volume is 0.2 fL attained by diffraction limited, confocal optics. The most common way to exceed the limits imposed by diffraction optics is to use an evanescent field (Ambrose et al., 1999). When light strikes a transparent surface at an angle that exceeds the critical angle essentially all of the light is reflected, this effect is called total internal reflection (TIR). However, a small fraction of light "leaks" into the media; this is the evanescent wave. The intensity of the evanescent wave decreases exponentially with an e^{-1} length in the range of 100–500 nm. Probe volumes smaller than 1 fL are attained by a combination of confocal detection, spatial confinement into volumes with dimensions smaller than the wavelength of light, and evanescent excitation (Foquet et al., 2002; Ha et al., 1999; Hassler et al., 2005; He et al., 2005; Mukhopadhyay et al., 2004; Schwille; 2003; Starr and Thompson, 2001). TIR techniques are well-suited for studying single molecules at interfaces such as surfaces and cellular membranes (Schwille, 2003; Starr and Thompson, 2001). Foquet et al. (2002) reported a probe volume of 0.05 fL by confocal excitation of a sample confined between two glass plates separated by 100 nm. Craighead's group recently combined spatial confinement and evanescent excitation; the sample was placed in a 200-nm inner-diameter quartz tube and excitation light was focused into the tube, resulting in an evanescent wave in the tube that gave a probe volume of 10^{-6} fL (Levene et al., 2003).

Two-photon excitation is another attractive way to define a small probe volume (Mertz et al., 1995). In two-photon spectroscopy, molecules are excited by the simultaneous absorption of two photons. Because the absorption is a two-photon process, the excitation probability is proportional to the square of the irradiance yielding probe volumes smaller than 1 fL (Schwille, 2001). Excitation of visible fluorophores is confined to the high irradiance region reducing both photobleaching of molecules outside of the probe volume and background luminescence from intracellular material.

1.4. STATISTICS OF SINGLE-MOLECULE MEASUREMENTS

1.4.1. Application to DNA Fingerprinting

This section describes our work on DNA fragment sizing (Ferris et al., 2005; Goodwin et al., 1993; Habbersett and Jett, 2004; Larson et al., 2000) that we include as a demonstration of the validity of extracting ensemble properties from less than ten molecules. Previously, we have measured the fluorescence intensity of individual DNA restriction fragments intercalated with a fluorescent dye to obtain a DNA fingerprint of a chromosome from a single bacterial cell or viral particle (Ferris et al., 2005). DNA intercalating dyes react stoichiometrically with DNA, resulting in a fluorescence signal that is proportional to the number of base pairs in the fragment. A histogram of the fluorescence intensity from individual restriction fragments is a DNA fingerprint.

1.4.1.1. Flow Cytometry

For analytical measurements, both an orderly delivery of analyte to the detection volume and efficient detection of analyte molecules are desired. One approach to this problem is detection in flowing sample streams. This method is based upon the use of hydrodynamic focusing to position sample streams in the center of a square-bore flow cell. When care is taken to ensure that the sample stream passes through the probe volume, such that all of the molecules experience similar excitation and optical detection efficiencies, the molecular detection efficiency is greater than 90% at a detection rate of approximately 100 molecules/s (Keller et al., 2002). Our flow approach to DNA sizing is roughly 100 times faster (from days to minutes), one million times more sensitive (from micrograms to femtograms), five times more certain (from 10% to 2%), and more accurate in the range of 125 bp to 450 kbp than gel electrophoresis (Larson et al., 2000). Figure 1.5 shows a DNA fingerprint from a *Hind*III digest of λ phage DNA (Habbersett and Jett, 2004). Fluorescence from approximately 1800 fragments is used to construct this histogram. How many were really needed?

Figure 1.6 illustrates how accurately the mean of a Gaussian distribution can be determined from the average of m measurements where $m = 2, 5, 10$, and 20 (Ferris et al., 2004). Only two measurements are required to reasonably estimate the mean and m greater than or equal to five does remarkably well for a Gaussian distribution. Figure 1.6 can be thought of as a graphical representation of the maximum likelihood theorem; that is, the most probable

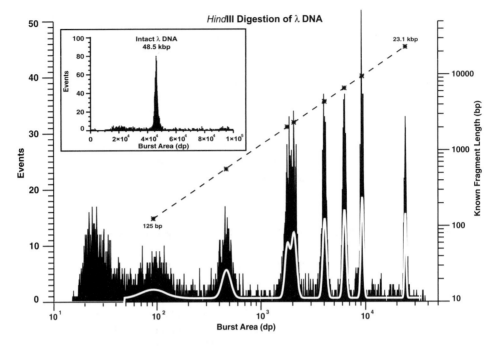

Figure 1.5. *Hind*III digest of λ phage DNA stained with SYTOX-orange and irradiated with 0.8 mW at 532 nm from continuous Nd:YAG laser. The *y* axis is the number of events, and the *x* axis is the burst area in detected photons. The histogram is constructed from a total of 1800 events recorded in 124 s from a 1–10 pM solution of fragments. The white curve is the sum of eight Gaussian distributions fit to the data; the amplitude of the curve is drawn at half-scale for visibility. The dashed line is a plot of the Gaussian centroids versus the known fragment lengths. The inset is a histogram of intact λ phage DNA with a CV of 1.6% on the major peak. [Reprinted from Habbersett and Jett (2004), with permission of Wiley-Liss.]

outcome for a random sequence of numbers from a Gaussian distribution is the average value of the distribution (Matthews and Walker, 1964).

Figure 1.7 shows the effect of reducing the sample size in an *S. aureus* restriction digest assay from 14,163 fragments to 570 fragments (Ferris et al., 2004). The sample size of 570 ensures that there are at least 10 replicate measurements of each fragment. The smaller-sized sample results in a loss of resolution, but still gives a representative fingerprint.

1.4.1.2. Optical Mapping

We use optical mapping techniques developed by David Schwartz and colleagues (Aston et al., 1999; Meng et al., 1995; Schwartz et al., 1993; Schwartz and Samad, 1997) to obtain DNA fingerprints from single viral genomes deposited on a surface (Ferris et al., 2005). Briefly, the experimental procedure involves embedding cells in an agarose matrix, lysing the cells, and processing to yield intact DNA. The DNA is then deposited on a derivatized glass substrate, which stretches out the genome. The elongated genome is digested with a restriction enzyme and stained with YOYO-1, an intercalating dye. The DNA relaxes upon cutting, leaving gaps at the restriction sites and intact DNA between restriction sites; after digestion, the order of the fragments is preserved. We use a concentration of 25 ng/mL

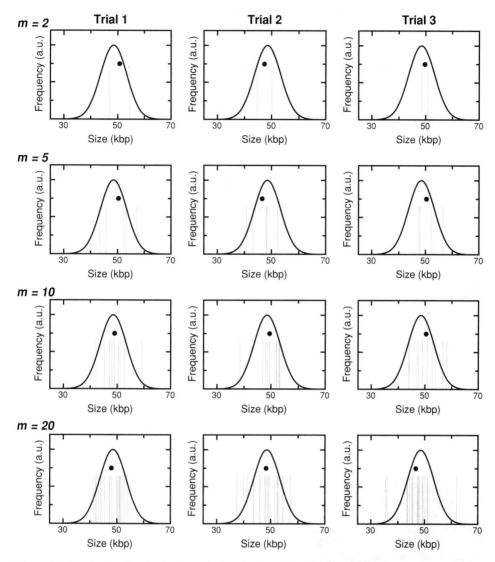

Figure 1.6. Distribution of randomly chosen single molecules within a Gaussian profile representative of an intact λ phage DNA fragment generated using a mean of 48.502 kbp and a CV of 10%. The solid curve represents the Gaussian distribution whereas the vertical lines denote the positions of the random picks and the filled circle denotes the average of the *m* picks. [Reprinted from Ferris et al. (2004), with permission of Wiley-Liss.]

of intact viral genomes to obtain one genome per 10^4 μm². Fluorescence from individual fragments is collected with a CCD camera mounted on a sensitive microscope.

Figure 1.8 shows fluorescence from a *Pme*I digest of a mixture of λ phage and bacteriophage T4, strain GT7. The fragment patterns are clearly different and easily distinguished from each other. We use the known sizes of T4 GT7 restriction fragments and a CV of 8%, characteristic of our measurements, to construct a virtual fingerprint of T4 GT7. DNA fragment lengths measured from seven T4 GT7 genomes, using λ phage DNA as an intensity standard, are plotted in Figure 1.8. Fragments from individual genomes are identified

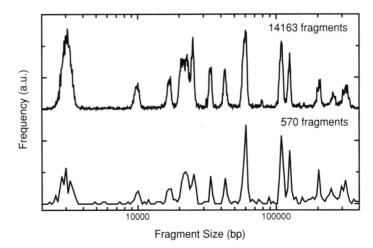

Figure 1.7. Comparison of the DNA fragment size distribution from a *S. aureus* Mu50 digested with *Sma*I derived from the measurement of 14000 fragments to a distribution derived from a measurement of 570 fragments. [Reprinted from Ferris et al. (2004), with permission of Wiley-Liss.]

by a color code. The fragment lengths from the individual cells are closely grouped, and a fingerprint from any one of the seven genomes represents the actual fingerprint.

Figure 1.9A shows a fluorescence image from a *Pme*I digest of a mixture of adenovirus, λ phage, and T4 GT7. Again the fingerprints are easily identified by their pattern. Figure 1.9B shows the fragment lengths measured in Figure 1.9A plotted on top of a virtual digest of adenovirus and T4 GT7. The λ digest is used to calibrate the fragment sizes. These data show that optical mapping techniques can identify the composition from a mixed sample. This is in contrast to restriction fragments analyzed by either gel electrophoresis or flow

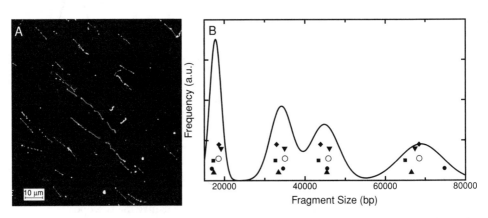

Figure 1.8. (A) A portion of a fluorescence image displaying stained DNA fragments of a *Pme*I digest of a mixture of λ phage DNA, falsely colored red, and T4 GT7, falsely colored green. The different genomes are easily identified by their respective patterns. See insert for color representation of figure A. (B) Seven distinct T4 GT7 genomes from the complete image shown in part A plotted on top of a virtual digest of T4 GT7. The data are offset from each other on the *y* axis for viewing. The two T4 GT7 genomes in part A are denoted by the square (■) and the open circle (O) in the plot. The fluorescence intensity of the T4 GT7 genomes is calibrated with data from the λ phage DNA. [Reprinted from Ferris et al. (2005), with permission of Elsevier.]

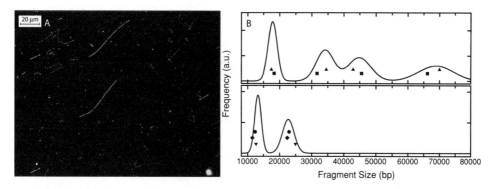

Figure 1.9. (A) *Pme*I digestion products of a mixture containing adenovirus (three genomes in blue), λ phage DNA (three genomes in red), and T4 GT7 DNA (two genomes in green). See insert for color representation of figure A. (B) Fragment sizes of T4 GT7 and adenovirus from part A are plotted on a virtual digests of the components using an 8% coefficient of variation. The lower plot in part B is adenovirus, whereas the upper plot is T4 GT7. [Reprinted from Ferris et al. (2005), with permission of Elsevier.]

cytometry that pass through a solution phase where fragment order is lost and therefore cannot be related to the species of origin.

Knowing the order of the restriction fragments on the genome provides another dimension in DNA fingerprinting that is a significant advantage when discriminating between two organisms or when matching the fingerprint of an unknown organism to a library entry. There are $N!$ ways of arranging N restriction fragments on a chromosome. Electrophoresis and flow cytometry provide a histogram of the fragment sizes but not the order of the fragments. This means there are $N!$ potential matches to other organisms in an electrophoretic or flow assay whereas an ordered map has only one match.

This point is illustrated in Figure 1.10A for the fingerprint of a pair of hypothetical genomes that contain three restriction fragments with two degenerate fragments. We have developed a "bingo card" display that incorporates both the fragment size and the fragment order to remove the degeneracy. The x axis reports the size of the restriction fragments, and the y axis indicates the order of the fragments.

1.5. VIRTUAL SORTING

It is difficult to sort single molecules by conventional flow cytometric techniques because the flow rates are too slow. Kapanidis et al. (2004) use fluorescence resonance energy transfer, herein referred to as FRET, to select emission from donor/acceptor-labeled biomolecules for "virtual sorting." In virtual sorting, the data stream is sorted instead of the sample stream. Consider the gedanken experiment shown in Figure 1.11. The sample contains four components: free dye, unlabeled target, labeled probe, and probe/target complex. One is interested in measuring the ratio of the number of bound and unbound probes containing a single tag. The chemistry is complex, and it is difficult to separate chemically free dye and probes containing more than one fluorescent tag from the solution. This can be important for analyses involving complex labeling schemes. (See Section 1.6.)

Theoretical data from a mixture of the three components is shown in Figure 1.11. The three components have different molecular weights and therefore different diffusion constants; consequently, they spend different times in the probe volume. The "sorter" directs the

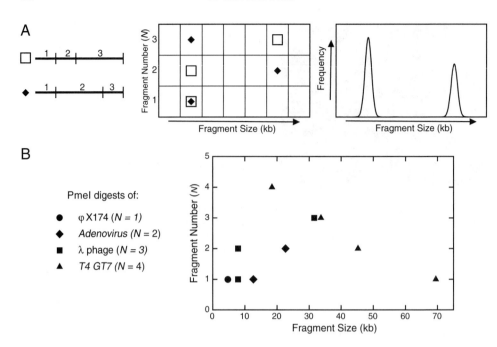

Figure 1.10. (A) A hypothetical example of a bingo card display for two different genomes; the two genomes are denoted by the open rectangle (□) and the filled diamond (♦). Both genomes contain identical fragments but with a different distribution. The middle of part A shows a bingo card display where each genome is distinguishable, whereas the rightmost plot in part A shows a histogram of the fragment sizes analogous to those from gel electrophoresis or flow cytometry. In the rightmost plot the fragments from the individual genomes are mixed together, eliminating the possibility of distinguishing between genomes. (B) Bingo card display for the genomes digested with *Pme*I in Figures 1.8 and 1.9. Fragments 1 and 2 of the λ digest, normally undetermined by gel electrophoresis and flow cytometry, are clearly visible in the bingo card presentation. [Reprinted from Ferris et al. (2005), with permission of Elsevier.]

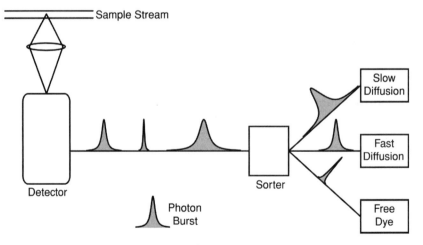

Figure 1.11. Schematic of an apparatus for sorting the data stream into bins corresponding to the molecule in the probe volume—that is, virtual sorting. The molecules are identified by their diffusion time, which is reflected by the length of the photon burst as the molecule diffuses through the probe volume.

data from the different species into the appropriate data bin. Hence, the ratio of the bound to unbound probes is obtained *without* ever purifying the mixture. Similar techniques can be used to determine FRET efficiencies without separating species containing only a donor or acceptor from the mixture.

1.6. STATISTICAL CHEMISTRY

Statistical chemistry is a new way of doing chemistry that would be a good application for virtual sorting. Figure 1.12 shows a synthetic pathway for the preparation of a molecule containing a FRET donor/acceptor pair for the study of conformation changes of large polymer molecules (e.g., proteins or DNA). Typically, two different reactive groups are added to the analyte (e.g., amine and carboxyl), and the donor and acceptor tags contain the appropriate complementary reactive groups. Consider Reaction A in Figure 1.12, which shows a labeling reaction for a protein with a donor and acceptor pair for a FRET experiment. DPA, which contains a single donor and acceptor, is the desired product; the other species are byproducts that contribute to the fluorescence background and perturb the answer. The amount of the various byproducts depends upon the selectivity of the reactive groups and the reaction conditions. Extensive column chromatography and other purification techniques can remove byproducts (i.e., unbound dyes) with only limited success. In fact, much of the data in the literature contain artifacts from byproducts. The autocorrelation function can show the presence of unwanted contaminants. For example, the presence of unreacted donors or acceptors would show up in the autocorrelation function as species with large diffusion coefficients.

A more challenging situation is where both reactive groups are identical—for example, either amine or carboxyl. In this case, virtual sorting can be a real aid. Consider Reaction B in Figure 1.12, where P has two reactive amines and both donors and acceptors contain a reactive SH group; both D and A are present at the same concentration. Again, the desired product is DPA, statistically 50% of the reaction yield. Extraction of these compounds from the reaction mixture by chromatography is exceedingly difficult, but separation of the data stream by selecting species that emit in both the red and green or emit in the red when excited in the green is relatively straightforward.

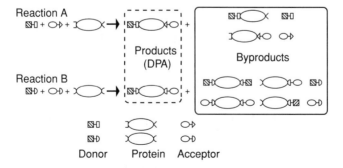

Figure 1.12. Reaction A is a donor/acceptor reaction for a protein with different binding sites for donor and acceptor. The desired product, DPA, is marked by a dashed rectangle, whereas the byproducts of the reaction (i.e., incompletely labeled protein, incorrectly labeled protein, and free dye) are marked by the solid rectangle. Reaction B is a donor/acceptor reaction for a protein with identical binding sites for donor and acceptor. The desired products and byproducts are denoted with a dashed and solid line rectangle, respectively.

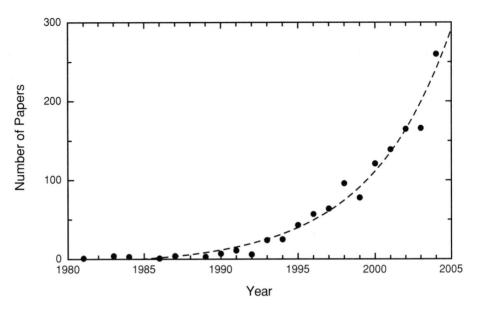

Figure 1.13. The approximate number of papers published concerning single-molecule studies; the number of papers doubles every three years. The dashed line is drawn only to indicate the trend. The number of papers published is found with SearchPlus™, a literature database.

1.7. THE FUTURE

It is appropriate to begin a discussion of the future with a brief summary of the past. Considerable progress has been made in analytic detection of organic molecules over the last 25 years. In 1983 the world record for the detection of fluorescent molecules in solution was 22,000 molecules in a 10-pL probe volume (Dovichi et al., 1983); in 1990, single-molecule detection in flow was achieved (Goodwin et al., 1995). Since 1990, the number of papers in single-molecule research has grown exponentially with no signs of a slowdown. Figure 1.13 shows a growth curve for single-molecule publications which indicates a promising future.

Currently, the majority of single-molecule research relies on fluorescence detection. In the near future, other intrinsic properties such as Raman spectra, absorption spectra, electrical properties, magnetic properties, and so on, will be increasingly used to characterize single molecules.

1.7.1. Medical Diagnostics

An area where single-molecule studies have great potential is medical diagnostics. Increasingly, modern medicine is becoming molecular medicine; diagnostics are beginning to be based upon single-molecule probe/target binding (e.g., antibody antigen). These diagnostics will become more important as probes become more selective, versatile, photostable, soluble, and compatible with intracellular environments. Indeed, one may even imagine mobile probes circulating through the body monitoring important functions and surveying for early indicators of a developing disease; these probes would broadcast reports to external receivers.

1.7.2. Single-Cell Chemistry and Biology

The ability to perform intracellular measurements for identification of species present, interaction between species, or tracking of species in response to stimulation is an emerging field in single-molecule detection. Future advancements will lead to a new understanding of basic biological processes in cells. Such information is not available from *in vitro* measurements. In fact, *in vitro* measurements are misleading because the pristine environment excludes interactions with surfaces and other cellular components that are important aspects of intracellular chemistry. A good starting point would be the encapsulation of selective intracellular species into lipid vesicles to mimic cells.

1.7.3. Sample Preparation and Detection

We expect a tremendous improvement in the handling of ultrasmall samples. One can detect single molecules, but it requires nanograms of sample to get a few molecules to the probe volume. New advances in nanotechnology permit handling of minute samples and delivery of such samples to an interrogation region.

1.7.3.1. Lab on a Chip

There is considerable activity in designing, constructing, and characterizing microfluidic devices of nanometer dimensions ("nanotechnology") for delivery of samples to small reaction chambers for the study of monomolecular and bimolecular chemistry of single species (e.g., single molecules, single DNA fragments, single cells, etc.). These devices incorporate valves, mixers, pumps, and detectors to form an integrated analysis system for nanochemistry—that is, a *"Lab on a Chip."*

In 2001, Chabinyc, Chiu, McDonald, Stroock, Christian, Karger, and Whitesides made considerable progress toward a total lab on a chip (Chabinyc et al., 2001). The chip incorporated a µAPD array, a fluidic system with a dedicated excitation/detection region, and fiber optics to deliver the excitation beam. The device had a detection sensitivity 25 nM fluorescein; future development will improve on the sensitivity.

Zare's group has developed a microfluidic device to study the reactions of single cells (Gao et al., 2004; Wheeler et al., 2003; Wu et al., 2004). The chip is constructed of soft polymethylsiloxane, has a reaction volume of 70 pL, and contains a metering system that can deliver picoliter volumes. In addition to ultrasensitive fluorescence detection, the device has an electrophoresis column and a fluidic trap incorporated into the chip. Briefly, samples are injected into a microchannel and a particular cell is selected and trapped in the reaction chamber. The cell is lysed and intracellular substances are released into the reaction chamber, where they combine with reagents introduced into the chamber. The products are then detected and characterized. In addition to a significant saving in time and reagents, intercellular heterogeneities are observed that would be obscured in the averaging that occurs in bulk measurements.

We anticipate a microchip incorporating many channels, where each channel contains (a) diode emitters for fluorescence excitation at a wavelength chosen by the operator, (b) pumps and valves for sample delivery, (c) an integrated single-molecule detector, (d) the ability to manipulate individual molecules, (e) a data processor, and (f) automated reporting.

1.7.3.2. Sample Preparation and Detection of Single Molecules on a Surface or in a Thin Layer of Agarose

By far, the easiest and most straightforward way to handle small samples is deposition on a surface or in a thin agarose gel (Ferris et al., 2005). Deposition on a surface minimizes sampling handling, allowing fragile samples to stay intact. In addition, it maintains spatial separation of the molecules in the sample, and thus any chemistry performed on the sample preserves this separation. Surface deposition analysis can easily be incorporated for automated analysis. We believe that surface deposition analysis will become a universal technique for single-molecule detection.

1.7.3.3. Analysis Rate

In order for single-molecule measurements to become a practical analytical tool, considerable improvement must be made in the analysis rate, currently 100 molecules/s. Capillary arrays are one approach that is being developed to increase the analysis rate. Currently, capillary arrays are used for DNA sequencing (Huang et al., 1992; Lu and Yeung, 1999; Mathies and Huang, 1992; Xue et al., 1999; Zagursky and McCormick, 1990; Zhang et al., 1999). However, capillary array detection is generally limited by a low duty cycle. We envision capillary arrays bundles becoming larger, having a larger throughput, increased sensitivity, and hyper-dimensional analysis. In fact, Dovichi has demonstrated two-dimensional analysis for DNA sequencing (Zhang et al., 2001).

1.7.3.4. Counting Individual Molecules and Sorting

Single-molecule detection and counting represent ultimate goals of analytical chemistry. If the signal from each molecule is observable, the molecular count rate is a digital process and directly measures the concentration. Current single-molecule detection efficiencies are between 80% and 90% (Goodwin et al., 1995; Lermer et al., 1997; Li and Davis, 1995; Zander et al., 1998), which will without doubt improve. In addition, probe volumes will become larger, allowing more sample to be analyzed at the same time.

Sorting techniques for detected and identified single molecules will be improved. This will enable separation and concentration of selected (rare) species from a heterogeneous sample for reanalysis, cloning, or further study. Reanalysis could involve increasing the accuracy/confidence of the initial analysis or identifying sorted species that contain another parameter (e.g., all DNA fragments of a given size that contain a specific sequence). Single-molecule manipulation techniques will enable selective chemical reactions and molecular assembly.

1.7.4. Simplified Apparatus or a Poor Man's Single-Molecule Detector

Presently, single-molecule detection, which includes the detection equipment, electronics, and light sources, occupies a substantial portion of a laboratory. We foresee a day where apparatus is less in both size and cost. "Off-the-shelf" components will be used to construct such apparatus. Single-molecule detectors will be come as commonplace as typical instruments in any analytical laboratory (see Figure 1.14) and will be found in medical, diagnostic, and analytical laboratories.

Figure 1.14. Single-molecule detection is becoming just "another tool" in an experimentalist's toolbox.

1.8. CONCLUSIONS

We constructed a model system for single and multiple molecules in a probe volume. We used simulated data to explore various methods of analysis: autocorrelation, averaging, and distribution analyses. Each method of analysis has its strengths and weaknesses; the most prudent method of analysis is often a combination of the various techniques to complement each other. Single-molecule properties (i.e., k and K_{eq}) can be extracted when there is no more than 10 molecules in the probe volume; synchronous starts are not necessary for this type of system.

Section 1.3 discusses the relationship between probe volumes and affinities for particular reaction kinetics. Insight to the reaction kinetics is helpful before beginning a single-molecule experiment to determine the "best" probe volume for a mechanism.

New methods of analysis can often simply experimental procedures. Statistical chemistry and virtual sorting can be used to analyze properties of selected molecules from a complex mixture without chemical separation and purification.

Lastly, the answer to our initial question, "Is one enough?" For accurate measurements, the answer is most likely *no*. However, we show that as few as two measurements are needed (see Figure 1.9); generally, 10–14 measurements are typically more than sufficient.

ACKNOWLEDGMENTS

We have been fortunate in having quality postdocs and colleagues, who have been largely responsible for our success. We thank the Department of Energy for 25 years of continued support. We also acknowledge the support of the NIH/NCRR funded National Flow Cytometry Resource, Grant RR-01315. Special thanks to Lawrence Pratt for informative discussions about data handling and subtleties in the statistics when studying the properties of a small number of molecules confined to a small volume.

REFERENCES

Ambrose, WP, Goodwin, PM, and Nolan, JP (1999). Single-molecule detection with total internal reflection excitation: Comparing signal-to-background and total signals in different geometries. *Cytometry B Clin Cytom* **36**:224–231.

Aragón, SR, and Pecora, R (1976). Fluorescence correlation spectroscopy as a probe of molecular dynamics. *J Chem Phys* **64**:1791–1803.

Aston, C, Mishra, B, and Schwartz, DC (1999). Optical mapping and its potential for large-scale sequencing projects. *Trends in Biotechnol* **17**:297–302.

Baldini, G, Cannone, F, and Chirico, G (2005). Pre-unfolding resonant oscillations of single green fluorescent protein molecules. *Science* **309**:1096–1100.

Cantor, CR, and Schimmel, PR (1980). *Biophysical Chemistry III: The Behavior of Biological Macromolecules*. Freeman, New York.

Chabinyc, ML, Chiu, DT, McDonald, JC, Stroock, AD, Christian, JF, Karger, AM, and Whitesides, GM (2001). An integrated fluorescence detection system in poly(dimethylsiloxane) for microfluidic applications. *Anal Chem* **73**:4491–4498.

Demas, JN (1983). *Excited State Lifetime Measurements*. Academic Press, New York.

Dovichi, NJ, Martin, JC, Jett, JH, and Keller, RA (1983). Attogram detection limit for aqueous dye samples by laser-induced fluorescence. *Science* **219**:845–847.

Elson, EL (1985). Fluorescence correlation spectroscopy, and photobleaching recovery. *Annu Rev Phys Chem* **36**:379–406.

Ferris, MM, Habbersett, RC, Wolinsky, M, Jett, JH, Yoshida, TM, and Keller, RA (2004). Statistics of single-molecule measurements: Applications in flow-cytometry sizing of DNA fragments. *Cytometry B Clin Cytom* **60A**:41–52.

Ferris, MM, Yoshida, TM, Marrone, BL, and Keller, RA (2005). Fingerprinting of single viral genomes. *Anal Biochem* **337**:278–288.

Foquet, M, Korlach, J, Zipfel, W, Webb, WW, and Craighead, HG (2002). DNA fragment sizing by single molecule detection in submicrometer-sized closed fluidic channels. *Anal Chem* **74**:1415–1422.

Gao, J, Yin, XF, and Fang, ZL (2004). Integration of single cell injection, cell lysis, separation and detection of intracellular constituents on a microfluidic chip. *Lab on a Chip* **4**:47–52.

Goodwin, PM, Wilkerson, CW, Ambrose, WP, and Keller, RA (1993). Ultrasensitive detection of single molecules in flowing sample streams by laser-induced fluorescence. *"Proc SPIE–Int Soc Opt Eng* **1895**:79–89.

Goodwin, PM, Affleck, RL, Ambrose, WP, Demas, JN, Jett, JH, Martin, JC, Reha-Krantz, LJ, Semin, DJ, Schecker, JA, Wu, M, and Keller, RA (1995). Progress toward DNA sequencing at the single molecule level. *Exp Tech Phys* **41**:279–294.

Ha, T, Laurence, TA, Chemla, DS, and Weiss, S (1999). Polarization spectroscopy of single fluorescent molecules. *J Phys Chem B* **103**:6839–6850.

Habbersett, RC, and Jett, JH (2004). An analytical system based on a compact flow cytometer for DNA fragment sizing and single-molecule detection. *Cytometry B Clin Cytom* **60A**:125–134.

Hassler, K, Anhut, T, Rigler, R, Gosch, M, and Lasser, T (2005). High count rates with total internal reflection fluorescence correlation spectroscopy. *Biophys J* **88**:L1–L3.

He, MY, Edgar, JS, Jeffries, GDM, Lorenz, RM, Shelby, JP, and Chiu, DT (2005). Selective encapsulation of single cells and subcellular organelles into picoliter- and femtoliter-volume droplets. *Anal Chem* **77**:1539–1544.

Hill, TL (1960). *Introduction to Statistical Thermodynamics*. Addison-Wesley, Reading, MA.

Huang, XC, Quesada, MA, and Mathies, RA (1992). DNA sequencing using capillary array electrophoresis. *Anal Chem* **64**:2148–2154.

Jia, YW, Talaga, DS, Lau, WL, Lu, HSM, DeGrado, WF, and Hochstrasser, RM (1999). Folding dynamics of single GCN4 peptides by fluorescence resonant energy transfer confocal microscopy. *Chem Phys* **247**:69–83.

Kapanidis, AN, Lee, NK, Laurence, TA, Doose, S, Margeat, EE, and Weiss SS (2004). Fluorescence-aided molecule sorting: analysis of structure and interactions by alternating-laser excitation of single molecules. *Proc Natl Acad Sci USA* **101**:8936–8941.

Keller, RA, Ambrose, WP, Arias, AA, Cai, H, Emory, SR, Goodwin, PM, and Jett, JH (2002). Analytical applications of single-molecule detection. *Anal Chem* **74**:316A–324A.

Krichevsky, O, and Bonnet, G (2002). Fluorescence correlation spectroscopy: The technique and its applications. *Rep Prog Phys* **65**:251–297.

Larson, E, Hakovirta, J, Cai, H, Jett, JH, Burde, S, Keller, RA, and Marrone, BL (2000). Rapid DNA fingerprinting of pathogens by flow cytometry. *Cytometry B Clin Cytom* **41**:203–208.

Laurence, TA, and Weiss, S (2003). How to detect weak pairs. *Science* **299**:667–668.

Leadbetter, JR (2003). Cultivation of recalcitrant microbes: Cells are alive, well and revealing their secrets in the 21st century laboratory. *Curr Opin Microbiol* **6**:274–281.

Lermer, N, Barnes, MD, Kung, CY, Whitten, WB, and Ramsey, JM (1997). High efficiency molecular counting in solution: Single-molecule detection in electrodynamically focused microdroplet streams. *Anal Chem* **69**:2115–2121.

Levene, MJ, Korlach, J, Turner, SW, Foquet, M, Craighead, HG, and Webb, WW (2003). Zero-mode waveguides for single-molecule analysis at high concentrations. *Science* **299**:682–686.

Li, LQ, and Davis, LM (1995). Rapid and efficient detection of single chromophore molecules in aqueous solution. *Appl Opt* **34**:3208–3217.

Lu, S, and Yeung, E (1999). Side-entry excitation and detection of square capillary array electrophoresis for DNA sequencing. *J Chromatogr* **853**:359–369.

Mathies, RA, and Huang, XC (1992). Capillary array electrophoresis—an approach to high-speed, high-throughput DNA sequencing. *Nature* **359**:167–169.

Matthews, J, and Walker, RL (1964). *Mathematical Methods of Physics*, 2nd edition. Addison-Wesley, New York.

Meng, X, Benson, K, Chada, K, Huff, EJ, and Schwartz, DC (1995). Optical mapping of lambda bacteriophage clones using restriction endonucleases. *Nat Genet* **9**:432–438.

Mertz, J, Xu C, and Webb, WW (1995). Single-molecule detection by two-photon-excited fluorescence. *Opt Lett* **20**:2532–2534.

Mukhopadhyay, A, Bae, SC, Zhao, J, and Granick, S (2004). How confined lubricants diffuse during shear. *Phys Rev Lett* **93**:236105–236114.

Park, M, Kim, HH, Kim, D, and Song, NW (2005). Counting the number of fluorophores labeled in biomolecules by observing the fluorescence-intensity transient of a single molecule. *Bull Chem Soc Jpn* **78**:1612–1618.

Pennisi, EE (2004). The biology of genomes meeting. Surveys reveal vast numbers of genes. *Science* **304**:1591–1591.

Rhoades, E, Gussakovsky, E, and Haran, G (2003). Watching proteins fold one molecule at a time. *Proc Natl Acad Sci USA* **100**:3197–3202.

Riley, KF, Hobson, MP, and Bence, SJ (1997). *Mathematical Methods for Physics and Engineering*. Cambridge University Press, Cambridge, UK.

Schenter, GK, Lu, HP, and Xie, XS (1999). Statistical analyses and theoretical models of single-molecule enzymatic dynamics. *J Phys Chem A* **103**:10477–10488.

Schwartz, DC, Li, X, Hernandez, LI, Ramnarain, SP, Huff, EJ, and Wang, YK (1993). Ordered restriction maps of *Saccharomyces cerevisiae* chromosomes constructed by optical mapping. *Science* **262**:110–114.

Schwartz, DC, and Samad, A (1997). Optical mapping approaches to molecular genomics. *Curr Opin Biotech* **8**:70–74.

Schwille, P (2001). Fluorescence correlation spectroscopy and its potential for intracellular applications. *Cell Biochem Biophys* **34**:383–408.

Schwille, P (2003). TIR-FCS: Staying on the surface can sometimes be better. *Biophys J* **85**:2783–2784.

Soper, SA, Nutter, HL, Keller, RA, Davis, LM, and Shera, EB (1993). The photophysical constants of several fluorescent dyes pertaining to ultrasensitive fluorescence spectroscopy. *Photochem Photobiol* **57**:972–977.

Starr, TE, and Thompson, NL (2001). Total internal reflection with fluorescence correlation spectroscopy: Combined surface reaction and solution diffusion. *Biophys J* **80**:1575–1584.

Talaga, DS, Lau, WL, Roder, H, Tang, JY, Jia, YW, DeGrado, WF, and Hochstrasser, RM (2000). Dynamics and folding of single two-stranded coiled-coil peptides studied by fluorescent energy transfer confocal microscopy. *Proc Natl Acad Sci USA* **97**:13021–13026.

Vazquez, MI, Rivas, G, Cregut, D, Serrano, L, and Esteban, M (1998). The vaccinia virus 14-kilodalton (a27l) fusion protein forms a triple coiled-coil structure and interacts with the 21-kilodalton (a17l) virus membrane protein through a c-terminal alpha-helix. *J Virol* **72**:10126–10137.

Verberk, R, and Orrit, M (2003). Photon statistics in the fluorescence of single molecules and nanocrystals: Correlation functions versus distributions of on- and off-times. *J Chem Phys* **119**:2214–2222.

Wheeler, AR, Throndset, WR, Whelan, RJ, Leach, AM, Zare, RN, Liao, YH, Farrell, K, Manger, ID, and Daridon, A (2003). Microfluidic device for single-cell analysis. *Anal Chem* **75**:3581–3586.

Wu, HK, Wheeler, A, and Zare, RN (2004). Chemical cytometry on a picoliter-scale integrated microfluidic chip. *Proc Natl Acad Sci USA* **101**:12809–12813.

Xue, G, Pang, HM, and Yeung, ES (1999). Multiplexed capillary zone electrophoresis and micellar electrokinetic chromatography with internal standardization. *Anal Chem* **71**:2642–2649.

Yang, H, and Xie, XS (2002a). Statistical approaches for probing single-molecule dynamics photon-by-photon. *Chem Phys* **284**:423–437.

Yang, H, and Xie, XS (2002b). Probing single-molecule dynamics photon by photon. *J Chem Phys* **117**:10965–10979.

Yang, SL, and Cao, JS (2001). Two-event echos in single-molecule kinetics: A signature of conformational fluctuations. *J Phys Chem B* **105**:6536–6549.

Zagursky, RJ, and McCormick, RM (1990). DNA sequencing separations in capillary gels on a modified commercial DNA sequencing instrument. *Biotechniques* **9**:74–79.

Zander, C, Drexhage, KH, Han, KT, Wolfrum, J, and Sauer, M (1998). Single-molecule counting and identification in a microcapillary. *Chem Phys Lett* **286**:457–465.

Zander, C, Enderlein, J, and Keller, RA, eds. (2002). *Single Molecule Detection in Solution: Methods and Applications*. Wiley-VCH, Berlin.

Zhang, J, Voss, KO, Shaw, DF, Roos, KP, Lewis, DF, Yan, J, Jiang, R, Ren, H, Hou, JY, Fang, Y, Puyang, X, Ahmadzadeh, H, and Dovichi, NJ (1999). A multiple-capillary electrophoresis system for small-scale DNA sequencing and analysis. *Nucleic Acids Res* **27**:e36–e36.

Zhang, JZ, Yang, MJ, Puyang, XL, Fang, Y, Cook, LM, and Dovichi, NJ (2001). Two-dimensional direct-reading fluorescence spectrograph for DNA sequencing by capillary array electrophoresis. *Anal Chem* **73**:1234–1239.

CHAPTER

2

DISSECTING CELLULAR ACTIVITY FROM SINGLE GENES TO SINGLE mRNAs

XAVIER DARZACQ, ROBERT H. SINGER, AND YARON SHAV-TAL

Fluorescent proteins have revolutionized cellular biology, introducing a time component in the spatial analysis of biological processes. However, it has been only recently that transcribed nucleic acids could be detected in live cells (Bertrand et al., 1998). Single-molecule detection of the constituents of gene expression is currently transforming our vision of molecular biology, allowing the focus to shift from cellular populations to single cells and eventually to the quantum units of single genes, the single mRNAs. Ultimately, we will be able to describe the dynamics of genes within live cells and analyze the synthesis and motion of single mRNAs traveling toward their translation sites.

2.1. LIVE-CELL SINGLE-LOCUS DETECTION OF DNA IN ITS GENOMIC CONTEXT

Chromosomes are the largest molecules present in cells. Human cells contain 46 chromosomes ranging from 58 megabases for chromosome Y (38 megatons/mol) to 243 megabases for chromosome 2 (160 megatons/mol). While most biological processes related to the regulation of genomic DNA have been biochemically or genetically dissected, it is only with the subcellular visualization of chromosomes—and more recently specific loci—that some aspects of replication, transcription, or chromosome segregation could be understood (Cremer and Cremer, 2001). Furthermore, our understanding that the different nuclear components are in constant motion (Phair and Misteli, 2000) has increased the need for the generation of new technical approaches that will enable the rapid imaging of nuclear dynamics in living cell systems. While examples of fluorescent proteins are known and have been routinely used in live cells (Tsien, 1998; Simpson et al., 2001; Zhang et al., 2002), nucleic acids possessing such properties have not been available until recently. Fluorescent thymidine analogs have been microinjected into nuclei of live cells and were incorporated into the cellular DNA. During subsequent cell cycles, the labeled DNA was replicated and segregated several times and the fluorescent nucleotide analogs were segregated to discrete chromosomes by semiconservative replication. This approach revealed that interphase chromosome territories are dynamic structures, the spatial position of which could be regulated (Zink and Cremer, 1998; Zink et al., 1998). The noninvasive (genetically encoded fluorescence) live cell detection of chromosomes could only be achieved using the affinity of DNA-interacting proteins to chromatin. The general affinity of chromatin interacting factors has been exploited to visualize

New Frontiers in Ultrasensitive Bioanalysis. Edited by Xiao-Hong Nancy Xu
Copyright © 2007 John Wiley & Sons, Inc.

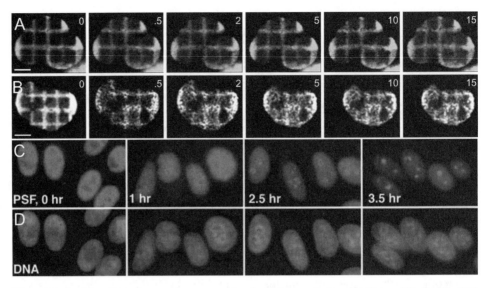

Figure 2.1. Nucleic acid binding proteins as tracers for chromatin organization. (A) Histone H2b fused to YFP was used to mark chromatin. Photobleaching of a grid on the nucleus led to the formation of a nuclear pattern that was used as a tracer for the spatial reorganization of chromatin in normal human cells or (B) in cells that were ATP depleted. (C) The formation of concave nucleolar caps is followed for time after actinomycin D treatment using the protein PSF as a marker. (D) Hoechst DNA staining shows the relationship of these caps to the nucleolar and perinucleolar chromatin. [A and B from (Shav-Tal et al., 2004a); C and D from Shav Tal et al. (2005).] See insert for color representation of this figure.

chromatin dynamics in interphase cells as well as individual chromosomes during mitosis (Lever et al., 2000; Phair and Misteli, 2000; Kimura and Cook, 2001; Forrest and Gavis, 2003; Gerlich et al., 2003; Walter et al., 2003; Phair et al., 2004; Shav-Tal et al., 2004a,b) (see Figure 2.1A, B). Proteins that interact with specific subchromosome regions have been fused to GFP and provided insights into the dynamic redistribution of centromeres (Shelby et al., 1996), nucleoli (Savino et al., 2001; Shav-Tal et al., 2005) (see Figure 2.1C,D), and telomeres (Hediger et al., 2002) within live cells.

Studying transcription and its regulation has proved to be a real challenge in live cells. Punctate nuclear structures are currently observed in fixed cells with phosphorylation site-specific antibodies directed to the RNA polymerase II catalytic subunit (CTD, C-terminal domain). These antibodies can designate the transcriptional state of the polymerase, which is known to correlate with specific sites of phosphorylation. However, this information is not available from live cells where GFP fusions of major components of the transcription apparatus do not carry information regarding transcriptional states (Iborra et al., 1996; Sugaya et al., 2000). In addition, the kinetic behavior of the polymerase alone is not sufficient to determine transcriptional kinetics since genes are dispersed and heterogeneous. What is needed is a specific, transcribing gene in order to interpret the flux of the transcription components. The first live-cell optical resolution of a single specific locus could be performed using the affinity of the lactose operon repressor protein (*lacI*) to its operon DNA sequence (*lacO*). Amplified repeats (256) of the *lacO* were sufficient to detect a single locus in a live cell using a lacI GFP fusion (Robinett et al., 1996) (see Figure 2.2). The *lacO* repeated sequences have been successfully integrated at specific loci by homologous recombination in

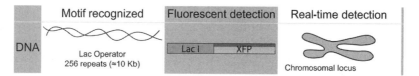

Figure 2.2. Live-cell single DNA locus detection using the *lacO/lacI* system. A tandemly repeated array (256 copies) of the lac operator (*lacO*) sequence is inserted into a chromosome. A fluorescent protein is then artificially tethered to this specific DNA sequence by fusion to the DNA-binding lac-repressor protein (*lacI*), which binds these repeats and thereby allows the detection of the integration locus in living cells. [Adapted from Shav-Tal et al. (2004b).] See insert for color representation of this figure.

yeast (Robinett et al., 1996; Straight et al., 1996; Heun et al., 2001), bacteria (Gordon et al., 1997), and viruses (Fraefel et al., 2004) or have been randomly integrated in mammalian cell genomes providing specific landmarks that can be subsequently mapped (Tsukamoto et al., 2000; Chubb et al., 2002). Other bacterial operons have been used, such as the tetracycline responsive element (TRE), providing the opportunity to label several loci in the same cell using different colors of fusion proteins (Michaelis et al., 1997). Finally, the engineering of an inducible transcription unit flanked by *lacO* repeats could reveal the dynamic interplay of chromatin-binding proteins and chromatin spatial organization during gene activation in a live cell (Tsukamoto et al., 2000; Janicki et al., 2004). The ability to follow transcription in live cells has opened a whole new field of investigation that addresses the time scale and dynamic plasticity of transcription regulatory mechanisms.

2.2. IN SITU DETECTION OF SINGLE RNA MOLECULES

2.2.1. Visualization of Single RNAs in Fixed Cells

The distribution of RNA molecules in fixed cells is visualized using the technique of fluorescent *in situ* hybridization (FISH). In this method the RNA of interest is detected and identified by exposing the fixed cell to an oligonucleotide probe consisting of the complementary nucleotide sequence to the RNA of interest. The probe is labeled with a fluorescent moiety and after hybridization of the probe with the RNA, and subsequent washing away of the unhybridized probe, the fluorescent signal can be detected by microscopy. This method can be utilized for quantifying the number of RNA molecules in fixed cells using sensitive quantitative digital microscopy with calibrated reagents (Femino et al., 1998; Fusco et al., 2003).

The RNA quantification procedure requires that a probe bind to only one unique sequence in the mRNA studied. These DNA probes are typically 50 nucleotides long and should contain at least four T positions for linking with fluorescent dyes. Also, a 50% G/C content for consistent hybridization kinetics is preferred. The RNA molecules throughout the cell are quantified by acquiring three-dimensional fluorescent image stacks of a region of interest from a cell that was hybridized with the specific probe, thus collecting the complete light output for each FISH probe, followed by deconvolution for the correction of optical blurring. After collecting an ideal three-dimensional digital volume of the cell, the cell is digitally divided into voxels, or three-dimensional pixels, in order to measure the number of fluorescent molecules throughout the cell (Femino et al., 2003). In parallel, the total fluorescent intensity (TFI) of one FISH probe is calculated by imaging known quantities of labeled probes. This information is then used to calculate the fluorescent signal in each voxel of the cell, and thus a quantification of the number and distribution of RNA molecules in a cell is obtained.

This technique was used to first detect the nuclear site of transcription of the β-actin gene and then to quantify the RNA molecules made at this locus. The β-actin gene is a serum-responsive gene. In order to detect the transcription site for β-actin mRNA, the gene was activated for increasing amounts of time using serum stimulation (Femino et al., 1998). Probes that hybridized to the 5' end of the gene detected nascent transcripts, whereas probes to the 3' untranslated region (UTR) detected almost completed transcripts, providing a means to address the transcription time and the kinetics of the transcriptional activation. Using the RNA quantification method, it was found that serum-starved cells contain 500 ± 200 β-actin mRNAs per cell. However, following serum induction this mRNA population increases to approximately 1500 copies per cell. As for transcription kinetics, the addition of serum resulted in the synchronous activation of transcription of one of the β-actin alleles in all cells of the population. This occurred in a matter of a few minutes. Some minutes later, both alleles were transcribing. This allowed the measurement of transcription rates, which corresponded to 1.1–1.4 kb/min. It could also be determined that the 3' end of the gene was more packed with polymerases than the 5' end, suggesting that termination was rate-limiting. This approach was therefore capable of producing high-resolution information of the endogenous transcriptional process.

This study was followed by single-cell analysis of gene expression for a number of active genes. Several oligonucleotide probes were conjugated with different fluorescent dyes, and mixtures of probes were designed to comprise a "spectral barcode" for each gene studied. Each designated gene was then labeled with a unique combination of dyes. The simultaneous detection of several transcription sites in single fixed cultured cells showed extensive cellular stochasticity in gene expression profiles (Levsky et al., 2002). This work emphasized the necessity of single-cell approaches to characterize gene expression profiles. For instance, the response of β-actin, γ-actin, c-jun, and c-fos genes to serum induction was readily detected on the single-gene level in single cells, and even minute incremental changes in gene expression could be observed, whereas microarray analysis could only detect a significant change in the c-fos gene.

2.2.2. Visualization of Single RNAs in Living Cells

Identification of RNA molecules in fixed cells can now be extended to the visualization of their travels within a living cell. While fluorescent fusion proteins have proven to be useful tools for following protein dynamics within a living cell, tagging of RNA for real-time studies requires a different approach. A number of protocols for RNA tagging and detection in living cells have been devised. Fluorescent *in vivo* hybridization (FIVH) uses a fluorescently labeled probe that will bind to the RNA of interest, similar to fixed cell FISH (Politz et al., 1999). While this hybridization approach allows the study of endogenous RNA molecules, double-stranded RNA complexes form that might trigger cellular degradation pathways and could affect the translational process. Moreover, the need for introduction of the probe into living cells by techniques such as microinjection or membrane permeabilization also impose stress on the cell and might indirectly affect the processes under study. Finally, since

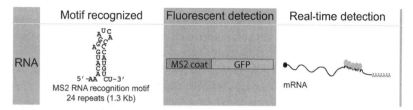

Figure 2.3. Live-cell detection of RNA molecules using the MS2 system. RNA molecules are detected using a fusion of green fluorescent protein (GFP) with the MS2 bacteriophage coat protein, which has an extremely high affinity for a short RNA-recognition motif that is derived from the phage genome. Single mRNA molecules that contain 24 MS2 RNA repeats can be detected. [Adapted from Shav-Tal et al. (2004b).] See insert for color representation of this figure.

one probe binds to one RNA molecule and an excess of probe is used in these techniques, the ability to detect a fluorescent signal from single mRNA molecules in living cells is confounded by the unhybridized probe.

In order to overcome limitations in signal detection, amplification of the fluorescent signal on a single RNA transcript is required. This has been addressed by the addition of specific protein-binding motifs to the RNA of interest. A phage RNA sequence that contains stem-loop binding sites (up to 24 sites) and that is specifically bound by the phage capsid protein MS2 is added to the DNA sequence of the gene of interest. This gene will now encode an RNA molecule with multiple binding sites for fluorescent molecules, which in this system are fluorescently tagged MS2 proteins (e.g., GFP-MS2) (see Figure 2.3). Since the binding of the MS2 protein (or GFP-MS2) to the stem-loop binding site is specific and extremely stable (K_d 39 nM), the GFP-MS2 protein binding occurs by dimerization; this site is multimerized 24 times, a strong signal above the diffuse GFP-MS2 background can be obtained, and this signal can be utilized for following the travels of specific RNAs in living cells.

This mRNA-tagging system was first used to follow the process of *ASH1* mRNA localization occurring in budding yeast (Bertrand et al., 1998). *ASH1* mRNA is one of many RNAs that are translocated from the mother yeast cell to the budding daughter cell and are found to concentrate at the bud tip (Long et al., 1997; Takizawa et al., 1997). The Ash1 protein is a nuclear DNA-binding protein required for control of mating-type switching in yeast (Darzacq et al., 2003). An mRNP complex containing an mRNA consisting of the 3′UTR of *ASH1* and MS2 repeats, bound by GFP-MS2 proteins, was imaged as it traveled from the mother cell to the bud tip. This movement was directional and exhibited speeds that correlated with that of the myosin V motor, known to be one of the required genes for successful localization.

While *ASH1* mRNP probably harbored several mRNA molecules, the behavior of single RNA molecules in living cells has also been studied using this live-cell approach (Fusco et al., 2003; Shav-Tal et al., 2004a). First, the RNA transcripts were quantified in fixed cells by sensitive FISH to confirm that the RNAs of interest traveled as single transcripts (Femino et al., 1998). The next step was visualization of these RNAs in living cells. The single mRNA molecules containing 24 MS2 stem-loop repeats were bound by GFP-MS2 proteins and formed RNA–protein complexes (mRNPs) that were amplified above background and tracked frame by frame in time-lapse movies made of living cells. Cytoplasmic mRNPs exhibited different types of movements that were characterized as diffusive, corralled, directed, or stationary. Switching between the different mobilities could be observed,

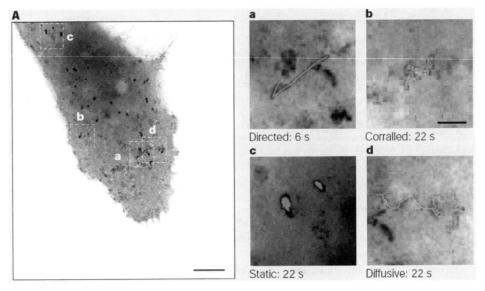

Figure 2.4. Live-cell analysis of cytoplasmic mRNPs. An mRNA transcript that contains the coding sequence of *lacZ*, 24 MS2 stem loops, and the 3′ untranslated region (UTR) of the human growth-hormone gene (hGH) was transiently expressed in COS cells together with the green fluorescent protein (GFP)–MS2 fusion protein. GFP–MS2-tagged RNA particles were followed, and different types of motility were detected in single living cells as indicated by the boxes in part A: directed (Aa), corralled (Ab), static (Ac), and diffusive (Ad). Bar, 2 μm The image in part A is a maximum-intensity image projection of 200 time frames. Bar, 10 μm. [From Fusco et al. (2003).] See insert for color representation of this figure.

and directional movements were demonstrated to follow cytoplasmic filaments (Fusco et al., 2003) (Figure 2.4).

Several studies have addressed the possible mechanisms for the nucleoplasmic travels of mRNAs in fixed cells (Zachar et al., 1993; Singh et al., 1999) and living cells (Politz et al., 1998; Politz et al., 1999; Calapez et al., 2002; Snaar et al., 2002; Molenaar et al., 2004). However, live-cell approaches have been limited to the tagging of the whole mRNA poly(A)$^+$ population in the nucleus, which could not provide a detailed analysis of a specific mRNA or of single mRNA molecules. Using the MS2 system, it is possible to probe the kinetics of single mRNA molecules as they traverse the nuclear environment (Shav-Tal et al., 2004a). In this study the travels of mRNPs containing single mRNAs, as they were released from the site of transcription and traveled to the nuclear pore, were examined in living cells using the MS2 system. Nuclear mRNPs were detected and analyzed by single-particle tracking and were found to exhibit random movements, following rules of simple diffusion (Figure 2.5). In contrast to the cytoplasm, directional movements of mRNPs were not seen in the nucleoplasm. While single-particle tracking proved useful for the analysis of mRNP movements in their nuclear microenvironments, FRAP analysis was used to follow the behavior of the population of these mRNPs and photoactivation was used to address the mobility of subpopulations of mRNPs at the time of their release from the site of transcription. These studies provided biophysical measurements for the diffusion coefficients of the mRNPs and consequently of nucleoplasmic viscosity.

Another approach for tagging of mRNA in living cells is by the use of a human U1A tag in experiments conducted in yeast cells. The U1A protein binds with high specificity

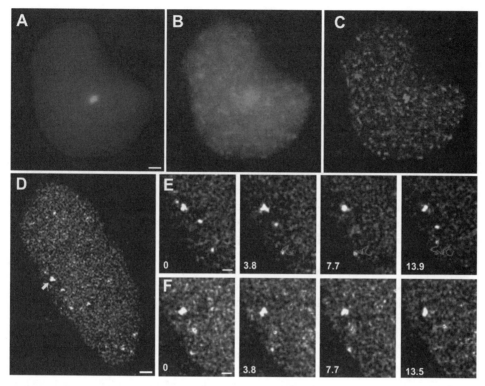

Figure 2.5. Live-cell analysis of nucleoplasmic mRNPs. (A) Image from a time-lapse movie acquired from transcriptionally induced cells containing a *lacO* cassette co-transfected with CFP-lac repressor (see Figure 2.2). (B) An inducible transcript tagged with MS2 repeats cotransfected with YFP-MS2 (see Figure 2.3). (C) Reduction of noise for tracking of RNA particles was obtained by deconvolution (Huygens software). Bar = 5 µm. (D) Tracking of particles in a transcriptionally active cell (arrow = transcription site) showed (E) diffusing particles, and (F) corralled particles. [From Shav-Tal et al. (2004a).] See insert for color representation of this figure.

and affinity to a small RNA hairpin, which does not exist in yeast. This tag is useful only in yeast since mammalian cells also have the endogenous U1A protein. In this system a GFP-U1A protein was used to follow the export pathway of different mRNAs that were engineered to contain the U1A hairpin (Brodsky and Silver, 2000). This approach was used for the analysis of the effects of UTRs, introns, RNA processing factors, nucleoporins, and transport factors on the export of the mRNAs and provided important insights on the connection between pathways of mRNA processing and mRNA export.

Most studies utilizing the MS2 or U1A RNA-tagging systems have been performed with exogenous genes since this strategy requires site-specific insertion of the hairpins into the gene of interest. Tracking RNAs produced from endogenous genes has been approached usually by indirect labeling of RNA with GFP-tagged RNA-binding proteins. Two examples are GFP-exuperantia, used to follow *bicoid* mRNA (Theurkauf and Hazelrigg, 1998) or GFP-polyA-binding protein II (PABP2) (Calapez et al., 2002). The former labels a specific RNA molecule, while the latter tags any RNA molecules transcribed by RNA polymerase II. However, the high abundance of these proteins and their lower affinity for RNA permits exchange and the few molecules of GFP bound per RNA do not allow single-molecule or single-mRNP studies. The introduction of multiple RNA-binding sites to a specific transcript is therefore an essential step in obtaining high-quality fluorescent signal over the background fluorescence.

The development of techniques that can provide a platform for the detection of mRNAs in living cells is of key interest when attempting to study mRNA kinetics of real-time processes occurring in living organisms. A perfect example in which the MS2 system has been utilized in an organism is the process of mRNA localization (Shav-Tal and Singer, 2005). Localized translation is controlled spatially and temporally in specified areas in *Drosophila* oocytes and embryos by the differential localization of a variety of mRNAs necessary for development. For instance, the nanos protein is required for the formation of the anterior–posterior body axis in the developing *Drosophila* oocyte (Tautz, 1988; Gavis and Lehmann, 1992). Some of *nanos* mRNA is localized to the posterior end of the oocyte during development, where it is stable and translated. This localization of *nanos* mRNA occurs in late stages of oogenesis. By tagging the nanos mRNA with a GFP-MS2 protein, it was possible to examine the dynamic aspects of the localization process. It was found that in contrast to other transport systems that require motor proteins for mRNA localization, *nanos* mRNA diffuses and reaches the posterior region through microtubule-dependent cytoplasmic movements, where it remains anchored by the actin cytoskeleton (Forrest and Gavis, 2003). Studies are now underway to introduce the MS2-binding sites into the genome of various organisms. This will ultimately allow the detection of the transcriptional activity of any endogenous gene and its mRNA complement in living cells and tissues.

2.2.3. Perspectives

Studying the kinetics of gene expression—from the locus of transcription, through nucleoplasmic travels, nuclear export, translation, and degradation—is of key importance for the understanding of fundamental processes taking place along the gene expression pathway. Major breakthroughs in the field of bioimaging have advanced our capability to probe these processes in living cells. However, at this time our technical abilities are mainly confined to addressing real-time processes occurring on exogenous gene constructs. We foresee a necessary step forward in the development of tools for the detection and analysis of endogenous gene expression. Such approaches will allow us to study the regulation of gene expression in the *in vivo* environment and ultimately even in the living organism.

ACKNOWLEDGMENTS

We would like to thank Shailesh M. Shenoy with his help with the submission and the figures. This work was supported by NIH grants GM57071 and EB2060 and DOE grant ER63056. Y.S. is a Stern–Lebell–Family Fellow in Life Sciences.

REFERENCES

Bertrand, E, Chartrand, P, Schaefer, M, Shenoy, SM, Singer, RH, and Long, RM (1998). Localization of ASH1 mRNA particles in living yeast. *Mol Cell* **2**:437–445.

Brodsky, AS, and Silver, PA (2000): Pre-mRNA processing factors are required for nuclear export. *RNA* **6**:1737–1749.

Calapez, A, Pereira, HM, Calado, A, Braga, J, Rino, J, Carvalho, C, Tavanez, JP, Wahle, E, Rosa, AC, and Carmo-Fonseca, M (2002). The intranuclear mobility of messenger RNA binding proteins is ATP dependent and temperature sensitive. *J Cell Biol* **159**:795–805.

Chubb, JR, Boyle, S, Perry, P, and Bickmore, WA (2002). Chromatin motion is constrained by association with nuclear compartments in human cells. *Curr Biol* **12**:439–445.

Cremer, T, and Cremer, C (2001). Chromosome territories, nuclear architecture and gene regulation in mammalian cells. *Nat Rev Genet* **2**:292–301.

Darzacq, X, Powrie, E, Gu, W, Singer, RH, and Zenklusen, D (2003). RNA asymmetric distribution and daughter/mother differentiation in yeast. *Curr Opin Microbiol* **6**:614–620.

Femino, AM, Fay, FS, Fogarty, K, and Singer, RH (1998). Visualization of single RNA transcripts *in situ*. *Science* **280**:585–590.

Femino, AM, Fogarty, K, Lifshitz, LM, Carrington, W, and Singer, RH (2003). Visualization of single molecules of mRNA *in situ*. *Methods Enzymol* **361**:245–304.

Forrest, KM, and Gavis, ER (2003). Live imaging of endogenous RNA reveals a diffusion and entrapment mechanism for nanos mRNA localization in *Drosophila*. *Curr Biol* **13**:1159–1168.

Fraefel, C, Bittermann, AG, Bueler, H, Heid, I, Bachi, T, and Ackermann, M (2004). Spatial and temporal organization of adeno-associated virus DNA replication in live cells. *J Virol* **78**:389–398.

Fusco, D, Accornero, N, Lavoie, B, Shenoy, SM, Blanchard, JM, Singer, RH, and Bertrand, E (2003). Single mRNA molecules demonstrate probabilistic movement in living mammalian cells. *Curr Biol* **13**:161–167.

Gavis, ER, and Lehmann, R (1992). Localization of nanos RNA controls embryonic polarity. *Cell* **71**:301–313.

Gerlich, D, Beaudouin, J, Kalbfuss, B, Daigle, N, Eils, R, and Ellenberg, J (2003). Global chromosome positions are transmitted through mitosis in mammalian cells. *Cell* **112**:751–764.

Gordon, GS, Sitnikov, D, Webb, CD, Teleman, A, Straight, A, Losick, R, Murray, AW, and Wright, A (1997). Chromosome and low copy plasmid segregation in E. coli: visual evidence for distinct mechanisms. *Cell* **90**:1113–1121.

Hediger, F, Neumann, FR, Van Houwe, G, Dubrana, K, and Gasser, SM (2002). Live imaging of telomeres: yKu and Sir proteins define redundant telomere-anchoring pathways in yeast. *Curr Biol* **12**:2076–2089.

Heun, P, Laroche, T, Shimada, K, Furrer, P, and Gasser, SM (2001). Chromosome dynamics in the yeast interphase nucleus. *Science* **294**:2181–2186.

Iborra, FJ, Pombo, A, Jackson, DA, and Cook, PR (1996). Active RNA polymerases are localized within discrete transcription 'factories' in human nuclei. *J Cell Sci* **109**(Pt 6):1427–1436.

Janicki, SM, Tsukamoto, T, Salghetti, SE, Tansey, WP, Sachidanandam, R, Prasanth, KV, Ried, T, Shav-Tal, Y, Bertrand, E, Singer, RH, and Spector, DL (2004). From silencing to gene expression: real-time analysis in single cells. *Cell* **116**:683–698.

Kimura, H, and Cook, PR (2001). Kinetics of core histones in living human cells: little exchange of H3 and H4 and some rapid exchange of H2B. *J Cell Biol* **153**:1341–1353.

Lever, MA, Th'ng, JP, Sun, X, and Hendzel, MJ (2000). Rapid exchange of histone H1.1 on chromatin in living human cells. *Nature* **408**:873–876.

Levsky, JM, Shenoy, SM, Pezo, RC, and Singer, RH (2002). Single-cell gene expression profiling. *Science* **297**:836–840.

Long, RM, Singer, RH, Meng, X, Gonzalez, I, Nasmyth, K, and Jansen, RP (1997). Mating type switching in yeast controlled by asymmetric localization of ASH1 mRNA. *Science* **277**:383–387.

Michaelis, C, Ciosk, R, and Nasmyth, K (1997). Cohesins: Chromosomal proteins that prevent premature separation of sister chromatids. *Cell* **91**:35–45.

Molenaar, C, Abdulle, A, Gena, A, Tanke, HJ, and Dirks, RW (2004). Poly(A)+ RNAs roam the cell nucleus and pass through speckle domains in transcriptionally active and inactive cells. *J Cell Biol* **165**:191–202.

Phair, RD, and Misteli, T (2000). High mobility of proteins in the mammalian cell nucleus. *Nature* **404**:604–609.

Phair, RD, Gorski, SA, and Misteli, T (2004). Measurement of dynamic protein binding to chromatin *in vivo*, using photobleaching microscopy. *Methods Enzymol* **375**:393–414.

Politz, JC, Browne, ES, Wolf, DE, and Pederson, T (1998). Intranuclear diffusion and hybridization state of oligonucleotides measured by fluorescence correlation spectroscopy in living cells. *Proc Natl Acad Sci USA* **95**:6043–6048.

Politz, JC, Tuft, RA, Pederson, T, and Singer, RH (1999). Movement of nuclear poly(A) RNA throughout the interchromatin space in living cells. *Curr Biol* **9**:285–291.

Robinett, CC, Straight, A, Li, G, Willhelm, C, Sudlow, G, Murray, A, and Belmont, AS (1996). *In vivo* localization of DNA sequences and visualization of large-scale chromatin organization using lac operator/repressor recognition. *J Cell Biol* **135**:1685–1700.

Savino, TM, Gebrane-Younes, J, De Mey, J, Sibarita, JB, and Hernandez-Verdun, D (2001). Nucleolar assembly of the rRNA processing machinery in living cells. *J Cell Biol* **153**:1097–1110.

Shav-Tal, Y, and Singer, RH (2005). RNA localization. *J Cell Sci* **118**:4077–4081.

Shav-Tal, Y, Darzacq, X, Shenoy, SM, Fusco, D, Janicki, SM, Spector, DL, and Singer, RH (2004a). Dynamics of single mRNPs in nuclei of living cells. *Science* **304**:1797–1800.

Shav-Tal, Y, Singer, RH, and Darzacq, X (2004b). Imaging gene expression in single living cells. *Nat Rev Mol Cell Biol* **5**:855–861.

Shav-Tal, Y, Blechman, J, Darzacq, X, Montagna, C, Dye, BT, Patton, JG, Singer, RH, and Zipori, D (2005). Dynamic sorting of nuclear components into distinct nucleolar caps during transcriptional inhibition. *Mol Biol Cell* **16**:2395–2413.

Shelby, RD, Hahn, KM, and Sullivan, KF (1996). Dynamic elastic behavior of alpha-satellite DNA domains visualized *in situ* in living human cells. *J Cell Biol* **135**:545–557.

Simpson, JC, Neubrand, VE, Wiemann, S, and Pepperkok, R (2001). Illuminating the human genome. *Histochem Cell Biol* **115**:23–29.

Singh, OP, Bjorkroth, B, Masich, S, Wieslander, L, and Daneholt, B (1999). The intranuclear movement of Balbiani ring premessenger ribonucleoprotein particles. *Exp Cell Res* **251**:135–146.

Snaar, SP, Verdijk, P, Tanke, HJ, and Dirks, RW (2002). Kinetics of HCMV immediate early mRNA expression in stably transfected fibroblasts. *J Cell Sci* **115**:321–328.

Straight, AF, Belmont, AS, Robinett, CC, and Murray, AW (1996). GFP tagging of budding yeast chromosomes reveals that protein–protein interactions can mediate sister chromatid cohesion. *Curr Biol* **6**:1599–1608.

Sugaya, K, Vigneron, M, and Cook, PR (2000). Mammalian cell lines expressing functional RNA polymerase II tagged with the green fluorescent protein. *J Cell Sci* **113**(Pt 15):2679–2683.

Takizawa, PA, Sil, A, Swedlow, JR, Herskowitz, I, and Vale, RD (1997). Actin-dependent localization of an RNA encoding a cell-fate determinant in yeast. *Nature* **389**:90–93.

Tautz, D (1988). Regulation of the *Drosophila* segmentation gene hunchback by two maternal morphogenetic centres. *Nature* **332**:281–284.

Theurkauf, WE, and Hazelrigg, TI (1998). *In vivo* analyses of cytoplasmic transport and cytoskeletal organization during *Drosophila* oogenesis: Characterization of a multi-step anterior localization pathway. *Development* **125**:3655–3666.

Tsien, RY (1998). The green fluorescent protein. *Annu Rev Biochem* **67**:509–544.

Tsukamoto, T, Hashiguchi, N, Janicki, SM, Tumbar, T, Belmont, AS, and Spector, DL (2000). Visualization of gene activity in living cells. *Nat Cell Biol* **2**:871–878.

Walter, J, Schermelleh, L, Cremer, M, Tashiro, S, and Cremer, T (2003). Chromosome order in HeLa cells changes during mitosis and early G1, but is stably maintained during subsequent interphase stages. *J Cell Biol* **160**:685–697.

Zachar, Z, Kramer, J, Mims, IP, and Bingham, PM (1993). Evidence for channeled diffusion of pre-mRNAs during nuclear RNA transport in metazoans. *J Cell Biol* **121**:729–742.

Zhang, J, Campbell, RE, Ting, AY, and Tsien, RY (2002). Creating new fluorescent probes for cell biology. *Nat Rev Mol Cell Biol* **3**:906–918.

Zink, D, and Cremer, T (1998). Cell nucleus: Chromosome dynamics in nuclei of living cells. *Curr Biol* **8**:R321–R324.

Zink, D, Cremer, T, Saffrich, R, Fischer, R, Trendelenburg, MF, Ansorge, W, and Stelzer, EH (1998). Structure and dynamics of human interphase chromosome territories *in vivo*. *Hum Genet* **102**:241–251.

CHAPTER 3

PROBING MEMBRANE TRANSPORT OF SINGLE LIVE CELLS USING SINGLE-MOLECULE DETECTION AND SINGLE-NANOPARTICLE ASSAY

XIAO-HONG NANCY XU, YUJUN SONG, AND PRAKASH NALLATHAMBY

3.1. INTRODUCTION

Every intracellular molecule possesses its specific cellular functions and plays a unique role in highly regulated cellular networks, such as gene expression, signal transduction, and membrane transport. Quantitative analysis of dynamic and kinetic parameters in the unitary processes of the cellular networks offers the possibility of depicting how these unitary processes are integrated into high-order functionalities. To this end, it is essential to detect and characterize dynamics behaviors and biochemical properties of individual molecules in single living cells in real time. Note that it is impossible to reliably measure the living cellular membrane transport kinetics using unhealthy or dead cells because the fluidity of phospholipid environment in the cell membrane is reduced or completely destroyed in dead cells.

Biomolecules—such as protein molecules, which play vital roles in cellular network—are remarkably complex and function as molecular machines. To study these elaborate molecular systems, conventional methods that study a particular function of bulk biomolecules of cells are quite often inappropriate. Since a multitude of complex processes in the living cell are continuously occurring at a molecular scale, ensemble methods are unable to depict the underlying mechanisms of cellular and subcellular function.

Single-molecule detection (SMD) offers the unique opportunity to investigate cellular and subcellular mechanisms, one molecule at a time, and is best suitable for probing living cellular pathway and function at the molecular level in real time. Note that bulk measurements cannot resolve the behavior of individual molecules in real time because bulk molecules of living cells are asynchronized. For example, each molecule in single living cells acts independently and hence is asynchronous. Unlike bulk measurement, SMD can distinguish the fluctuations and statistical distributions of dynamic and kinetic behaviors of individual molecules in real time, which is essential to investigate heterogeneity of living cellular mechanism and function. Thus, SMD offers the possibility of quantitative analysis of living cellular events (e.g., membrane transport and signaling transduction pathways) at the single-molecule level in real time.

New Frontiers in Ultrasensitive Bioanalysis. Edited by Xiao-Hong Nancy Xu
Copyright © 2007 John Wiley & Sons, Inc.

To detect single molecules, one will need to (i) design detection volume in which a single molecule is present statistically and (ii) ensure sufficient high signal-to-noise ratio (S/N), allowing single molecules to be detected individually. These are two primary criteria for successful detection of single molecules. To effectively enhance the S/N ratio, it is even more important to reduce the noise rather than enhancing the signal, because the signal of single molecules is always limited by chemical and physical properties of molecules of interest. For example, the fluorescence intensity of a given fluorophor is fundamentally limited by its quantum efficiency and photodecomposition lifetime. To reduce the background noise, it is crucial to decrease the detection volume, which will dramatically low the noise generated by the matrix, such as water, in which a single molecule is present. Other approaches that have been used to reduce the background noise include preventing the excitation beam from reaching detector and using time-gated detection method.

A variety of approaches have been reported for SMD in solution (Dovichi et al., 1984; Ambrose et al., 1999; Barnes et al., 1995; Nie and Zare, 1997; Yeung, 1997; Xu and Yeung, 1997; Weiss, 1999; Xu et al., 2001; Zander et al., 2002) and in living cells (Byassee et al., 2000; Xu et al., 2003; Sako and Yanagida, 2003; Tinnefeld and Sauer, 2005; Xie et al., 2006; Pramanik, 2004). In this chapter, we will primarily focus on describing the basic experimental design and approaches of using optical and fluorescence microscopy and spectroscopy to detect a single molecule (SM) in single living cells in real time, especially SMD of cellular membrane transport of single living cells. Chapter 2 in this book focuses on describing the study of cellular activity at single genes and single mRNAs level, while Chapter 1 centers on addressing how many measurements of a single molecule is needed to represent bulk measurement at single-molecule resolution.

Membrane transport plays a leading role in a wide spectrum of cellular and subcellular pathways, including multidrug resistance (MDR), cellular signaling, and cell–cell communication. Multi-substrate extrusion systems are being reported for both prokaryotes and eukaryotes. Microbial cells, however, have membrane transport mechanisms that are substantially distinguished from those in mammalian cells. For example, current evidence shows that endocytosis, pinocytosis, and exocytosis do not exist in microbial cells (prokaryotes), whereas these processes are widely observed in mammalian cells (eukaryotes).

Membrane permeability and active extrusion systems in many living organisms, such as bacteria, yeast, molds, and mammalian cells, play a crucial role in controlling the accumulation of specific intracellular substances, leading to the cellular self-defense mechanism that can resist incoming noxious compounds (Cole et al., 1992; Nakae, 1997; Poole, 2001; Ryan et al., 2001). For instance, the efflux pump of *Pseudomonas aeruginosa* (gram-negative bacteria) can extrude a broad spectrum of structurally and functionally diverse substrates, which lead to multidrug resistance (Nakae, 1997; Poole, 2001). It is very likely that membrane proteins are triggered by pump substrates to assemble membrane transporters optimizing for the extrusion of encountered substrates. Therefore, direct measurements of the sized change of efflux pump in real time at the molecular level are crucial to better understand such universal extrusion cellular defense mechanisms. Currently, the sizes of membrane transporters are determined solely by x-ray crystallography measurements, which are limited by the difficulties of crystallization of membrane proteins. In addition, the x-ray crystallography cannot be used for the study of real-time dynamics of membrane transport. Therefore, despite extensive studies over decades, the mechanisms of action and resistance of antimicrobial agents by efflux pump still remain unclear.

3.2. RESEARCH OVERVIEW

3.2.1. Detection Configurations

Fluorescence microscopy and spectroscopy has been used as a popular tool for SMD in solution and in living cells (Zander et al., 2002; Byassee et al., 2000; Xu et al., 2003a; Sako and Yanagida, 2003; Tinnefeld and Sauer, 2005; Xie et al., 2006; Pramanik, 2004; Ichinose and Sako, 2004). High-quantum-efficient fluorophors (e.g., R6G), fluorescent proteins (e.g., GFP), and fluorescence quantum dots have been used to tag the molecules of interest and have been used as a probe to trace the cellular pathways using fluorescence microscopy. The change of fluorescence intensity of a particular fluorophor (e.g., EtBr) based upon their interactions with intracellular DNA has been used to measure the cellular transport mechanism and measure transport rate of individual membrane transporters of living bacterial cells (Xu et al., 2003a).

Currently, five detection configurations of fluorescence microscopy and spectroscopy have been used for single molecule study of single live cells (Byassee et al., 2000; Xu et al., 2003a; Sako and Yanagida, 2003; Tinnefeld and Sauer, 2005; Xie et al., 2006). These configurations include total internal reflection fluorescence microscopy (TIR-FM) (Funatsu et al., 1995; Xu and Yeung, 1997), confocal fluorescence microscopy (CFM) (Nie and Zare, 1997; Byassee et al., 2000; Xie and Lu, 1999; Tinnefeld and Sauer, 2005), thin-layer total internal reflection microscopy (TL-TIR-FM) (Xu and Yeung, 1998; Xu et al., 2001; Xu et al., 2003a), fluorescence correlation spectroscopy (FCS) (Rigler, 1995; Xie and Lu, 1999; Vukojevic et al., 2005; Min et al., 2005), and spectrally resolved fluorescence lifetime imaging microscopy (SFLIM) (Tinnefeld and Sauer, 2005).

TIR-FM and TL-TIR-FM have been widely used for SMD in solution (Funatsu et al., 1995; Xu and Yeung, 1997), at liquid–solid interface (Xu and Yeung, 1998), and in living cells (Sako and Yanagida, 2003; Xu et al., 2003a).

Total internal reflection creates an evanescent field at the interface. The intensity of evanescent wave decays exponentially as it penetrates into the medium away from the interface, generating a penetration depth of about half-wavelength of excitation source (Axelrod et al., 1984), a submicrometer-depth detection zone ($d = 0.25$ μm), as illustrated in Figure 3.1A. By coupling with the intensified CCD camera that offers a pixel array detector with a single pixel resolution (area of each pixel = 0.2×0.2 μm^2 for a pixel of 0.2 μm), TIR-FM provides an array of 10 aL (10^{-17} L) detection volume ($V = $ area $\times d = 10^{-17}$ L) (Xu and Yeung, 1997). Thus, a solution of 1.7×10^{-7} M will allow a single molecule to statistically occupy a detection volume of 10 aL. Note that a single molecule in a solution of 10 aL equals $(1/6.02 \times 10^{23})/10^{-17}$ (mol/L) = 1.7×10^{-7} M.

TIR-FM has unique advantages over other detection configurations, offering low detection volume and high S/N by using an evanescent wave excitation, which deviates the excitation beam from entering the CCD detector and effectively reduces the background noise. Since the evanescent wave only penetrates into the interface at the depth of half-wavelength of excitation source, TIR-FM is most suitable to study the events occurring at the interface. However, as molecules move away, out of the evanescent-wave field, these molecules will not be able to visualize and study by TIR-FM (Figure 3.1A).

TIR-FM was first used as a detection configuration to directly visualize single myosin, individual ATP turnover, and movement of single kinesin molecules along microtubules in 1995–1996 (Funatsu et al., 1995). Myosin, ATP, and kinesin are labeled with a fluorescent

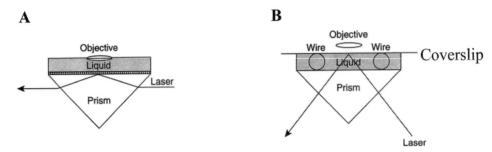

Figure 3.1. Schematic optical configurations of SMD: (A) total internal reflection (TIR) with evanescent wave excitation; (B) thin-layer total internal reflection (TL-TIR) fluorescence microscopy. The unfocused output of a purified Ar^+ laser beam is directed, below the critical angle, into a fused-silica right-angle prism located on the microscope stage. The excitation zone is imaged by a microscope through a 100× objective (numerical aperture = 1.3) immersed directly in the solution (A), or through a 0.08-mm cover slip without immersion oil (B). Fluorescence of single molecules is detected with ICCD mounted on top of a microscope. In part A, the estimated 0.15-μm thickness of the evanescent field defines a SMD volume of 10 aL (an attoliter equals 10^{-18} liter). In part B, the excitation depth is defined by placement of two 4-μm-thick tungsten wires on the prism surface. One drop of sample is sandwiched between the prism surface and a cover slip with the wires as a spacer. The laser penetrates the solution before being totally reflected at the cover slip surface, providing a lower background than epifluorescence arrangements. [Reprinted from *Science* **275**:1106–1109 (1997), with permission of American Association for the Advancement of Science.]

molecule (Cy3) and are detected at the liquid and coverglass surface. In the following years, TIR-FM has been used to study the diffusion of protein molecules trapped at the liquid–solid (fused silica) interface and to better understand the interaction of protein molecules with the fused silica surface (Xu and Yeung, 1998). This study demonstrates the possibility of investigating the chromatographic interactions at the liquid–solid interface in real time at the molecule level, offering a powerful tool to investigate the fundamental mechanism of partition and retention at the molecule level and opening up the new opportunity to design more effective separation means.

TIR-FM has also been created by carefully aligning the excitation beam via objective, so-called objective-mode TIR-FM (Tokunaga et al., 1997). This experimental design has been used to (a) study the turnover of individual ATP molecules labeled with fluorophor on the surface of a glass cover slide and (b) visualize individual Cy3-cAMP molecules on the surface of living cells sitting on glass slides (Sako and Yanagida, 2003). However, the objective-mode TIR-FM shares the same limitation of TIR-FM, which only permits the molecules at the TIR interface within evanescent-wave field to be investigated. Consequently, TIR-FM cannot be used to study the molecules in the bulk solution and track the intracellular pathway.

To overcome such limitation, TL-TIR-FM has been developed (Xu and Yeung, 1998; Xu et al., 2001; Xu et al., 2003a), as shown in Figure 3.1B. In this configuration, an evanescent wave is no longer used as an excitation source, but instead to deviate the excitation beam from reaching the ICCD detector and hence reduce the background. The thin-layer channels created using a micrometer spacer (e.g., 4-μm wire, or 6-μm-thick membrane) provide a detection volume that is about 20 or 30 times larger than that generated by evanescent-wave excitation, respectively. The molecules in the microchannel are illuminated with excitation beam all the time. Thus, these molecules can be continuously monitored in solution, allowing one to measure the diffusion distance and photodecomposition lifetime of single molecules in real time (Xu and Yeung, 1997). On the basis of random-walk theory, the diffusion distance is related to the mass and size of individual molecules (Bard and Faulkner, 1980). Thus, by measuring the diffusion distance of individual molecules, Xu and her co-workers have demonstrated the possibility of screening the different size (molecular weight) of individual tumor markers (e.g., PAS-complex, PSA-free) (Xu et al., 2001) and probing the membrane transport of single living cells using single-molecule diffusion microcopy (TL-TIR-FM) (Xu et al., 2003a).

Confocal fluorescence microscopy (CFM) interfaced with avalanche photodiodes (APDs) as detectors has also been widely used for SMD in aqueous solution (Nie and Zare, 1997) and has been demonstrated for SMD in living cells (Byassee et al., 2000). Using the confocal volume (10^{-15} L) as a detection volume—along with the optical design of a confocal microscope, a pinhole (50–100 μm in diameter), to reject the scattering light from out-of-focal planes and hence effectively reduce the background—CFM conveniently creates a SMD configuration. As confocal microscope became commercially available, CFM has widely been used for SMD in solution (Weiss, 1999), such as the study of photophysics of single fluorophors (Willets et al., 2005) and conformational dynamics of enzyme and DNA (Min et al., 2005; Xie and Lu, 1999). The depth of confocal volume is about 1 μm or less (Nie and Zare, 1997), which is about five times larger than evanescent excitation. Thus, it creates larger individual detection volume than those generated by TIR-FM. Nevertheless, it is still easy for molecules to move out of confocal volume, prohibiting one from tracing the same molecules over an extended period of time. In addition, confocal images are created by individual point measurements, offering low temporal resolution and eliminating the possibility of monitoring multiple molecules simultaneously.

Fluorescence correlation spectroscopy (FCS) and spectrally resolved fluorescence lifetime imaging microscopy (SFLIM) have been carried out using optical design, similar to those of CFM and on the basis of the same principle of using confocal volume as SM detection volume and using a pinhole to reject the background. However, the substantial modification of excitation and detection system has been applied in SFLIM to achieve the simultaneous measurement of spectra, intensity, and lifetime of individual molecules. FCS has been widely used in SMD (Rigler, 1995; Vukojevic et al., 2005) and has recently been demonstrated for the study of individual ligand–receptor interactions on live cells (Pramanik, 2004). Sauer and his co-workers recently introduced SFLIM and have used it for detecting single molecules (fluorescently labeled mRNA molecules) in living cells (Tinnefeld and Sauer, 2005).

3.2.2. Detection Probes

Fluorescent probes including high-quantum-yield fluorescent dyes and fluorescent proteins have been widely used for SMD in aqueous solution, as overviewed by Chapters 1 and 2 (Zander et al., 2002). In recent years, fluorescent probes, such as R6G and fluorescent proteins, have been successfully used to probe individual molecules in single living cells and depict cellular pathway (e.g., membrane transport and gene expression) in single living cells at the single-molecule level (Femino et al; 1998; Byassee et al., 2000; Xu et al., 2003a; Janicki et al., 2004; Yu et al., 2006), as illustrated in Chapter 2.

Fluorescence probes provide limited lifetime for tracing and quantitative analysis of single molecules in living cells, because of photodecomposition of fluorophor. The second disadvantage of fluorescence probes is an uneven labeled ratio of molecules of interest with fluorophor, which makes the quantitative analysis impossible. In addition, the excitation beam that is needed for lighting up the fluorescence probes for detection creates a certain degree of damage to the living cells. Furthermore, as fluorophores photodecompose, they may generate free radicals, such as superoxide ions, which are very active species and can react with any nearby cellular substances in live cells, causing significant damage to the interior of cells and lead to cell death because cells can only tolerate a limited amount of free radicals (Tsien and Waggoner, 1995).

Single green fluorescence protein (GFP) molecules immobilized in polyacrylamide gels were first detected by Dickson and coworkers, showing on-and-off blinking behavior (Dickson, 1997). Green fluorescence protein and its derivatives have become popular fluorescent probes for cellular imaging, in which the molecules of interest are genetically encoded with fluorescent proteins as probes (Sako et al., 2000; Iino et al., 2001; Singer, 2004; Tinnefeld and Sauer, 2005; Yu et al., 2006). Chapter 2 illustrates the examples of using fluorescent proteins to dissecting the cellular activity in living cells at the single gene level. Recently, Xie and his co-workers have used a similar approach to determine gene expression in individual living cells at single protein resolution (Yu et al., 2006). Xie and his co-workers have successfully detected the production of single-fusion protein molecules of YFP in living bacterial cells using an intense pulse laser of 300 μs as an excitation source to capture the sharp images of fast motion of individual protein molecules before the molecules diffuse out of optical detection limit (Yu et al., 2006).

Fluorescent proteins are bright and relatively biocompatible and possess a well-defined labeled ratio with protein molecules of interest. Thus, fluorescence proteins are a superior choice of fluorescent probes for studying the dynamics of individual proteins. However, fluorescence proteins still suffer from photodecomposition and cannot be used for the study of dynamics of individual molecules for an extended period of time (Harms et al., 2001). The blinking (on-and-off) phenomena of individual fluorescence protein molecules (e.g., GFP) also prevent fluorescence proteins from being used to continuously track the kinetics of individual molecules. The advent of the relatively stable GFP makes it possible to fluorescently label proteins of interest in living cells with no need of protein purification (Tsien, 1998; Haraguchi, 2002). Nevertheless, utilization of GFP for single-molecule detection in living cells is still obstructed by its low detection ratio with respect to the flavinoid fluorescence, so-called auto-fluorescence (Harms et al., 2001). Therefore, brighter photostable probes with high detection ratio are still highly sought for single-molecule analysis of single live cells.

Recent development in nanoscience and nanotechnology provides a new kind of optical nanoprobes for SMD in single live cells, such as semiconductor quantum dots (QDs) and noble metal nanoparticles (Bruchez et al., 1998; Chan and Nie, 1998; Xu et al., 2002, 2004; Kyriacou et al., 2004; Xu and Patel, 2005; Agrawal et al., 2006). Quantum dots have unique optical properties in comparison with fluorescence dyes and proteins, such as tunable narrow emission spectrum, broad excitation spectrum, high photostability, and long fluorescence lifetime. The functional QDs are water-soluble and have been used as tags to conjugate with biomolecules of interest for ultrasensitive analysis in living cells, as described in Chapter 4. These features make them invaluable probes for the study of dynamics events in living cells at a single-molecule level for an extended period of time (Dahan et al., 2003). Nevertheless, QD still suffers from a certain degree of photodecomposition. Furthermore, it remains a challenge to prevent intracellular QDs from aggregation and to generate a uniform labeled ratio of QDs with molecules of interest, which hinders quantitative analysis of molecules of interest using QDs as probes. Moreover, the biocompatibility of QDs still remains under investigation despite the fact that preliminary studies show the insignificant cytotoxicity of low concentration of QDs (Derfus et al., 2004; Gao et al., 2004). More thorough and extensive studies are needed to address this issue.

Noble metal nanoparticles (Ag, Au, and their alloys) have unique optical properties, such as surface plasmon resonance, showing the dependence of optical properties on its size, shape, surrounding environment, and dielectric constant of the embedding medium (Mie, 1908; Bohren and Huffman, 1983; Kreibig and Vollmer, 1995; Mulvaney, 1996). Recent research demonstrates the feasibility of (a) design and preparation of monodisperse desired size and shape of noble metal nanoparticles (Haes and Van Duyne, 2002) and (b) using intrinsic optical properties of the nanoparticles for imaging single living cells in real time with sub-100-nm spatial resolution and millisecond to nanosecond time resolution (Xu et al., 2002, 2004; Kyriacou et al., 2004). The quantum yield of Rayleigh scattering of 2-nm Ag nanoparticles is about 10^4 times higher than that of a single fluorescent dye molecule. The scattering intensity of noble metal nanoparticles is proportional to the volume of nanoparticles (Mie, 1908; Bohren and Huffman, 1983; Kreibig and Vollmer, 1995; Mulvaney, 1996). Thus, these nanoparticles are extremely bright and can be directly observed using dark-field microscopy equipped with CCD camera, digital camera, or bare eyes. Unlike fluorescent dyes, fluorescent proteins, and quantum dots, these noble metal nanoparticles do not suffer photodecomposition and can be used as a probe to continuously monitor dynamic events in living cells for an extended period of time. Furthermore, the optical properties (color) and localized surface plasmon resonance (LSPR) spectra of nanoparticles show size-dependence (Mie, 1908; Bohren and Huffman, 1983; Kreibig and Vollmer, 1995). Thus, one can use the color (LSPR spectra) index of these multicolor nanoparticles as a nanometer-size index to directly measure membrane transports of nanoparticles and to determine the change of membrane pores at the nanometer scale in real time (Xu et al., 2002, 2004; Kyriacou et al., 2004).

A distinguished advantage of noble metal nanoparticles over QDs is non-photodecomposition. Nevertheless, noble metal nanoparticles share the similar drawbacks of QDs for single-molecule analysis of single living cells. For example, it is still impossible to effectively control the labeled ratio of nanoparticles with the molecules of interest, such as achieving one-to-one labeled ratio. In general, individual nanoparticles tend to bind to several molecules simultaneously. This problem can be overcome by reducing the size of nanoparticles. Thus, the surface of an individual nanoparticle can only accommodate a single molecule of interest. However, as the size of nanoparticles decreases, the surface area of nanoparticles will increase. Thus, the nanoparticles may become even more unstable and subject to the aggregation. In addition, the scattering intensity of nanoparticles will decrease because the scattering intensity of individual nanoparticles is proportional to its size.

Like QDs, the cytotoxicity and genotoxicity of noble metal nanoparticles awaits further study (Xu et al., 2002). The preliminary study has showed that the cytotoxicity of silver nanoparticles to bacterial cells (e.g., *P. aeruginosa*) highly depends on nanoparticle concentration. At low nanoparticle concentration (< 5 pM), the cells appear to grow at the same rate as those in the absence of nanoparticles, and no significant cytotoxicity is observed. As nanoparticle concentration increases, the aggregation of nanoparticles on the cell membrane at high concentration leads to cell death (Xu et al., 2002). This problem may be able to overcome by carefully designing of functional group attached onto nanoparticle surface, which will prevent the aggregation of nanoparticles and make nanoparticles more biocompatible. The research of this kind opens up the entire new research area of the study of cytotoxicity and genotoxicity of nanomaterials and offers the possibility of rational design of biocompatible nanomaterials for a wide spectrum of biological applications.

3.3. RESEARCH HIGHLIGHTS

Single-molecule detection and single-nanoparticle tracking have been widely used to trace the diffusion of biomolecules in cellular membrane and intracellular medium, aiming to understand its dynamics associated with biological function. For example, studies of SM diffusion in phospholipid membrane (Schmidt et al., 1996) and free solution (Keller et al., 1996; Nie et al., 1994; Xu and Yeung, 1997; Xu et al., 2001) have been demonstrated. Detection of single molecules labeled with fluorophors in single living cells and on cellular membrane (Byassee et al., 2000; Sako et al., 2000; Xu et al., 2003a; Sako and Yanagida, 2003; Pramanik, 2004; Tinnefeld and Sauer, 2005; Xie et al., 2006), and single-molecule analysis of chemotactics signaling in living cells (Ueda et al., 2001) have also been reported.

Here, we will (a) focus on the topics that are directly related to probing individual molecules transporting through living cellular membrane and (b) select a research project that has been carried out in Xu research group to illustrate the experimental approach of SMD and its potential application of better understanding the transport mechanism of living cellular pathway.

Direct monitoring of influx and efflux of a single molecule through living cellular membrane remains enormously challenge. The key challenges of such an endeavor include (i) how to detect a single molecule in and out of a living cell and distinguish the difference between them; (ii) how to develop a sufficient rapid mean to follow the molecule while it is passing through the living cellular membrane, and (iii) how to select the biocompatible molecules that will not perturb the cellular function.

3.3.1. Probing Efflux Pump Machinery of Single Living Bacterial Cells

Our research group has demonstrated the possibility of real-time measurement of membrane pump machinery of single live cells at the single-molecule level (Xu et al., 2003a). The MexAB-OprM membrane pump in WT of *P. aeruginosa* is used as a working model, and the pump efficiency of single EtBr molecules transported by WT and its mutants is investigated using single-molecule fluorescence microscopy and spectroscopy.

Pseudomonas aeruginosa is a ubiquitous gram-negative bacterium and has emerged to be the leading cause of nosocomial infections in cancer, transplantation, burn, and cystic fibrosis patients. These infections are impossible to eradicate in part because of the bacterial intrinsic resistance to a wide spectrum of structurally and functionally unrelated antibiotics (Ryan et al., 2001; Poole 2001; Nakae 1997; Ma et al., 1994). Several efflux systems including MexAB-OprM, MexCD-OprJ, MexEF-OprN, and MexXY-OprM have been reported in *P. aeruginosa* (Ma et al., 1994; Masuda et al., 2000; Maseda et al., 2000; Morshed et al., 1995). The MexAB-OprM pump is the major efflux pump in wild-type (WT) cells. This pump consists of two inner membrane proteins (MexA and MexB) and one outer membrane protein (OprM) (Maseda et al., 2000; Morshed et al., 1995). The MexB protein consists of 1046 amino acid residues and is assumed to extrude the xenobiotics by utilizing the proton motive force as the energy source (Ma et al., 1994; Masuda et al., 2000). Studies show that the inner-membrane subunits appear to specify the substrates to be transported (Maseda et al., 2000; Morshed et al., 1995). Several models have been proposed in an attempt to describe the possible mechanism of the pump (Maseda et al., 2000; Morshed et al., 1995; Lee et al., 2000; Germ et al., 1999; Nakae, 1995; Lomovskaya et al., 2001). Time-course fluorescence spectroscopy has been used as a popular tool for real-time monitoring of the substrates accumulated in bulk bacterial cells (Mortimer and Piddock, 1991). Such bulk measurements have significantly advanced the understanding of efflux machinery in both prokaryotes and eukaryotes (Ryan et al., 2001; Mortimer and Piddock, 1991). However, these bulk measurements show the accumulated kinetics of substrates in the cells and represent the average behavior of a massive number of pumps in cells. SMD allows one to look beyond the ensemble average and provides the new opportunity to study the pump machinery at the molecular level. Thus, it is essential to apply SMD for the study of influx (passive diffusion into the cell) and efflux kinetics of substrates at the molecular level.

EtBr emits a weak fluorescence in aqueous solution (outside the cells) and becomes strongly fluorescence as it enters the cells and intercalates with DNA. Thus, EtBr is especially suitable for tracing the transport of molecules in and out of living cells (Ocaktan et al., 1997; Yoneyama et al., 2000; Kyriacou et al., 2002; Xu et al., 2003a,b).

Time courses of fluorescence intensity of 40 μM EtBr in the cell suspension of WT, nalB-1, and ΔABM (Figure 3.2A) show that the mutant (nalB-1) of overexpress of MexAB-OprM accumulates EtBr much more slowly than WT and ΔABM (mutant devoid of MexAB-OprM), demonstrating efflux function of MexAB-OprM in *P. aeruginosa*. The fluorescence intensity of EtBr in the buffer solution remains almost unchanged over time, showing the photostability of EtBr. Taken together, the result suggests that the change of fluorescence intensity in cell suspension could be used to follow the accumulation of EtBr in cells. The result also clearly illustrates that time courses of fluorescence intensity of EtBr depends on substrate (EtBr) concentrations. As substrate concentration decreases, the distinguished extrusion kinetics in three strains decreases.

In 40 μM EtBr, the fluorescence intensity of EtBr in all three strains increases with time, indicating the accumulation of EtBr in cells (Figure 3.2A). The fluorescence intensity of ΔABM (mutant devoid of MexAB-OprM) and nalB-1 (mutant with over-expression pump) increases with time most and least rapidly, respectively. This result suggests that nalB-1 extrudes EtBr out of cells much more effectively than WT and ΔABM because of the overexpression of pump proteins. In contrast, ΔABM is unable to extrude EtBr using MexAB-OprM and thus accumulates EtBr more rapidly than WT and nalB-1. Unlike those observed in 40 μM EtBr, time courses of the fluorescence intensity in 0.010 μM EtBr for all three strains shows the similarity that the fluorescence intensity remains almost unchanged over time (Figure 3.2B). This result suggests that no detectable change of fluorescence intensity is observed by fluorescence spectrometer. In order to probe the pump machinery in the low substrate concentration, single-molecule fluorescence microscopy and spectroscopy (TL-TIR) has been developed and applied to measure the influx and efflux kinetics of EtBr in live cells, as shown in Figure 3.3.

3.3.1.1. Sizes of Single Living Cells as SM Detection Volume

To detect single-substrate molecules (EtBr), sizes of single live cells are used as individual detection volumes and thin-layer total-internal reflection (TL-TIR) fluorescence microscopy is used to reduce the background noise of excitation beam and hence increase the signal-to-noise ratio (Xu et al., 2001, 2003a). High-resolution optical images of WT of *P. aeruginosa* acquired using optical dark-field microscopy show that the sizes of optical images of cells are around 2 μm in length and 0.5 μm in width and in height (Figure 3.4A), which are similar to the sizes of ΔABM and nalB-1 and have been confirmed by transmission electron microscope (TEM). Thus, the detection volume of the single cell is 5×10^{-16} L. Cell concentration

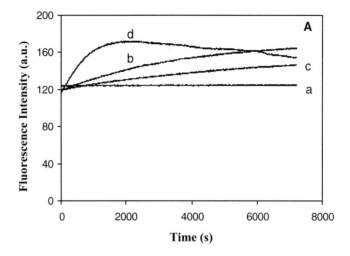

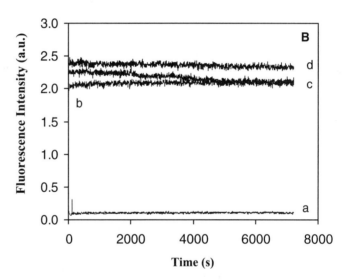

Figure 3.2. Accumulation kinetics of EtBr in intact cells of *P. aeruginosa* shows concentration and mutant dependence: time courses of fluorescence intensity of EtBr at 590 nm (a) in the PBS buffer and in the presence of intact cells of *P. aeruginosa*, (b) WT, (c) nalB-1, and (d) ΔABM, acquired directly from a 3.0 mL cell solution containing (A) 40 μM and (B) 0.010 μM (10 nM) EtBr using fluorescence spectroscopy with a time-drive mode at a 3-second data acquisition interval and 488-nm excitation. [Reprinted from *Biochem Biophys Res Commun* **305**:941–949, with permission of Academic Press.]

(20 × dilution of $A_{600\ nm} = 0.1$) and 0.2–0.4 nM EtBr are used to ensure statistically a single cell with about a single EtBr molecule or less.

The location of a single cell in the solution is identified using dark-field optical microscopy. The subframe that contains a single cell (solid-line square) (Figure 3.4A) is selected, and then sequence fluorescence images are acquired using SM fluorescence microscopy and spectroscopy. Integrated fluorescence intensity of a detection zone in the part of the image including a single cell (solid-line square) and the other part of image excluding

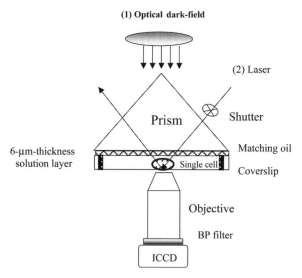

Figure 3.3. Schematic of single-molecule microscopy and spectroscopy for study of membrane transport of single living cells: (1) dark-field optical microscopy for imaging single living bacterial cells, *P. aeruginosa*, and (2) thin-layer total-internal reflection (TL-TIR) fluorescence microscopy for detection of single molecules in and out of single living cells. Dark-field microcopy uses halogen lamp (100 W) as an illumination source, whereas the TL-TIR microscopy uses the unfocused filtered Ar^+ laser beam at 514.5 nm as the excitation source. The laser penetrates the solution before being totally reflected at the interface of the coverslip and air, providing a lower background than epifluorescence arrangement. A mechanical shutter placed in front of the laser beam is synchronized with ICCD shutter to limit exposure of the sample to the laser and to prevent the sample from photobleached. Exposure time of ICCD and shuttle delay time are monitored using an oscilloscope to ensure the precise measurement of temporal resolution. [Reprinted from *Biochem Biophys Res Commun* **305**:941–949, with permission of Academic Press.]

the cell (solution only, dash-line square) is measured in the same solution using the same laser power and recorded in the same frame. The former serves as a signal for monitoring of the influx and efflux of a single EtBr molecule by a single living cell. The latter offers the real-time measurement of a single EtBr molecule in the solution that serves as a reference and background for the signal measurement. The subtracted fluorescence intensity of the latter from the former represents a single EtBr molecule in or out of a single living cell, as illustrated in Figure 3.4B. Using this novel approach, the fluctuation of ICCD dark noise, laser beam, buffer solution, and photodecomposition of EtBr molecules is reduced to the minimum because the fluorescence intensity of EtBr in and out of the cell is measured simultaneously in the same frame from the same solution at the same laser power. This allows one to distinguish single molecules in and out of single living cells and measure influx and efflux kinetics at the molecular level.

3.3.1.2. Membrane Pump Efficiency of WT

Two hundred sequence images of WT cells in 0.2 nM EtBr are recorded using ICCD with 70-ms temporal resolution and 99.84 ms exposure time. Seventeen sets of such sequence images recorded over 80 min show snap shots of influx and efflux of a single EtBr by a single WT cell. An area (13 × 13 pixel) that contains an individual cell (solid-line square)

A.

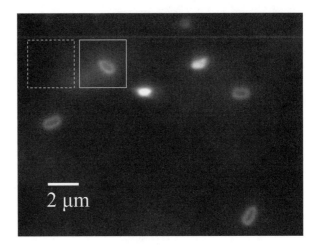

B.

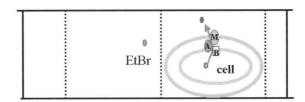

Figure 3.4. Illustration of using a single living cell as SM detection volume: (A) Sub-frame high-resolution optical image of single living bacterial cells, *P. aeruginosa* (WT), acquired using dark-field microscopy, showing the size of single cells. Solid-line and dash-line squares represent the selected pixel area for integrated fluorescence intensity analysis of EtBr in a single cell and in solution, respectively. (B) Schematic illustration of the selected detection volume containing a single EtBr molecule in the solution (left) and with a single live cell (right) in the microchannel. EtBr enters the cell using passive diffusion and is extruded out of the cell by MexAB-OprM. [Reprinted from *Biochem Biophys Res Commun* **305**:941–949, with permission of Academic Press.] See insert for color representation of these figures.

and an equal sized pixel area that contains no cells (solution only) (dash-line square) in the same frame (Figure 3.4A) are selected. The integrated fluorescence intensity of area (dash-line square) where no cell is presence (EtBr in solution) is subtracted from that of the cell is presence (solid-line square) (EtBr in a single cell). Representative plots of subtracted integrated fluorescence intensity versus time Figure 3.5A clearly demonstrate that the change of fluorescence intensity of a single EtBr molecule with a single WT cell is above the fluctuation of background emission of a single cell and ICCD noise (baseline). Unlike eucaryote cells, the auto-fluorescence of the bacterial cells is not observed. This might be attributable to its tiny size ($2 \times 0.5 \times 0.5$ μm).

As EtBr molecules enter the cell and intercalate with DNA, the fluorescence intensity of a single EtBr molecule in the cell increases over the intensity of EtBr in solution. Thus, the subtracted integrated intensity becomes positive. In contrast, as EtBr molecules are transported out of the cell, the fluorescence intensity of a single cell decreases below the intensity of EtBr in solution. Thus, the subtracted integrated intensity becomes negative. EtBr concentration in solution is higher than in a live cell. Thus, the molecule must be extruded out

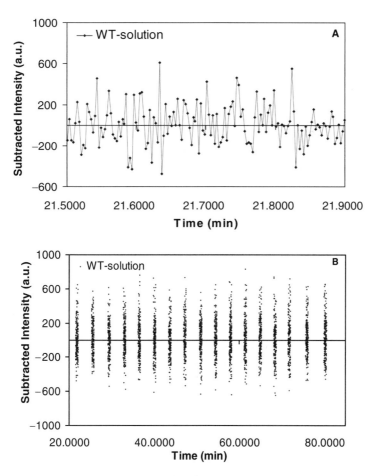

Figure 3.5. SMD of membrane pump efficiency of single living cells. Representative plots of (A) subtracted integrated fluorescence intensity of a single EtBr molecule in solution (dash-line square) from that in a single live cell (WT) (solid-line square) versus time (min) at 21.5000–21.9000 min; (B) seventeen sets of the 200 consecutive images are taken during 80 min. A zoom-in display of (B) at 21.5000–21.9000 min is shown in (A). Every set of 200 sequence images is taken with a 70-ms temporal resolution and 99.84-ms CCD exposure time. [Reprinted from *Biochem Biophys Res Commun* 305:941–949, with permission of Academic Press.]

of the cell through the active efflux mechanism. Taken together, the subtracted fluorescence intensity is positive or negative, representing the location of a single EtBr molecule, in or out of a single living cell, respectively. The frequency of change sign of the subtracted fluorescence intensity from negative to positive or from positive to negative represents the rates of influx and efflux of a single EtBr molecule by a single living cell. Seventeen sets of 200 sequence images recorded in Figure 3.5B demonstrates the stochastic influx and efflux rate of a single EtBr molecule with the average at (2.86 ± 0.12) s^{-1}. The average of influx and efflux rate remains nearly constant over time, which is in an excellent agreement with the bulk measurement of accumulated EtBr in WT cells (Figure 3.2B).

The study of the dependence of efflux rate on temporal resolution, EtBr (substrate) concentration, and mutants indicates that the efflux rate of two mutants (nalB-1, ΔABM) in the presence of 0.2–0.4 nM EtBr at the temporal resolution of 60 ms and 70 ms is the same. The results suggest that the efflux rate is at least 10 ms, indicating that the influx and efflux rate at the single molecule level is indeed measured because the pump efficiency is independent on substrate concentration only at the single-molecule level and is dependent on substrate concentration at the multimolecule level (Kyriacou et al., 2002; Xu et al., 2003a,b). Furthermore, the result suggests that the influx and efflux rates of a single-substrate molecule (EtBr) are independent on number of pumps expressed per cell. This is consistent with the unique feature of single-molecule measurements. The observation of influx and efflux of EtBr by ΔABM suggests that other pump proteins (e.g., MexCD-OprJ) with the low expression level in ΔABM may be responsible, which is another distinguished feature of SMD, allowing rare phenomena to be observed at the SM level.

3.3.2. Sizing the Membrane Transport of Single Living Cells in Real Time

While fluorophors (e.g., EtBr) can be used to trace the influx and efflux rate of membrane pump, it does not provide the size information of membrane pores. It is very likely that membrane proteins can specifically recognize an array of structurally unrelated substrates (e.g., chemotoxics) and assemble membrane transporters optimized for the extrusion of specific encountered substrates. Such fascinating smart sensing and transport mechanisms occur at the nanoscale regime. Thus, studies of the mechanism and assembly of the efflux pump will offer new insights into the function of efflux pump and will provide new knowledge that is essential for the design of self-assembly of smart molecular pumps.

Currently, the sizes of membrane transporters are determined solely by x-ray crystallography measurements, which are limited by the difficulties of crystallization of membrane proteins. In addition, x-ray crystallography cannot be used to study real-time dynamics of self-assembly of pump proteins in living cells (Tate et al., 2001). Despite extensive study over decades (Poole et al., 1993; Nakae, 1995; Ryan et al., 2001; Li and Nikaido, 2004), the structure, mechanisms, and function of extrusion transport remain unclear.

To address some of these questions, our research group has developed a new tool that uses silver (Ag) nanoparticles as nanometer probes to determine the sizes of substrates that can be transported through the membrane of living microbial cells and to measure accumulation kinetics of the substrates in real time at single-cell resolution (Xu et al., 2002, 2004). The challenges of such a study include: (i) how to overcome the rapid motion of tiny bacterial cells with size of $2 \times 0.5 \times 0.5$ μm in suspension so that the living cells can be confined and continuously monitored in suspension for hours; (ii) how to simultaneously monitor a group of individual cells so that one can obtain statistical information of bulk cells at the single-cell resolution; and (iii) how to develop a new imaging tool that can measure the nanometer probes moving in and out of living cells for real-time monitoring change of membrane permeability and efflux kinetics at the nanometer (nm) resolution.

Our group has overcome these challenges by developing and utilizing a microchannel system to confine living bacterial cells in suspension with no need to immobilize them, which allows living bacterial cells to be continuously and simultaneously monitored and imaged at the single-cell resolution for hours using a CCD camera coupled with dark-field optical microscope (Kyriacou et al., 2002; Xu et al., 2003; Xu et al., 2004). Furthermore, Ag nanoparticles as nanometer probes are developed to directly image the changes of membrane pore sizes and permeability at the nanometer (nm) and millisecond (ms) resolution using optical microscopy (Xu et al., 2002; Kyriacou et al., 2004; Xu et al., 2004).

3.3.2.1. Surface Plasmon Resonance of Single Silver Nanoparticles

Optical properties, such as localized surface plasmon resonance spectra (LSPRS) of Ag nanoparticles, depend on size and shape of nanoparticles and dielectric constant of its embedded medium (Mie, 1908; Bohren and Huffman, 1983; Kreibig and Vollmer, 1995; Mulvaney, 1996; Lamprecht, 2000; Haynes and Van Duyne, 2001).

Unlike bulk material, individual nanoparticles have a much higher surface-to-volume ratio. The surface properties contribute significantly to optical properties of nanoparticles. Unlike the bulk plasmon, the surface plasmon of nanoparticles can be directly excited by propagating light waves (electromagnetic waves), leading to the selective absorption and scattering of particular wavelength of light.

When noble metal nanoparticles are illuminated by light (electromagnetic wave), due to the small size of nanoparticles, the electromagnetic field generates polarized charges on the nanoparticle surface, which creates a linear restoring force, leading to subsequent charges on the surface of nanoparticles (Raether, 1988; Bohren and Huffman, 1983; Lamprecht, 2000). Thus, the conduct electrons in a nanoparticle act as an oscillator system, leading to a resonance behavior of the electron plasmon oscillation, as illustrated in Figure 3.6. The frequency of oscillation (surface plasmon resonance spectra) depends upon the energy (wavelength) of light and geometry of nanoparticles. The phenomena of LSPR of nanoparticles have been described by Mie theory (Mie, 1908). With an assumption of quasi-static regime, nanoparticle radius $(R) \ll \lambda$ (wavelength of the light), the electromagnetic excitation field is assumed to be nearly constant for each location of the entire volume of a spherical nanoprticle. With such an assumption, the polarizability of nanoparticles shows the dependence of shape, aspect ratio and frequency-dependent dielectric function of nanoparticles, and dielectric constant of embedding medium of nanoparticles, as described in the following equation (Mie, 1908; Bohren and Huffman, 1983, Kreibig and Vollmer, 1995; Lamprecht, 2000):

$$\alpha_i(\omega) = \varepsilon_0 V \frac{\varepsilon(\omega) - \varepsilon_m}{\varepsilon_m + [\varepsilon(\omega) - \varepsilon_m]L_i},$$

where $\alpha_i(\omega)$, ω, L_i, V, ε_0, ε_m, and $\varepsilon(\omega)$ represent frequency-dependent electric polarizability, frequency, a geometrical depolarization factor (dependent on the shape and aspect ratio of nanoparticle), volume of a nanoparticle, dielectric constant of vacuum and of the embedding medium, and frequency-dependent dielectric function of the nanoparticle.

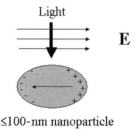

≤100-nm nanoparticle

Figure 3.6. Illustration of interaction of light (electromagnetic wave) with a noble metal nanoparticle, creating subsequent charges on the surface of a nanoparticle and thereby leading to a resonance behavior of the electron plasmon oscillation, a phenomenon known as localized surface plasmon resonance (LSPR) of a nanoparticle.

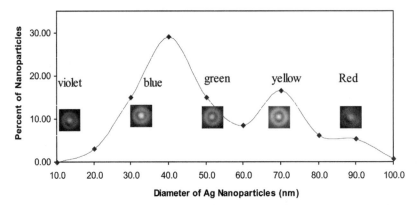

Figure 3.7. Dependence of optical properties (LSPRS, color) of nanoparticles on its size. Correlation of color index and size index of Ag nanoparticles is determined using cumulative histograms of the number of Ag nanoparticles versus nanoparticle sizes. The color index of Ag nanoparticles obtained from 100 nanoparticles determined by the optical microscope and spectroscopy is compared with the size index of the nanoparticles measured by TEM, showing that color index of Ag nanoparticles with violet, blue, green, and red is correlated with the size index of 30 ± 10, 50 ± 10, 70 ± 10, and 90 ± 10 nm, respectively. [Reprinted from *Biochemistry* **43**:140–147, with permission of American Chemical Society.] See insert for color representation of this figure.

Therefore, the colors (LSPRS) of Ag nanoparticles are correlated with the sizes of nanoparticles, while the shapes of nanoparticles and their embedded medium remain unchanged (Figure 3.7) (Xu et al., 2002, 2004; Kyriacou et al., 2004). Such unique size-dependent optical properties allow us to use the color index of nanoparticles (violet, blue, green, red) as the nanometer size index probes (30 ± 10, 50 ± 10, 70 ± 10, 90 ± 10 nm) for real-time sizing the change of living cellular membrane permeability and porosity at the nanometer scale.

Individual Ag nanoparticles are extremely bright under dark-field optical microscope and can be directly imaged by a digital or CCD camera through the dark-field microscope. Unlike fluorescence dyes, these nanoparticles do not suffer photodecomposition and can be used as a probe to continuously monitor dynamics and kinetics of membrane transport in living cells for an extended period of time (hours). In addition, these nanoparticles can be used as nanometer probes to determine the sizes of substrates that are transporting in and out of the living cellular membrane at the nanometer scale in real time.

As described above, the quantum yield of Rayleigh scattering of 10 nm of Ag nanoparticles is about six orders of magnitude higher than the fluorescent quantum yield of a single fluorescence dye molecule (R6G). The scattering intensity of Ag nanoparticles increases proportionally as the volume of nanoparticles increases. Unlike fluorescence dyes, Ag nanoparticles resist photobleaching. Therefore, these Ag nanoparticles are extremely bright under the dark-field microscope and can be used for real-time monitoring of membrane transport in living cells for an extended time.

These new tools have been used to study the function of aztreonam (AZT) at the nanometer scale and single living bacteria (*P. aeruginosa*) resolution (Kyriacou et al., 2004). AZT is a monocyclic β-lactam antibiotic. It is well known that the mode of action of AZT is the disruption of the cell wall (Greenwood, 1997). Thus, AZT is used to validate Ag nanoparticles as nanometer assays for real-time sizing membrane porosity and permeability as AZT disrupts the cell walls (Kyriacou et al., 2004). The complete disruption of the cell wall by a high concentration of AZT (31.3 µg/mL) causes the overflowing of intracellular nanoparticles, leading to the aggregation of intracellular Ag nanoparticles.

The validated nanoparticle assay is then used to study the membrane transport mechanisms of living microbial cells (*P. aeruginosa*) in the absence of antibiotics and in the presence of an antibiotic (chloramphenicol), which is neither a β-lactam nor aminoglycoside antibiotic. The primary target of chloramphenicol in microbial cells is to inhibit the ribosomal peptidyl transferase rather than disrupt cell wall (Greenwood, 1997). The dependence of membrane transport upon the dosages of chloramphenicol is investigated using single-nanoparticle optics and single-living-cell imaging. An efflux pump (MexAB-OprM) of living microbial cells (*P. aeruginosa*) is selected as a working model for probing the role of the efflux pump in controlling the accumulation of substrates (nanoparticles, EtBr, chloramphenicol) in *P. aeruginosa*.

As these nanoparticles transport through the living microbial cells, their sizes have been measured at the nanometer scale in real-time using single-nanoparticle optics, demonstrating that single-Ag-nanoparticle optics and single-living-cell imaging can be used to monitor the modes of action of antibiotics in real time and probe the new function of antibiotics at single-living-cell resolution with temporal (ms) and size (nm) information (Xu et al., 2004). TEM has also been used to determine the size and location of the intracellular Ag nanoparticles at subnanometer resolution, confirming the size measured using single-nanoparticle optics. Such new tools offer an important new opportunity to advance our understanding of membrane transport kinetics and MDR mechanism in living cells in real time.

3.3.2.2. Real-Time Sizing Membrane Permeability Using Single-Nanoparticle Optics

Silver nanoparticles with diameters 10–100 nm are prepared by reducing $AgNO_3$ aqueous solution with freshly prepared sodium citrate aqueous solution as described previously (Lee and Meisel, 1982; Kyriacou et al., 2004; Xu et al., 2004). Ag nanoparticle concentration is determined by dividing the moles of nanoparticles by the volume of the solution. The moles of Ag nanoparticles are determined by dividing the number of Ag nanoparticles with Avogadro's constant (6.02×10^{23}). These Ag nanoparticles are characterized using UV-vis spectroscopy, dark-field optical microscopy and spectroscopy (LSPRS), and TEM (Kyriacou et al., 2004; Xu et al., 2004).

The color distribution of single nanoparticles in 0.4 nM solution measured using dark-field microscopy and spectroscopy is used to compare with the size distribution of single nanoparticles from the same solution determined by TEM. The results indicate that the color index of violet, blue, green, and red is correlated with size index of 30 ± 10, 50 ± 10, 70 ± 10, and 90 ± 10 nm, respectively, as shown in Figure 3.7 (Kyriacou et al., 2004). The solution contains approximately 23% of violet (30 ± 10 nm), 53% of blue (50 ± 10 nm), 16% of green (70 ± 10 nm), and 8% of red (90 ± 10 nm) nanoparticles.

These multicolored nanoparticles are then used as nanometer-sized probes to directly measure the sizes of substrates that can transport through living microbial membrane, aiming to determine the change of membrane permeability and pore sizes in real time. To minimize the possible effects of competitive transports of nanoparticles with the substrates of interest (e.g., chloramphenicol) and prevent the possible aggregation of nanoparticles, a low concentration of nanoparticles (1.3 pM) is used to incubate with the living cells ($OD_{600\ nm} = 0.1$) for real-time monitoring of membrane transport at the nanometer scale.

Many living individual cells and single nanoparticles in the microchannel are monitored simultaneously using a CCD camera through a dark-field microscope, showing that more nanoparticles are observed in the cells as chloramphenicol concentration increases. Representative images of single cells selected from the full images illustrate that the cells contain nanoparticles in detail, indicating that more nanoparticles are present in the cells as chloramphenicol concentration increases (Figure 3.8A). Normalized histograms of the number of nanoparticles with the cells versus sizes of the nanoparticles (Figure 3.8B) indicate that a greater number of larger Ag nanoparticles are with the cells as chloramphenicol concentration increases, suggesting that the permeability and porosity of the cellular membrane increase as chloramphenicol concentration increases. Note that a higher percent of 50 ± 10-nm nanoparticles (53% of total nanoparticles) are present in the solution. Therefore, more 50 ± 10-nm nanoparticles are observed in the cells. The relative numbers of specific sizes of intracellular nanoparticles in the presence of 0, 25, and 250 μg/mL chloramphenicol are compared to determine the change of membrane permeability and porosity as chloramphenicol concentration increases.

The integrated scattering intensity of selected individual nanoparticles in and out of cells is measured and subtracted from the background in the same image, indicating that the intensity of intracellular nanoparticles is about 10% less than extracellular nanoparticles in the solution and about 20% less than those extracellular nanoparticles on the membrane. The scattering intensity of nanoparticles decreases slightly ($\sim$10%) as nanoparticles enter the cellular membrane because the cellular membrane and matrix absorb the microscope illuminator light and reduce its intensity. The decreased intensity of intracellular nanoparticles appears to depend upon the location of nanoparticles inside the cells, such as the depth below the cellular membrane. Quantitatively, study of the dependence of scattering intensity of intracellular nanoparticles upon their locations inside the cells become impossible because optical diffraction limit defines the spatial resolution ($\sim$200 nm) and makes the nanoparticles appear larger than their actual sizes. Nevertheless, intracellular and extracellular nanoparticles can be qualitatively determined based upon their intensity changes (Xu et al., 2004).

The quantum yield of Rayleigh scattering of intracellular Ag nanoparticles is smaller than extracellular Ag nanoparticles because intracellular Ag nanoparticles are surrounded by biomolecules (e.g., proteins, lipid) that reduce the reflection coefficient of Ag nanoparticles (Mie, 1908; Bohren and Huffman, 1983, Kreibig and Vollmer, 1995). Therefore, the nanoparticles outside the cellular membrane appear to radiate more sharply and brightly, whereas nanoparticles inside the cells look blurry and dim as shown in Figure 3.9. This feature allows us to determine whether the nanoparticles are inside or outside the cells using optical microscopy. The size of the nanoparticles (<100 nm) and the thickness of cell membrane (36 nm including 8-nm inner or outer membrane and $\sim$20 nm between inner and outer membrane of gram-negative bacteria) are below the optical diffraction limit ($\sim$200 nm). The scattering intensity of nanoparticles is much higher than that of the cellular membrane because of the higher intrinsic optical dielectric constant of Ag nanoparticles. Therefore, the nanoparticles are much brighter than the membrane, appearing to be accumulated on the membrane (Figure 3.9A). However, these nanoparticles are blurry and are dimmer than the extracellular nanoparticles (Figure 3.9B), indicating that these blurry nanoparticles are inside the cells. Taken together, the results indicate that, in the absence of chloramphenicol, the majority of red nanoparticles (> 80 nm) appear to stay outside the cells, whereas violet, blue, and green nanoparticles (< 80 nm) can enter the cells.

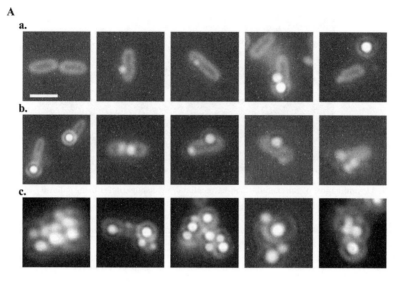
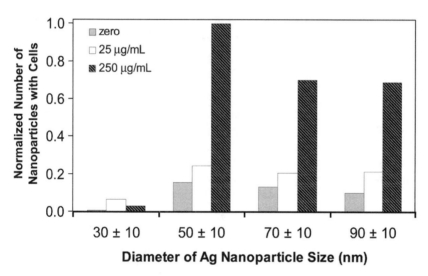

Figure 3.8. Real-time sizing membrane permeability and pore sizes in single living cells (WT), induced by chloramphenicol in a dose-dependent manner, using single nanoparticle optics: (A) Representative optical images (a–c) of single cells selected from ∼60 cells in the full-frame images. The solutions contain the same concentration of cells ($OD_{600\,nm} = 0.1$) and Ag nanoparticles (1.3 pM), but different concentration of chloramphenicol: (a) 0, (b) 25, and (c) 250 μg/mL chloramphenicol. Each solution is prepared in a vial and imaged in the microchannel at 15-min intervals for 2 h. The scale bar represents 2 μm. (B) Representative normalized histograms of the number of Ag nanoparticles with the cells versus sizes of nanoparticles from the solutions in (A), showing that more larger nanoparticles are accumulated inside the cells as chloramphenicol concentration increases. [Reprinted from *Biochemistry* **43**:10400–10413, with permission of American Chemical Society.] See insert for color representation of this figure.

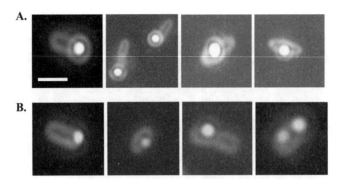

Figure 3.9. Determination of intracellular and extracellular Ag nanoparticles using intensity of nanoparticles: optical images of Ag nanoparticles accumulated (A) on membrane (outside the cell) and (B) inside single live cells (*P. aeruginosa*) recorded by the CCD with 100-ms exposure time and digital color camera through the dark-field optical microscope. Extracellular nanoparticles appear to radiate more sharply and brightly than intracellular nanoparticles because the scattering intensity of nanoparticles decreases about 10% as nanoparticles enter the cells, owing to light absorption of cellular membrane and decreased of reflection coefficient of intracellular Ag nanoparticles. The scale bar represents 2 μm. [Reprinted from *Biochemistry* **43**:10400–10413, with permission of American Chemical Society.] See insert for color representation of this figure.

To determine the sizes and locations of intracellular nanoparticles at the subnanometer (Å) level, ultrathin sections (70–80 nm) of the fixed cells that have been incubated with 1.3 pM nanoparticles and chloramphenicol (0, 25, and 250 μg/mL) for 2 hours are prepared and imaged using TEM as described above. TEM images in Figure 3.10 unambiguously demonstrate that Ag nanoparticles with a variety of sizes (20–80 nm in diameter) are inside the cells in the absence of chloramphenicol. The majority of the nanoparticles are located in the cytoplasmic space of the cells, whereas a few nanoparticles are just underneath the cellular membrane. The cells with embedded triangular nanoparticles (Figure 3.10) are particularly selected for easy identification of Ag nanoparticles. The TEM images show that the nanoparticles with sizes ranging up to 80 nm are embedded inside the cells in the absence of chloramphenicol, which agrees well with those measured using single-nanoparticle optics, suggesting that, even in the absence of chloramphenicol, the nanoparticles with size

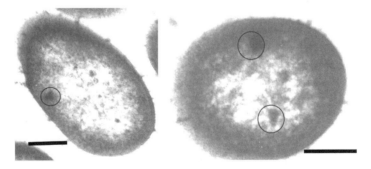

Figure 3.10. TEM images show sizes and locations of Ag nanoparticles inside the cells (WT), which are acquired from ultra-thin sections (70–80 nm) of fixed WT cells sliced using a diamond knife. The cells (OD$_{600\,nm}$ = 0.1) have been incubated with 1.3 pM Ag nanoparticles in the absence of chloramphenicol for 2 h before fixation. The scale bars represent 220 nm, and the circles are used to highlight the nanoparticles. [Reprinted from *Biochemistry* **43**:10400–10413, with permission of American Chemical Society.]

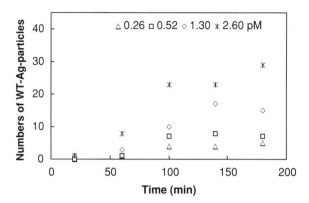

Figure 3.11. Dependence of accumulation kinetics of single living cells (WT) upon nanoparticle concentration and incubation time. Representative plots of the number of cells with silver-enhanced gold nanoparticles versus incubation time for the cells that are incubated with 0.26, 0.52, 1.30, and 2.60 pM nanoparticles. The 400 cells in total are analyzed for each data point to ensure sufficient statistics of single-cell measurement that will represent the bulk cells at the single-cell resolution. [Reprinted from *Nano Letters* **2(3)**:175–182, with permission of American Chemical Society.]

up to 80 nm can transport through the outer and inner cell membrane of *P. aeruginosa*, a size which is about 50 times larger than conventional antibiotics. As chloramphenicol concentration increases to 25 and 250 μg/mL, larger Ag nanoparticles are observed inside the cells, suggesting that chloramphenicol increases the membrane permeability and sizes of membrane pores.

To ensure that such an event of uptake and efflux is not a random event, the accumulation rate of intracellular nanoparticles upon the nanoparticle concentration and incubation time has been studied (Xu et al., 2002). Representative plots of number of *P. aeruginosa* cells with nanoparticles versus incubation time at nanoparticle concentration of 0.26, 0.52, 1.30, and 2.60 pM demonstrates that the nanoparticles accumulated by cells increase as incubation time and nanoparticle concentration increase prior to reaching the equilibrium of influx and efflux (Figure 3.11). The uptake and efflux dynamics show mutant-dependence (Xu et al., 2002, 2004).

It is important to note that cells with single nanoparticles at such a low nanoparticle concentration could still grow and divide with each individual nanoparticle remaining well isolated and color unchanged, suggesting that single nanoparticles may not create significant disturbance on cellular physiological environments. Nevertheless, further studies will be required to better understand the cytotoxicity and genotoxicity of Ag nanoparticles.

To investigate the role of the efflux pump proteins (MexAB-OprM) in controlling the accumulation of intracellular nanoparticles and function of chloramphenicol, the membrane permeability of two mutants, nalB-1 (the mutant with the overexpression level of MexAB-OprM) and ΔABM (the mutant devoid of MexAB-OprM), has been investigated. The study shows that the accumulation of intracellular Ag nanoparticles is highly dependent upon expression level of pump protein (MexAB-OprM) and chloramphenicol concentration.

The results show that the number of Ag nanoparticles accumulated in all three strains increases as chloramphenicol concentration increases and as incubation time increases, suggesting that chloramphenicol increases membrane porosity and permeability (Xu et al., 2004). In addition, the results indicate that accumulation kinetics of intracellular nanoparticles is associated with the expression levels of MexAB-OprM. The mutant with overexpression level of MexAB-OprM (nalB-1) accumulates the least number of intracellular Ag nanoparticle, whereas the mutant devoid of MexAB-OprM (ΔABM) accumulates the greatest number of Ag nanoparticles, suggesting that MexAB-OprM plays a critical role in controlling of accumulation of intracellular Ag nanoparticles.

3.3.2.3. Validation of Single-Nanoparticle Probes

To rule out the possibility of steric and size effect of using Ag nanoparticle probes for real-time sizing membrane permeability and porosity induced by chloramphenicol (antibiotic), a small molecule (EtBr) is used as a fluorescence probe to measure accumulation kinetics of EtBr by bulk WT, nalB-1, and ΔABM in the presence of 0, 25, and 250 μg/mL chloramphenicol. EtBr and chloramphenicol enter the cells through passive diffusion and are extruded out of the cells by the efflux pumps in *P. aeruginosa* (Yoneyama et al., 1997, 1998, 2000). The EtBr concentration at 10 μM is deliberately selected to be much lower than chloramphenicol concentration, 25 μg/mL (77 μM) and 250 μg/mL (770 μM), allowing chloramphenicol to dominate the passive diffusion pathway and enter the cells. This approach minimizes the possible effect of the presence of EtBr on the accumulation of chloramphenicol, ensuring the function of chloramphenicol to be studied using EtBr.

Transients of time-dependent accumulation of intracellular Ag nanoparticles and EtBr by the cells (WT) in the absence and presence of chloramphenicol (0, 25, and 250 μg/mL) are shown in Figure 3.12. The changes of fluorescence intensity over time are used to real-time monitor the accumulation kinetics of EtBr in the cells because EtBr enters the cells and intercalates with DNA, leading to 10-fold increased fluorescence intensity. The accumulation rate of intracellular Ag nanoparticles and EtBr in 0, 25, and 250 μg/mL chloramphenicol are measured and compared to determine the dose effect of chloramphenicol upon the membrane permeability. Note that in these experiments the cells are incubated in the PBS buffer solution at room temperature and the cells do not grow and divide significantly in such an environment. Thus, the cell concentration remains unchanged over 2-hour incubation.

In the absence of chloramphenicol (Figure 3.12A: a), a very small number of intracellular Ag nanoparticles are observed in WT and the number of accumulated intracellular Ag nanoparticles remains almost unchanged over time. Similar phenomena are observed using EtBr as a molecular probe, indicating the low fluorescence intensity of EtBr in WT, and the fluorescence intensity remains constant over time (Figure 3.12B, plot a).

As chloramphenicol concentration increases to 25 μg/mL (Figure 3.12A, plot b), the number of intracellular Ag nanoparticles in WT increases with time at a rate of 4×10^{-5} s^{-1}, which are about 10-fold higher than in the absence of chloramphenicol. Similar transient is observed using EtBr, showing that the fluorescence intensity increases with time at a rate of 9×10^{-4} s^{-1}, which is slightly (1.8-fold) higher than those observed in the absence of chloramphenicol (Figure 3.12B, plot b).

As chloramphenicol concentration increases further to 250 μg/mL (Figure 3.12A, plot c), the number of intracellular Ag nanoparticles in WT increases with time at a rate of 1×10^{-4}, which are about 2.5-fold higher than in 25 μg/mL chloramphenicol. The influx of large amount of intracellular nanoparticles overwhelms the capacity of individual cells, leading to the aggregation and overlapping of intracellular nanoparticles, which makes it impossible to determine the number of individual intracellular nanoparticles. A similar result is acquired using EtBr, illustrating a huge increase of fluorescence intensity over time at 9×10^{-3} s^{-1}, which is 18 times more rapid than in 25 μg/mL chloramphenicol.

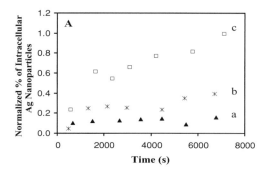

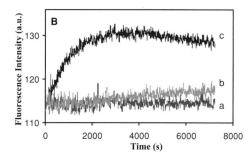

Figure 3.12. Validation of single nanoparticle probes using a fluorophor (EtBr). (A) Direct observation of accumulation kinetics of living cells (WT) using single Ag nanoparticles as probes, showing that membrane porosity and permeability increase as chloramphenicol concentration increases. Plots of normalized number of intracelluar Ag nanoparticles with the cells (WT) versus time from solution containing the cells ($OD_{600\ nm} = 0.1$), 1.3 pM Ag nanoparticles, and (a, ▲) 0, (b, *) 25, and (c, □) 250 μg/mL chloramphenicol, respectively. The 600 cells in total are analyzed for each data point to ensure sufficient statistics of single-cell measurements that will represent the bulk cells at the single-cell resolution. (B) Real-time monitoring of EtBr accumulation kinetics in bulk living cells (WT). Representative time courses of fluorescence intensity of EtBr at 590 nm from the solutions containing 10 μM EtBr, the WT cells ($OD_{600\ nm} = 0.1$), and (a) 0, (b) 25, and (c) 250 μg/mL chloramphenicol, respectively. Times courses are acquired from the 3.0 mL solution in a quartz cuvette using fluorescence spectroscopy with time-drive mode at a 3-s data acquisition interval and 488-nm excitation. [Reprinted from *Biochemistry* **43**:10400–10413, with permission of American Chemical Society.]

Taken together, the results measured using EtBr agree qualitatively with those measured using single-nanoparticle optics and single-living-cell imaging (Figure 3.12), validating that single Ag nanoparticles can be used as nanometer probes for real-time sizing membrane transport in living cells and eliminating the concern of possible size and steric effects of nanoparticles in the absence and low concentration of chloramphenicol. Unlike the accumulation kinetics measured using fluorescence probes (EtBr), the results acquired by single nanoparticles and single-living-cell images offer the new insights into the change of membrane permeability and pore sizes at the nanometer scale (Figures 3.8 and 3.9). For example, Ag nanoparticles with a size beyond 80 nm (red nanoparticle) appear to be difficult to transport through the membrane of *P. aeruginosa* in the absence of chloramphenicol. In contrast, a greater number of larger nanoparticles are accumulated in living cells as chloramphenicol concentrate increases, suggesting that chloramphenicol increases the membrane permeability and pore sizes. The study of mutant-dependence has also been carried out using both single nanoparticles and EtBr, showing the mutant-dependence with similar accumulation kinetics of both probes (Xu et al., 2004).

3.4. SUMMARY AND PROSPECTIVE

In summary, SMD is a unique and powerful tool for the study of living cellular membrane transport, demonstrating the potential for tracing the transport mechanism of living cells at the SM level in real time. A single-nanoparticle assay offers the possibility of imaging membrane transport of living cells with unprecedented spatial and temporal resolution, providing the new opportunity of measuring the kinetic and size transformation of individual membrane pores of living cells in real time. These initial studies present the new way to explore living cellular imaging and living cellular membrane transport in real time. Similar approaches can be applied for the study of living subcellular membrane transport, such as nuclear membrane pores.

Note that the single-nanoparticle assay presented here is not the same as single-nanoparticle tracking, because single-nanoparticle tracking uses individual nanoparticles solely as a tag to label and trace the motion of molecules of interest, which does not use nanoparticles as nanometer-size probes (Saxton and Jacobson, 1997; Kusumi, 2005). In some studies, the sizes of nanoparticles were characterized in fixed cells using TEM, which could not offer the temporal resolution of cellular events (cellular or subcellular membrane transport) in living cells because TEM imaging has to be performed under the vacuum, leading to the cell death.

There is no doubt that SMD and single-nanoparticle assay will advance our understanding of an array of cellular and subcellular mechanisms of living cells with high temporal and spatial resolution at the single-molecule level. The primary research focuses and advances in this particular field will include the following topics:

1. Development of biocompatible, ultrasensitive, and photostable single-molecule optical probes (tags) that can be conjugated with molecules of interest (e.g., protein) with well-defined label ratio for single-molecule detection and sensing. Such idea probes and sensors will open up the new possibility of probing an array of biochemical pathways and mechanisms in individual living cells at the single-molecule resolution. For example, using such probes, one can trace the transport of a wide variety of functional molecules and drugs in and out of individual living cells and subcellular compartments (e.g., nuclei) in real time; one can tag protein molecules that are involved in the assembly of membrane transporters and measure their dynamics and interactions at nanometer resolution in real time; using the similar approaches, one can map single ligands and receptors and measure their binding affinity and kinetics on single living cells for better understanding their vital roles in cellular function; and one can also conjugate crucial molecules (e.g., signaling molecules) for the study of signaling transduction pathways and cell–cell communication. Such experiments are expected to provide new insights into cellular and subcellular function at single-living-cell and single-molecule level with nanometer resolution in real time. Such a powerful molecular probe will surely become an extremely valuable tool to address numerous analytical, biochemical, and biomedical problems associated with analysis of single living cells. The development of this type of single-molecule probes and sensors will generate a wide variety of new knowledge and technique, which will be essential to the design of biocompatible molecular probes and sensors for molecular imaging and diagnosis *in vivo*. Such new knowledge will advance an array of research fields, including biomedical, chemical, and material sciences and engineering.

2. Development of high-speed, high-resolution, and ultrasensitive imaging tools that will be able to directly track multiple individual molecular probes in single living cells simultaneously. The design of new detection and imaging scheme to further improve signal-to-noise ratio and reduce the detection volume for single-molecule analysis of single living cells will be actively pursued. The new technology for the development of new type of imaging microscopy and detector, which can offer high temporal, spatial, and spectra (color) resolution of molecules of interest, will be in high demand, aiming to simultaneously analyze and characterize multiple analytes of interest in single living cells in real time. Such powerful tools will be absolutely essential for unraveling an array of cellular and subcellular pathways and mechanisms, including membrane transport, signaling transduction, gene expression, and cell–cell communication, in living cells at single-molecule level in real time.

3. It is essential to develop advanced detection means for quantitative and qualitative analysis of multiple individual molecules of interest, especially unknown molecules, in individual living cells in real time, in order to map the network of the molecules that are constantly changing in living cells with sufficient temporal and spatial resolution. Unfortunately, such crucial tools are not yet available, which hinders the progress of better understanding of cellular and subcellular function. Currently, mass spectrometry is the only tool that can be used to characterize unknown molecules in cells as described in Chapter 12. Regrettably, imaging cells using mass spectrometry is currently performed under vacuum, which leads to the cell death. The spatial and temporal resolution and sensitivity of current mass spectrometry are still unable to offer meaningful results for tracing cellular and subcellular pathways and mechanism at the molecular level in single living cells. Development of new tools to solve these technique problems is not a trivial task, and one may meet the fundamental limitation. Thus, it is essential and urgent to develop groundbreaking innovative tools that gear toward quantitative and qualitative analysis of individual molecules of interest and probing their interactions and regulation dependence in single living cells with high spatial and temporal resolution.

4. Single-nanoparticle assay will be further developed to serve as single-molecule nanosensors, which can be used for imaging single living cells with nanometer spatial resolution in real time. Multiple colors of nanoparticles will be further developed to prepare an array of nanosensors, which can meet the challenges of detecting multiple molecules of interest in single living cells simultaneously. The current challenges of development of such single-molecule nanosensors include how to prepare uniform surface and optical properties of nanoparticles, how to prevent the aggregation of nanoparticles in a wide variety of medium, how to design the functional surface of nanoparticles that can be conjugated with the molecules of interest (e.g., protein) with desire ratio, and how to design biocompatible nanoparticles that will not perturb and destroy intrinsic biomolecular machinery.

In conclusion, quantitative and qualitative imaging and analysis of single living cells at the single-molecule level with high spatial (nm) and temporal resolution (ns) will remain an extremely active and exciting field for years to come. Advances in this research field will heavily rely upon research breakthrough in an array of research fields, especially interdisciplinary research approaches. Consequently, outcomes of such research effort will create enormous impacts to a wide variety of research fields, including biological, biomedical, chemical, material, and environmental sciences and engineering and will certainly lead to a wide spectra of technology advances.

ACKNOWLEDGMENTS

We are indebted to all students of Xu group for their contribution of the development of SMD and nanoparticle assays for the study of cellular function in live cells. The support of this work in part by NIH (R21 RR15057; R01 GM0764401), NSF (BES 0507036; DMR 0420304), DOE (DE-FG02-03ER63646), and Old Dominion University is gratefully acknowledged.

REFERENCES

Agrawal, A, Zhang, C, Byassee, T, Tripp, RA, and Nie, S (2006). Counting single native biomolecules and intact viruses with color-coded nanoparticles. *Anal Chem* **78**(4):1061–1070.

Ambrose, WP, Goodwin, PM, Jett, JH, Orden, AV, Werner, JH, and Keller, RA (1999). Single molecule fluorescence spectroscopy at ambient temperature. *Chem Rev* **99**:2929–2956 and references therein.

Axelrod, D, Burghardt, TP, and Thompson, NL (1984). Total internal reflection fluorescence. *Ann Rev Biophys Bioeng* **13**:247–268.

Bard, AJ, and Faulkner, LR (1980). *Electrochemical Methods Fundamentals and Applications*. Wiley, New York, pp. 488–510.

Barnes, MD, Whitten, WB, and Ramsey, JM (1995). Detection single molecules in liquids. *Anal Chem* **67**:418A and references therein.

Bohren, CF, and Huffman, DR (1983). *Absorption and Scattering of Light by Small Particles*. Wiley, New York, pp. 287–380 and references therein.

Bruchez, M, Moronne, M, Gin, P, Weiss, S, and Alivisatos, AP (1998). Semiconductor nanocrystals as fluorescent biological labels. *Science* **281**:2013–2016.

Byassee, TA, Chan, W, and Nie, S (2000). Probing single molecules in single living cells. *Anal Chem* **72**:5606–5611.

Chan, WCW, and Nie, S (1998). Quantum dot bioconjugates for ultrasensitive nonisotopic detection. *Science* **281**:2016–2018.

Cole, SP, Bhardwaj, G, Gerlach, JH, Mackie, JE, Grant, CE, Almquist, KC, Stewart, AJ, Kurz, EU, Duncan, AM, and Deeley, RG (1992). Overexpression of a transporter gene in a multidrug-resistant human lung cancer cell line. *Science* **258**:1650–1654.

Dahan, M, Levi, S, Luccardini, C, Rostaing, P, Riveau, B, and Triller, A (2003). Diffusion dynamics of glycine receptors revealed by single-quantum dot tracking. *Science* **302**:442–445.

Derfus, AM, Chan, WCW, and Bhatia, SN (2004). Probing the cytotoxicity of semiconductor quantum dots. *Nano Lett* **4**:11–18.

Dickson, RM, Cubitt AB, Tsien RY, and Moerner WE (1997). On/off blinking and switching behaviour of single molecules of green fluorescent protein. *Nature* **388**:355–358.

Dovichi, NJ, Martin, JC, Jett, JH, Trkula, M, and Keller, RA (1984). Laser-induced fluorescence of flowing samples as an approach to single-molecule detection in liquids. *Anal Chem* **56**(3):348–354.

Femino, AM, Fay, FS, Fogarty, K, and Singer, RH (1998). Visualization of single RNA transcripts *in situ*. *Science* **280**:585–590.

Funatsu, T, Harada, Y, Tokunagida, M, and Yanagida, SK (1995). Imaging of single fluorescent molecules and individual ATP turnovers by single myosin molecules in aqueous solution. *Nature (London)* **374**:555–559.

Gao, X, Cui, Y, Levenson, RM, Chung, LW, and Nie, S (2004). *In vivo* cancer targeting and imaging with semiconductor quantum dots. *Nat Biotechnol* **22**:969–976.

Germ, M, Yoshihara, E, Yoneyama, H, and Nakae, T (1999). Interplay between the efflux pump and the outer membrane permeability barrier in fluorescent dye accumulation in *Pseudomonas aeruginosa*. *Biochem Biophys Res Commun* **261**(2):452–455.

Greenwood, D (1997). Modes of action in antibiotic and chemotherapy. In: O'Grady, F, Lambert, HP, Finch, RG, and Greenwood, D (eds.), *Antibiotic and Chemotherapy: Anti-infective Agents and Their Use in Therapy*, 7th edition. Churchill Livingstone, New York, pp. 10–21.

Haes, AJ, and Van Duyne, RP (2002). A nanoscale optical biosensor: Sensitivity and selectivity of an approach based on the localized surface plasmon resonance spectroscopy of triangular silver nanoparticles. *J Am Chem Soc* **124**:10596–10604 and references therein.

Haraguchi, T (2002). Living cell imaging: Approaches for studying protein dynamics in living cells. *Cell Struct Functi* **27**:333–334.

Harms, GS, Cognet, L, Lommerse, PH, Blab, GA, and Schmidt, T (2001). Autofluorescent proteins in single-molecule research: Applications to live cell imaging microscopy. *Biophys J* **80**:2396–2408.

Haynes, C, and Van Duyne, R (2001). Nanosphere lithography: A versatile nanofabrication tool for studies of size-dependent nanoparticle optics. *J Phys Chem B* **105**:5599–5611 and references therein.

Ichinose, J, and Sako, Y (2004). Single-molecule measurement in living cells. *Trends Anal Chem* **23**: 587–594.

Iino, R, Koyama, I, and Kusumi, A (2001). Single molecule imaging of green fluorescent proteins in living cells. *Biophys J* **80**:2667–2677.

Janicki, SM, Tsukamoto, T, Salghetti, SE, Tansey, WP, Sachidanandam, R, Prasanth, KV, Ried, T, Shav-Tal, Y, Bertrand, E, Singer, RH, and Spector, DL (2004). From silencing to gene expression: Real-time analysis in single cells. *Cell* **116**(5):683–698.

Keller, RA, Ambrose, WP, Goodwin, PM, Jett, JH, Martin, JC, and Wu, M (1996). Single-molecule fluorescence analysis in solution. *Appl Spectrosc* **50**:12A–32A.

Kreibig, U, and Vollmer, M (1995). *Optical Properties of Metal Clusters*. Springer, Berlin, pp. 14–123 and references therein.

Kusumi, A, Nakada, C, Ritchie, K, Murase, K, Suzuki, K, Murakoshi, H, Kasai, RS, Kondo, J, and Fujiwara, T (2005). Paradigm shift of the plasma membrane concept from the two-dimensional continuum fluid to the partitioned fluid: High-speed single-molecule tracking of membrane molecules. *Annu Rev Biophys Biomol Struct* **34**:351–378 and references therein.

Kyriacou, S, Brownlow, W, and Xu, XH (2004). Nanoparticle optics for direct observation of functions of antimicrobial agents in single live bacterial cells. *Biochemistry* **43**:140–147.

Kyriacou, S, Nowak, M, Brownlow, W, and Xu, XH (2002). Single live cell imaging for real-time monitoring of resistance mechanism in *Pseudomonas aeruginosa*. *J Biomed Opt* **7**(4):576–586.

Lamprecht, B (2000). *Ultrafast Plasmon Dynamics in Metal Nanoparticles*. Karl-Franzens-University of Graz, Dissertation, Germany.

Lee, A, Mao, W, Warren, MS, Mistry, A, Hoshino, K, Okumura, R, Ishida, H, and Lomovskaya, O (2000). Interplay between efflux pumps may provide either additive or multiplicative effects on drug resistance. *J Bacteriol* **182**(11):3142–3150.

Lee, PC, and Meisel, D (1982). Adsorption and surface-enhanced Raman of dyes on silver and gold sols. *J Phys Chem* **86**(17):3391–3395.

Li, XZ, and Nikaido, H (2004). Efflux-mediated drug resistance in bacteria. *Drug* **64**:159–204 and references therein.

Lomovskaya, O, Warren, MS, Lee, A, Galazzo, J, Fronko, R, Lee, M, Blais, J, Cho, D, Chamberland, S, Renau, T, Leger, R, Hecker, S, Watkins, W, Hoshino, Ishida, KH, and Lee, VJ (2001). Identification and characterization of inhibitors of multidrug resistance efflux pumps in *Pseudomonas aeruginosa*: Novel agents for combination therapy. *Antimicrob Agents Chemother* **45**(1):105–116.

Ma, D, Cook, DN, Hearst, JE, and Nikaido, H (1994). Efflux pumps and drug resistance in gram-negative bacteria. *Trends Microbiol* **2**(12):489–493.

Masuda, N, Sakagawa, E, Ohya, S, Gotoh, N, Tsujimoto, H, and Nishino, T (2000). Substrate specificities of MexAB-OprM, MexCD-OprJ, and MexXY-oprM efflux pumps in *Pseudomonas aeruginosa*. *Antimicrob Agents Chemother* **44**(12):3322–3327.

Maseda, H, Yoneyama, H, and Nakae, T (2000). Assignment of the substrate-selective subunits of the MexEF-OprN multidrug efflux pump of *Pseudomonas aeruginosa*. *Antimicrob Agents Chemother* **44**(3):658–664.

Mie, G (1908). Beiträg zur Optik trüber Medien, speziell kolloidaler Metrallösungen. *Ann Phys* **25**:377.

Min, W, English, BP, Luo, G, Cherayil, BJ, Kon, SC, and Xie, XS (2005). Fluctuating enzymes: lessons from single-molecule studies. *Acc Chem Res* **38**(12):923–931 and references therein.

Morshed, SR, Lei, Y, Yoneyama, H, and Nakae, T (1995). Expression of genes associated with antibiotic extrusion in *Pseudomonas aeruginosa*. *Biochem Biophys Res Commun* **210**(2):356–362.

Mortimer, PG, and Piddock, LJ (1991). A comparison of methods used for measuring the accumulation of quinolones by Enterobacteriaceae, *Pseudomonas aeruginosa* and *Staphylococcus aureus*. *J Antimicrob Chemother* **28**(5):639–653.

Mulvaney, P (1996). Surface plasmon spectroscopy of nanosized metal particles. *Langmuir* **12**:788–800.

Nakae, T (1995). Role of membrane permeability in determining antibiotic resistance in *Pseudomonas aeruginosa*. *Microbiol Immunol* **39**(4):221–229 and references therein.

Nakae, T (1997). Multiantibiotic resistance caused by active drug extrusion in *Pseudomonas aeruginosa* and other gram-negative bacteria. *Microbiologia* **13**(3):273–284 and references therein.

Nie, S, and Zare, RN (1997). Optical detection of single molecules. *Annu Rev Biophys Biomol Struct* **26**:567 and references therein.

Nie, S, Chiu, DT, and Zare, RN (1994). Probing individual molecules with confocal fluorescence microscopy. *Science* **266**:1018–1021.

Ocaktan, A, Yoneyama, H, and Nakae, T (1997). Use of fluorescence probes to monitor function of the subunit proteins of the MexA-MexB-OprM drug extrusion machinery in *Pseudomonas aeruginosa*. *J Biol Chem* **272**(35):21964–21969.

Poole, K (2001). Multidrug efflux pumps and antimicrobial resistance in *Pseudomonas aeruginosa* and related organisms. *J Mol Microbiol Biotechnol* **3**(2):255–264.

Poole, K, Krebes, K, McNally, C, and Neshat, S (1993). Multiple antibiotic resistance in *Pseudomonas aeruginosa*: Evidence for involvement of an efflux operon. *J Bacteriol* **175**(22):7363–7372.

Pramanik, A (2004). Ligand–receptor interactions in live cells by fluorescence correlation spectroscopy. *Curr Pharm Biotechnol* **5**:205–212.

Raether, H (1988). *Surface Plasmons on Smooth and Rough Surfaces and on Gratings*. Springer, Berlin, and references therein.

Rigler, R (1995). Fluorescence correlations, single molecule detection and large number screening. Applications in biotechnology. *J Biotechnol* **41**:177–186.

Ryan, BM, Dougherty, TJ, Beaulieu, D, Chuang, J, Dougherty, BA, and Barrett, JF (2001). Efflux in bacteria: what do we really know about it? *Expert Opin Invest Drugs* **10**(7):1409–1422 and references therein.

Sako, Y, and Yanagida, T (2003). Single-molecule visualization in cell biology. *Nat Rev Mol Cell Biol* Suppl SS1–5.

Sako, Y, Minoghchi, S, and Yanagida, T (2000). Single-molecule imaging of EGFR signalling on the surface of living cells. *Nat Cell Biol* **2**:168–172.

REFERENCES

Saxton, MJ, and Jacobson, K (1997). Single-particle tracking: Applications to membrane dynamics. *Annu Rev Biophys Biomol Struct* **26**:373–399 and references therein.

Schmidt, T, Schutz, G, Baumgartner, W, Gruber, H, and Schindler, H (1996). Imaging of single molecule diffusion. *Proc Natl Acad Sci USA* **93**:2926–2929.

Singer, RH (2004). Dynamics of single mRNPs in nuclei of living cells. *Science* **304**:1797–1800.

Tate, CG, Kunji, ERS, Lebendiker, M, and Schuldiner, S (2001). The projection structure of EmrE, a proton-linked multidrug transporter from *Escherichia coli*, at 7 Å resolution. *EMBO* **20**: 77–81.

Tinnefeld, P, and Sauer, M (2005). Branching out of single-molecule fluorescence spectroscopy: Challenges for chemistry and influence on biology. *Angew Chem Int Ed* **44**:2642–2671.

Tokunaga, M, Kitamura, K, Saito, K, Iwane, AF, and Yanagida, T (1997). Single molecule imaging of fluorophores and enzymatic reactions achieved by objective-type total internal reflection fluorescence microscopy. *Biochem Biophys Res Commun* **235**:47–53.

Tsien, RY (1998). The green fluorescent protein. *Annu Rev Biochem* **67**:509–544.

Tsien, RY, and Waggoner, AS (1995). Fluorophores for confocal microscopy: Photophysics and photochemistry. In: Pauley, J (ed.), *Handbook of Confocal Microscopy*, 2nd ed. Plenum Press, New York, pp. 267–279 and references therein.

Ueda, M, Sako, Y, Tanaka, T, Devreotes, P, and Yanagida, T (2001). Single-molecule analysis of chemotactic signaling in *Dictyostelium* cells. *Science* **294**:864–867.

Vukojevic, V, Pramanik, A, Yakovleva, T, Rigler, R, Terenius, L, and Bakalkin, G (2005). Study of molecular events in cells by fluorescence correlation spectroscopy. *Cell Mol Life Sci* 62(5):535–550.

Weiss, S (1999). Fluorescence spectroscopy of single biomolecules. *Science* **283**:1676–1683 and references therein.

Willets, KA, Nishimura, SY, Schuck, PJ, Twieg, RJ, and Moerner, WE (2005). Nonlinear optical chromophores as nanoscale emitters for single-molecule spectroscopy. *Acc Chem Res* **38**(7):549–556 and references therein.

Xie, XS, and Lu, HP (1999). Single-molecule enzymology. *J Biol Chem* **274**(23):15967–15970 and references therein.

Xie, XS, Yu, J, and Yang, WY (2006). Living cells as test tubes. *Science* **312**:228–230 and references therein.

Xu, XH, and Patel, RN (2004). Nanoparticles for live cell dynamics. In: Nalwa HS (ed.), *Encyclopedia of Nanoscience and Nanotechnology*, Vol. 7. American Scientific Publishers, Stevenson Ranch, CA, pp. 189–192.

Xu, XH, and Patel, RN (2005). Imaging and assembly of nanoparticles in biological systems. In: Nalwa, HS (ed.), *Handbook of Nanostructured Biomaterials and Their Applications in Nanobiotechnology*, Vol. 1. American Scientific Publishers, Stevenson Ranch, CA, pp. 435–456.

Xu, XH, and Yeung, ES (1997). Direct measurement of single-molecule diffusion and photodecomposition in free solution. *Science* **275**:1106–1109.

Xu, XH, and Yeung, ES (1998). Long-range electrostatic trapping of single protein molecules at a liquid/solid interface. *Science* **281**:1650–1653.

Xu, XH, Jeffers, R, Gao, J, and Logan, B (2001). Novel solution-phase immunoassays for molecular analysis of tumor markers. *Analyst* **126**:1285–1292.

Xu, XH, Chen, J, Jeffers, RB, and Kyriacou, S (2002). Direct measure of sizes and dynamics of single living membrane transporters using nanooptics. *Nano Letters* **2**(3):175–182.

Xu, XH, Brownlow, W, Huang, S, and Chen, J (2003a). Real-time measurements of single membrane pump efficiency of single living *Pseudomonas aeruginosa* cells using fluorescence microscopy and spectroscopy. *Biochem Biophys Res Commun* **305**:79–86.

Xu, XH, Wan, Q, Kyriacou, S, Brownlow, W, and Nowak, M (2003b). Direct observation of substrate induction of resistance mechanism in *Pseudomonas aeruginosa* using single live cell imaging. *Biochem Biophys Res Commun* **305**:941–949.

Xu, XH, Brownlow, W, Kyriacou, S, Wan, Q, and Viola, J (2004). Real-time probing of membrane transport in living microbial cells using single nanoparticle optics and living cell imaging. *Biochemistry* **43**(32):10400–10413.

Yeung, ES (1997). Single molecule spectroscopy. In: Hill, M (ed.), *Yearbook of Science and Technology*, pp. 433–435; also see references therein. McGraw-Hill, New York.

Yoneyama, H, Ocaktan, A, Masataka, T, and Nakae, T (1997). The role of Mex-gene products in antibiotic extrusion in *Pseudomonas aeruginosa*. *Biochem Biophys Res Commun* **233**:611–618.

Yoneyama, H, Ocaktan, A, Gotoh, N, Nishino, T, and Nakae, T (1998). Subunit swapping in the Mex-extrusion pumps in *Pseudomonas aeruginosa*. *Biochem Biophys Res Commun* **244**:898–902.

Yoneyama, H, Maseda, H, Kamiguchi, H, and Nakae, T (2000). Function of the membrane fusion protein, MexA, of the MexA, B-OprM efflux pump in *Pseudomonas aeruginosa* without an anchoring membrane. *J Biol Chem* **275**(7):4628–4634.

Yu, J, Xiao, J, Ren, X, Lao, K, and Xie, XS (2006). Probing gene expression in live cells, one protein molecule at a time. *Science* **311**:1600–1603.

Zander, C, Enderlein, J, and Keller, RA (2002). *Single Molecule Detection in Solution*, 1st edition. Wiley-VCH, Berlin, Germany.

CHAPTER

4

NANOPARTICLE PROBES FOR ULTRASENSITIVE BIOLOGICAL DETECTION AND IMAGING

AMIT AGRAWAL, TUSHAR SATHE, AND SHUMING NIE

4.1. INTRODUCTION

The development of high-sensitivity and high-specificity probes is of considerable interest in many areas of cancer research, ranging from basic tumor biology to *in vivo* imaging and early detection. The process to develop new imaging probes, however, remains a slow and expensive undertaking. In this context, research on quantum nanostructures and self-assembled materials has attracted much attention because it could lead to a new generation of nanoparticle contrast agents for *in vivo* tumor imaging at high sensitivity and specificity. Indeed, recent advances have led to the development of functional nanoparticles that are covalently linked to biological molecules such as peptides, proteins, and nucleic acids (Akerman et al., 2002; Bruchez et al., 1998; Chan et al., 2002; Chan and Nie, 1998; Dahan et al., 2003; Dubertret et al., 2002; Ishii et al., 2003; Jaiswal et al., 2003; Larson et al., 2003; Lidke et al., 2004; Mattoussi et al., 2000; Medintz et al., 2003; Wu et al., 2003). Due to their size-dependent properties and dimensional similarities to biomacromolecules (Alivisatos, 1996; Henglein, 1989; Schmid, 1992), these nanobioconjugates are well-suited as contrast agents for *in vivo* magnetic resonance imaging (MRI) (Bulte and Brooks, 1997; Bulte et al., 2001; Josephson et al., 1999a), as controlled carriers for drug delivery, and as structural scaffolds for tissue engineering (Curtis and Wilkinson, 2001; Gref et al., 1994). Furthermore, several classes of colloidal nanoparticles are under intense development for potential uses in combinatorial chemistry (Boal et al., 2000; Li et al., 1999; Templeton et al., 2000), medical diagnostics (Gao and Nie, 2003; Han et al., 2001), and multiplexed optical encoding (Klarreich, 2001; Mitchell, 2001).

In comparison with traditional contrast agents such as radioactive small molecules, gadolinium compounds, and labeled antibodies, quantum dots and related nanoparticles provide several unique features and capabilities. First, their optical and electronic properties are often dependent on size, and they can be tuned continuously by changing the particle size (Alivisatos, 1996). This "size effect" provides a broad range of nanoparticles for simultaneous detection of multiple cancer biomarkers. Second, nanoparticles have more surface areas and functional groups that can be linked with multiple diagnostic (e.g., radioisotopic or magnetic) and therapeutic (e.g., anticancer) agents. This opens the opportunity to design multifunctional or "smart" nanoparticle reagents for multi-modality imaging

---

New Frontiers in Ultrasensitive Bioanalysis. Edited by Xiao-Hong Nancy Xu
Copyright © 2007 John Wiley & Sons, Inc.

or simultaneous imaging and treatment. In the following, we discuss recent developments in nanoparticle probes and their applications in ultrasensitive detection and imaging.

4.2. PROBE PREPARATION

Several types of nanoparticles have been prepared using magnetic, semiconductor, metal, or organic materials. A typical process involves chemical synthesis of monodisperse nanoparticles, water solubilization, and functionalization with targeting ligands such as peptides, proteins, or nucleic acids (see Figure 4.1). In this section, we briefly discuss preparation procedures for semiconductor quantum dots, metal, and magnetic nanoparticles and also describe common procedures for nanoparticle solubilization and bioconjugation.

Semiconductor Quantum Dots. High-quality QDs are typically prepared at elevated temperatures in organic solvents, such as TOPO and HDA, both of which are high-boiling-point solvents containing long alkyl chains. These hydrophobic organic molecules not only serve as the reaction medium, but also coordinate with unsaturated metal atoms on the QD surface to prevent formation of bulk semiconductors. As a result, the nanoparticles are capped with a monolayer of the organic ligands and are soluble only in nonpolar hydrophobic solvents such as chloroform. For biological applications, these hydrophobic dots can be solubilized by using amphiphilic polymers that contain both a hydrophobic segment or side-chain (mostly hydrocarbons) and a hydrophilic segment or group (such as polyethylene glycol or multiple carboxylate groups). A number of polymers have been reported including octylamine-modified low-molecular-weight polyacrylic acid, PEG derivatized phospholipids, block copolymers, and polyanhydrides (Akerman et al., 2002; Henglein, 1989; Savic et al., 2003; Soo et al., 2002). The hydrophobic domains strongly interact with tri-n-octylphosphine oxide (TOPO) on the QD surface, whereas the hydrophilic groups face outward and render QDs water-soluble. Note that the coordinating organic ligands (TOP or TOPO) are retained on the inner surface of QDs, a feature that is important for maintaining the optical properties of QDs and for shielding the core from the outside environment.

Magnetic Nanoparticles. Similar to the method for semiconductor nanoparticle synthesis, Sun et al. (2000) developed a procedure to make uniform FePt magnetic nanoparticles by simultaneous reduction of platinum acetylacetonate Pt(acac)$_2$ with a long-chain diol. The procedure involves thermal decomposition of iron pentacarbonyl in the presence of oleic acid and oleylamine, which acts as reaction medium and stabilizing ligand. This procedure was taken further in the size-dependent synthesis of magnetite particles by reducing Fe(acac)$_3$ with the same diol and the same stabilizing ligands. The size could be tuned by simply varying the reaction solvent. For example, using benzyl ether as a solvent (bp $\sim$300°C) results in 6-nm particles, whereas using phenyl ether (bp $\sim$265°C) results in 4-nm particles.

Metal Nanoparticles. The procedure for metal nanoparticle synthesis involves reduction of metal compounds with an electron-donating species in hot aqueous solution. For example, gold nanoparticles can be synthesized by adding 7.5 mL of 1% sodium citrate solution to 500 mL of 0.01% HAuCl$_4$ maintained under hot boiling conditions (Grabar et al., 1995). Within 95 seconds, the reaction is complete as indicated by color change to red-orange. After this the heating is continued for 10 minutes. The heat is removed later and solution is stirred for 15 minutes. In this procedure, 18-nm gold nanoparticles were obtained. Particle diameter can be increased by reducing sodium citrate concentration and vice versa.

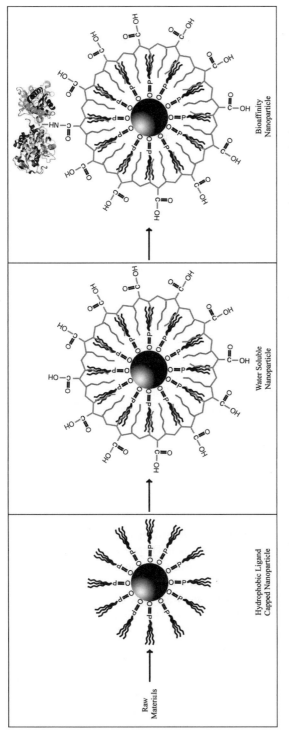

Figure 4.1. Preparation of bioconjugated nanoparticle probes. First a high-temperature organic phase synthesis is carried out to produce highly monodisperse and uniform nanoparticles. Then, QDs are rendered water-soluble by using hydrophobic interactions of the surface ligand with an amphiphilic polymer or by ligand exchange with water-soluble ligands. Finally, water-soluble QDs are conjugated with biomolecule of interest using well-established bioconjugation procedures. See insert for color representation of this figure.

Solubilization and Bioconjugation. Because most synthesis methods that produce highly monodisperse, homogenous nanoparticles use organic solvents, the resulting particles need to be rendered water-soluble for biological applications. Specifically, four key requirements should be met: (i) increased stability of nanoparticles in water over a long period of time; (ii) presence of sterically accessible functional groups for bioconjugation; (iii) biocompatibility and nonimmunogenicity of particles in living systems; and (iv) lack of interference with the nanoparticles' native properties (Brooks et al., 2001; Gao et al., 2004). Two types of procedures have been used to solubilize semiconductor quantum dots. The first procedure is a ligand-exchange method in which the surface ligands are exchanged with a thiol ligand such as mercaptoacetic acid (Chan and Nie, 1998) or polysilanes (Bruchez et al., 1998) (Figure 4.2a). Another procedure involves interactions between the hydrophobic ligand on the nanoparticle surface and hydrophobic part of an amphiphilic polymer construct (Figure 4.2a). Other water solubilization procedures such as encapsulation in phospholipids micelles (Nitin et al., 2004; Palmacci and and Josephson, 1993) or coating the nanoparticle with a polysaccharide layer (Palmacci and and Josephson, 1993) have been used for magnetic nanoparticles.

For bioconjugation, several types of electrostatic, hydrophobic, and covalent binding have been developed for linking nanoparticles to biomolecules (Figure 4.2b). Electrostatic schemes depend on charge–charge interactions between oppositely charged entities such as a positively charged peptide and a negatively charged polymer-coated nanoparticle. Methods based on hydrophobic interactions utilize the entropic factors that force hydrophobic parts of the coating on the nanoparticle and the biomolecule to interact stably with each other. Finally, covalent conjugation techniques use bifunctional linkers to attach biomolecules with the nanoparticle. With a broad variety of bioconjugation methods available (Hermanson, 1996), it is now possible to conjugate nanoparticles with ligands, peptides, carbohydrates, nucleic acids, proteins, lipids, and polymers. Probe design plays a critical role when the probes are used in biological solution, inside cells, or *in vivo*. However, the probes need to be delivered to the site of the target. In the following, we describe recent advances in optimizing the interaction of bioaffinity nanoparticle probes with their intended targets.

4.3. DELIVERING AND TARGETING

For dynamic intracellular observation of molecular processes, one needs to deliver individual, functional nanoparticles into cells. Several strategies have been explored for delivering nanoparticle cargos into living cells (Stephens and Pepperkok, 2001). These methods can be broadly divided into three categories: (a) delivery by transient cell permeabilization, (b) carrier mediated delivery, and (c) direct physical delivery. As shown in Figure 4.3, the transient permeabilization method utilizes bacterial toxins called cytolysins (Palmer, 2001). These toxins are protein molecules that are incorporated in cholesterol-containing cell membranes and generate pores with a diameter of 30–40 nm (Bhakdi et al., 1985, 1993; Palmer et al., 1998). At low cytolysin concentrations, the cells are permeabilized reversibly and nanoparticles can diffuse through the pores into the cell. The main advantage of this approach is that nanoparticles are delivered in a concentration-dependent fashion. For nanoparticles larger than 30 nm, the delivery occurs in a time-dependent fashion. This approach is promising for delivery of functional, individual nanoparticles into living cells. A key disadvantage is the lack of selectivity in delivery; that is, it is difficult to deliver nanoparticles into a selected group of cells.

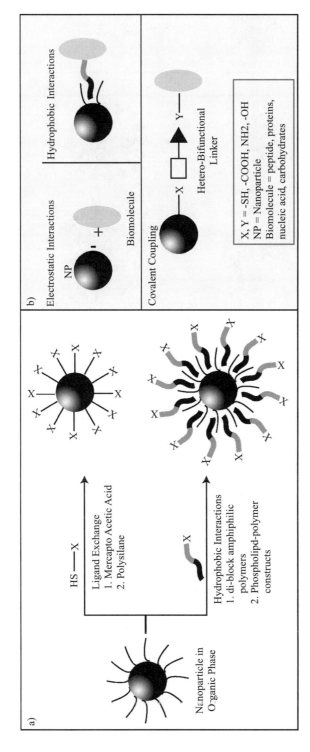

Figure 4.2. Procedures for solubilizing nanoparticles in water solution and for conjugation to biological molecules. See insert for color representation of this figure.

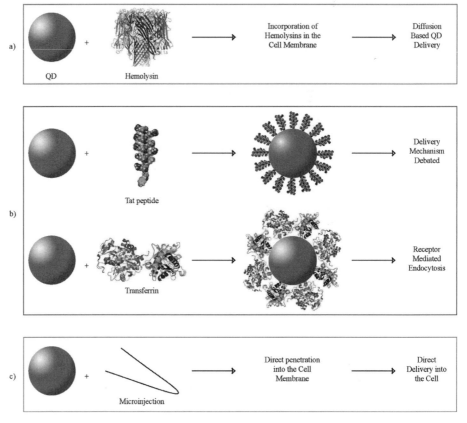

Figure 4.3. Schematic illustration of strategies for delivering nanoparticle probes into living cells. See insert for color representation of this figure.

In contrast to permeabilization-mediated delivery, carrier-mediated delivery relies on receptor mediated endocytosis. A classic example is the delivery of transferrin-conjugated QDs via transferrin-receptor-mediated endocytosis (Chan and Nie, 1998; Derfus et al., and Weissleder, 2004). Note that receptor-mediated endocytosis takes place in less than 30 min at 37°C, and it is suppressed and nanoparticles are not delivered into the cell at 4°C. When cell-penetrating peptides such as HIV-TAT peptide are used, the nanoparticles are able to escape the endosomal compartment and enter the nucleus or retain function for intracellular staining of actin filaments (Agrawal et al., 2003). The mechanism of HIV-TAT-peptide-mediated delivery is still a matter of debate, but recent research has shown that macropinocytosis (Wadia et al., 2004) and endocytosis followed by initial ionic interaction (Vives, 2003; Vives et al., 2003) are important steps in tat-peptide-mediated delivery. A major problem with this approach is that the nanoparticles are delivered as aggregates and may not be available for target binding. Physical delivery methods include microinjection and electroporation. In microinjection, nanoparticles are delivered into a cell using a micro needle, but this procedure is time-consuming (one cell at a time) and requires considerable training and skills. Electroporation was explored by Defrus et al. and has been found to result in nanoparticle aggregation (Derfus et al., 2004b). At present, a perfect method is still not available for delivery and targeting of nanoparticle probes inside living cells. This area will require considerable research effort in the next few years.

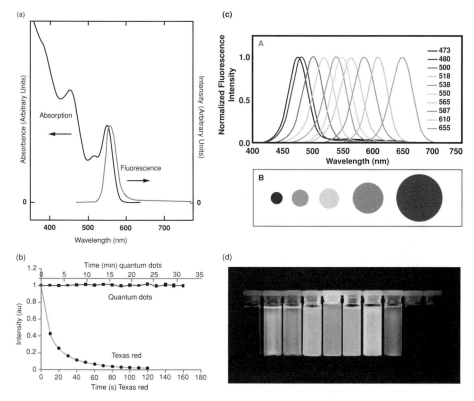

Figure 4.4. Optical properties of semiconductor QDs. (a) Absorption and fluorescence spectra of a 3.5-nm CdSeQD. [Reprinted from Murray et al. (1993), with permission of American Chemical Society.] (b) Photostability of QDs compared with Texas Red. [Reprinted from Gao et al. (2005) in *Current Opinion in Biotechnology*.] (c,d) Size tunable emission properties of CdSe QDs. [Reprinted from Smith and Nie, *Analyst* **129**:672–7 (2004) and Chan et al. (2002) in *Current Opinion in Biotechnology*.] See insert for color representation of parts (c) and (d) of this figure.

4.4. *IN VITRO* DETECTION AND IMAGING

Quantum Dots. Semiconductor quantum dots have received considerable attention due to their novel optical and electronic properties (Figure 4.4). By varying the size and composition, QDs with visible and near-IR fluorescence [400–2000 nm (Bruchez et al., 1998)] have been produced (Bailey and Nie, 2003; Kim et al., 2003; Wehrenberg et al., 2002; Zhong et al., 2003a, 2003b). Due to their broad absorption profile, the same light source can excite QDs emitting at different wavelengths, which makes them ideal for multiplexed biological detection. QDs are also photostable (Gao et al., 2005) and have large excitation coefficients: 10^5–10^6 $M^{-1}cm^{-1}$ (Chan et al., 2002). A new development is the ultrasensitive detection and molecular analysis of single intact viruses, especially the human respiratory syncytial virus (RSV), which causes serious lower respiratory infections in children and the immune-compromised (Agrawal et al., 2005). In this application, two antibodies to different surface proteins on RSV particles were attached with two nanoparticles that emit light at different wavelength when excited with blue light. When these bioaffinity nanoparticle probes bind to the same viral particle, they pass together through a small confocal probe volume, thereby producing coincident photons (Figure 4.5a). Furthermore, if a virus binds to more

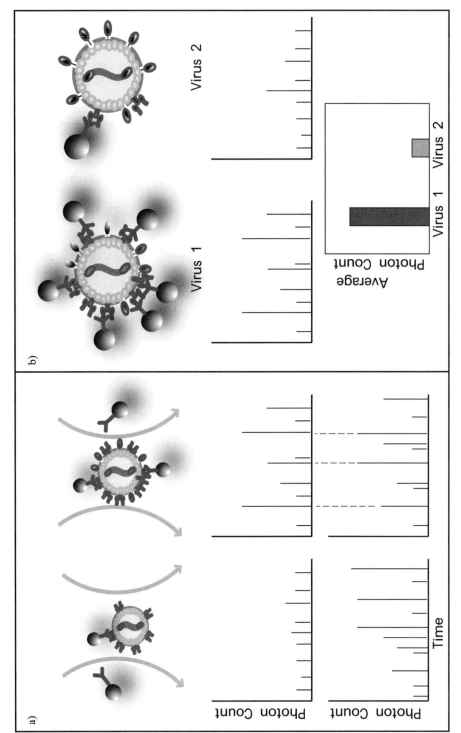

Figure 4.5. Principle of (a) real-time viral detection and (b) viral surface protein estimation using two-color nanoparticle probes. [Reprinted from Agrawal et al. (2005), with permission of the American Society of Microbiology.] See insert for color representation of this figure.

nanoparticles, it is likely to produce a higher number of photons (Figure 4.5b) in a fixed period of time. This principle has been used to probe protein expression levels on a viral surface.

QDs have shown immense promise in ultrasensitive, live-cell, and solution-based biological detection assays. However, it has not matched the sensitivity reported by Mirkin and co-workers in using metal nanoparticle for nucleic acid (Taton et al., 2000) and protein (Nam et al., 2003) detection, as discussed below.

Metal Nanoparticles. Colloidal gold–protein complexes have long been used in electron microscopy for biological studies (Dickson et al., 1981; Faulk and Taylor, 1971). However, the light-scattering properties of metal nanoparticles were not exploited in biological detection assay until 1995 when Stimpson et al. demonstrated the use of selenium nanoparticles for DNA detection assay on a chip (Stimpson et al., 1995). The authors used a 2-D optical waveguide to observe scattering from nanoparticles attached to the waveguide surface. They demonstrated nucleic acid detection limit of up to 0.4 nM. In 1997 Nie and Emery demonstrated extremely large Raman enhancement factors for R6G molecules adsorbed on silver and gold nanoparticles (Nie and Emery, 1997). These enormous enhancement factors indicate that surface-enhanced Raman scattering (SERS) (Jeanmaire and Vanduyne, 1977) can be used for high-sensitivity biological detection.

Gold nanoparticles are known to scatter light at different wavelengths in a size-dependent manner when illuminated with a white light (van de Hulst, 1981). Clustering of metal nanoparticles can also lead to change in the color (Kreibig and Genzel, 1985). Based on this observation, Elghanian et al. (1997) demonstrated a highly sensitive colorimetric assay for detection of polynucleotides. The authors coupled gold nanoparticles with complementary polynucleotides and allowed aggregation to occur via DNA hybridization. Using color change as an indicator, the authors achieved a detection limit as low as 10 femtomoles. However, light scattering depends on the particle size, shape, and orientation, and the orientation of nanoparticles is especially difficult to control; also, the light-scattering approach makes it difficult to calibrate the signal from the detection system (Rosi and Mirkin, 2005). SERS-based detection can overcome this problem and can enable ultrasensitive multiplexed detection of biological targets. Haes and Van Duyne (2002) reported protein detection by measuring the wavelength shift in the surface plasmon resonance of nanoparticles upon binding with a protein. In a series of three articles (Elghanian et al., 1997; Nam et al., 2003; Taton et al., 2000), Mirkin's group developed innovative schemes for sensitive detection of polynucleotides and protein molecules. Using what the authors call bio-bar codes, they demonstrated detection of prostate-specific antigen (PSA) at 3–30 femtomolar concentration. The design of the assay is shown in Figure 4.6. The novelty of the method lies in the use of gold nanoparticles which were coupled to PSA polyclonal antibody and are also functionalized with a specific DNA sequence. A complementary DNA sequence (the bio-bar code) is first allowed to hybridize to the gold nanoparticle probe. Then, a magnetic microparticle functionalized with PSA monoclonal antibody is used to capture the target (PSA) in the solution. Then, the gold nanoparticle probe is allowed to bind to the target immobilized on the magnetic particle surface. After purification using a magnet, the magnetic particles with target and nanoparticle probe are incubated in a denaturing buffer to release the hybridized DNA bio-bar-code molecules. Finally, these DNA molecules are detected via a scanometric approach or via gel electrophoresis after amplification of DNA bio-bar codes using a polymerase chain reaction (PCR). The scanometric approach involves detection of individual gold nanoparticles hybridized to a glass surface using a flat-bed scanner (Taton

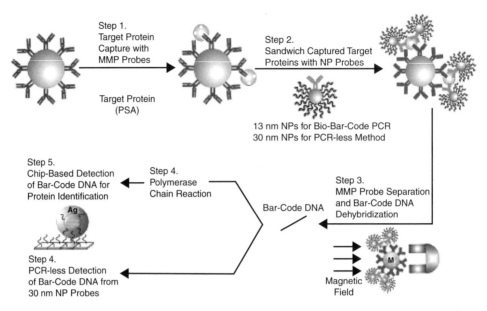

Figure 4.6. Schematic illustration of using nanoparticle biobarcodes for direct protein detection. [Reprinted from Nam et al. (2003), with permission of American Association for the Advancement of Science.] See text for details. See insert for color representation of this figure.

et al., 2000). The authors achieved single nanoparticle sensitivity by depositing silver on single gold nanoparticles (called silver enhancement).

Magnetic Nanoparticles. Magnetic nanoparticles are a class of contrast agents for *in vivo* cellular and molecular imaging (Lewin et al., 2000), tracking (Lewin et al., 2000), cell (Josephson et al., 1999b; Melville et al., 1975), molecular enrichment (Nam et al., 2003), and magnetic drug targeting. In particular, iron oxide nanoparticles have attracted considerable interest because of their superparamagnetic properties. When multidomain ferrimagnetic bulk material such as iron oxide is reduced in size, thermal energy is able to disrupt the fixed alignment of electron spins so that the spins are randomly aligned and the net magnetic moment is effectively nullified. However, in the presence of an external magnetic field, the spins align themselves parallel to the field and the net magnetic moment is much greater. This switchable magnetic property is typically displayed by nanometer-sized particles (<25 nm for iron oxides).

Iron oxide (Fe_3O_4) nanoparticles were first employed as a negative MR contrast agent (Duda et al., 1994) due to their ability to darken the area of surrounding tissue as a result of a change in T2 relaxation time of nearby water protons. Unlike positive T1 contrast agents (e.g., gadolinium chelates), which light up its near environment, iron oxide nanoparticles are able to alter the water-rich tissue environment. Furthermore, these superparamagnetic iron oxide nanoparticles (SPION) are less toxic than gadolinium and can be metabolized and stored by the liver. Physicians and radiologists later observed that the human body's reticuloendothelial system (RES) was very efficient at clearing the nanoparticles into the liver, spleen, and lymph nodes, and so these organs were naturally the first to be imaged using MRI with iron oxide as a contrast agent.

Weissleder and co-workers first demonstrated *in vivo* targeting and imaging with SPION derivatized with dextran and cross-linked with epichlorohydrin (CLIO) to allow conjugation of targeting agents and improve circulation time in blood (Josephson et al., 1999a; Shen et al., 1993). By conjugating HIV-TAT peptide (Josephson et al., 1999a), the same iron oxide nanoparticles were shown to internalize into several cell types (Lewin et al., 2000) such as human CD4+ lymphocytes and mouse neural progenitor cells. It was also possible to track these cells by MRI (Zhao et al., 2002). Weissleder and co-workers further observed a unique phenomenon associated with magnetic nanoparticle aggregation (Perez et al., 2002). The group observed extensive changes (lowering) in the spin–spin relaxation time (T2) of surrounding water protons when target binding caused magnetic nanoparticles' aggregation. The authors detected this change with a relaxometer when magnetic nanoparticles formed aggregates as a result of binding to nucleic acid molecules, protein molecules, and viruses (Perez et al., 2003).

4.5. INTRACELLULAR AND *IN VIVO* APPLICATIONS

Quantum Dots. Latex beads (Meier et al., 2001) and dye molecules have long been used for tracking cell surface receptor dynamics. However, large nanoparticles can overwhelm the native physiology of the cell, and smaller dye molecules suffer from photobleaching. Quantum dots fall in the mesoscopic size range and do not suffer from photobleaching. Realizing the benefits of these unique properties of QDs, Dahan et al. (2003) have used quantum dots to study the diffusion dynamics of glycine receptors on the neuronal membrane of living cells. Glycine is an inhibitory neurotransmitter found in the central nervous system (Vannier and Triller, 1997). GlyR, the receptor for glycine, is present in clusters in the postsynaptic face of a nerve terminal (Triller et al., 1985). Gephyrin is a scaffolding protein known to stabilize these clusters or microdomains via mictotubule attachment of the GlyR receptors (Kirsch et al., 1993). The movement of these receptors within the cell membrane is critical in understanding synapse plasticity (Choquet and Triller, 2003). Dahan et al. first

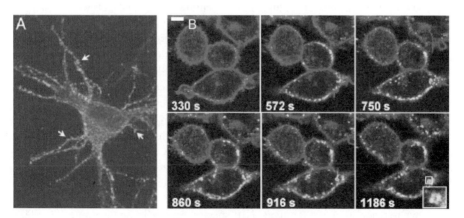

Figure 4.7. Single QDs for live cell imaging. (A) Red QDs bound to GlyR receptor on neuron cell surface. Microtubule associated protein protein-2 is in green. [Reprinted from Dahan et al. (2003), with permission of American Association for the Advancement of Science.] (B) EGF-QDs endocytosed into CHO cells in a time dependent fashion. GFP-EGFR is in green. [Reprinted from Lidke et al. (2004), with permission of Nature Publishing Group.] See insert for color representation of this figure.

incubated spinal cultured neurons with primary antibodies against the GlyR receptor on the neuron surface. Next, biotinylated secondary antibody fragments were allowed to bind to the primary antibodies on the cell surface, followed by addition of streptavidin-coated quantum dots. Single QDs were imaged with an exposure of 75 ms for approximately 60 s (Figure 4.7a). Even at such short exposure time, a signal-to-noise ratio of about 50 was obtained, which allowed single QD tracking on the cell surface with a lateral resolution of 5 nm. It was found that the GlyR receptors were distributed in three distinct regions: synaptic, perisynaptic, and extrasynaptic. Since GlyR receptors are clustered in the synaptic region, lower QD diffusion rates were observed in the synaptic and perisynaptic regions. Furthermore, the diffusion coefficients observed in the extrasynaptic region with the QDs were much higher than those observed with 500-nm latex in a previous study (Meier et al., 2001), confirming that large latex beads retarded the receptor diffusion in the membrane of the cell. Since QDs are electron-dense, the authors were also able to perform electron microscopy to confirm their findings. Following this report, Lidke et al. (2004) demonstrated use of QDs in probing the mechanism of erbB/HER receptor-mediated signal transduction. Briefly, CHO cells were incubated with streptavidin-coated QDs coupled with biotinylated epidermal growth factor (EGF). Binding of QD-EGF with the EGF receptor on the cell surface allowed observation of single QDs on the cell membrane and inside the cell after endocytosis (Figure 4.8b). Using the data obtained, the authors detected homodimer formation of the epidermal growth factor receptors on the cell surface.

For *in vivo* studies using animal models, Akerman et al. (2002) linked QDs with peptide ligands that were specific for murine lung cells. After injection of the QD-peptide construct through the tail vein of a mouse, the animal was sacrificed and its tissue was examined. The QDs were found to localize specifically in the lungs. This result was reinforced by similar experiments targeting blood and lymphatic vessels in tumors. Furthermore, PEG molecules (MW = 5000) on the surface improved circulation time and enabled the QD probe to evade the RES (reticuloendothelial system), which is the body's foreign particle filtration system. One step further, Dubertret et al. (2002) developed a more robust coating and demonstrated the biocompatibility and lack of toxicity for QD probes under *in vivo* conditions. A phospholipids-PEG block copolymer coating was used to stabilize the dots in an aqueous environment and render them biocompatible. The coated QDs were then injected into *Xenopus* embryos, which were used due to their sensitivity to toxicity and because small cellular disturbances could be revealed in biological phenotypes. Surprisingly, even when injected with 2 billion QDs per cell, the embryos showed little change in phenotype, although abnormal traits were evident at concentrations above 5 billion QDs per cell.

In a recent paper, Gao et al. (2004) reported *in vivo* cancer imaging and targeting with a new class of bioconjugated quantum dots. Antibodies specific to a prostate cancer marker called PSMA were conjugated to QDs coated with a PEGylated ABC triblock copolymer. The probes were injected into nude mice implanted with human prostate cancer cells and *in vivo* imaging was performed. The results showed QDs accumulated specifically at the implanted tumor site as shown in Figure 4.8. Due to high autofluorescence from the mouse skin, a spectral unmixing algorithm was applied to subtract the background fluorescence. This would not be necessary if NIR QDs are used because autofluorescence emission has been shown to taper off at ∼800 nm and NIR spectral region lies in the range 700–900 nm. To that effect, another paper reported the use of NIR QDs for sentinel lymph node mapping (Kim et al., 2004). By injecting NIR QDs into pigs and mice and then imaging with an NIR imaging setup, a surgeon was able to identify the lymph node without interference from autofluorescence. Although this paper marks the first *in vivo* application of NIR QDs

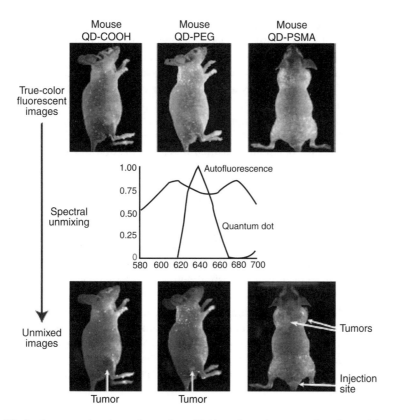

Figure 4.8. *In vivo* cancer imaging and targeting with bioconjugated quantum dots. By applying a spectral unmixing algorithm, it is possible to subtract background autofluorescence and visualize visible-range emitting QD aggregation at target sites. Also illustrated is evidence that actively targeted QD probes (QD-PSMA) reach the implanted tumor more efficiently than passive targeting (QD-PEG) and no targeting (QD-COOH). [Reprinted from Gao et al. (2004) in *Nature Biotechnology*.]

that can be detected, considerable work still needs to be done. The quality of NIR QDs needs to be improved so that they are bright, stable, and tunable. In fact, Kim et al. have predicted that the two optimum spectral windows for *in vivo* QD imaging are 700–900 nm and 1200–1600 nm. The ultimate goal of this technology is its application for clinical use in human subjects. In order to realize this, toxicity issues need to be addressed. As noted earlier, QDs are typically composed of CdSe, of which Cd^{2+} ions are especially toxic to living cells (Derfus et al., 2004a). The experiments discussed in this section so far indicate that there are no significant short-term toxic effects of QDs if they are stable and protected with a ZnS layer. However, detailed and long-term studies need to be performed on how QDs can affect the body upon degradation.

Magnetic Nanoparticles. Magnetic nanoparticles include paramagnetic gadolinium-DTPA molecules, magnetodendrimers, and superparamagnetic iron oxide nanoparticles (SPION). We shall focus on superparamagnetic nanoparticles and their applications in molecular and cellular imaging *in vivo*. There are several advantages of SPION that make it a good candidate for *in vivo* imaging. First, iron oxide is nontoxic in low dosage since it can be metabolized, and iron is essential for normal cellular growth (Arbab et al., 2003). Second,

SPION have been developed as an MRI negative contrast agent for liver imaging and are thus familiar to radiologists. Third, SPION with the appropriate coating have been shown to circulate in the blood for 24–36 hours (Weissleder et al., 2000).

In an effort to incorporate SPION into cells and track them *in vivo*, Lewin et al. (2000) designed a multifunctional SPION by using both a delivery peptide (Tat) and a fluorophore (FITC). Compared to untagged particles, SPION-Tat particles uptake was improved by nearly 100-fold in human CD4+ lymphocytes (Lewin et al., 2000). Similar improvements were also obtained for human CD34+ progenitor cells, mouse neural progenitor cells, and mouse splenocytes. A major finding was that upon introduction of these SPION-loaded cells into mice, MRI detected homing of these cells to the bone marrow.

Application of SPION to imaging of tumors and angiogenesis also holds great promise. In particular, labeling deep tissue tumor to determine their physical parameters such as tumor mass, volume, degree of neovascularization, and stage of metastasis can be accomplished with SPION. An even more exciting prospect is the ability to deliver SPION conjugated to anti-angiogenic therapeutic agents. In fact, targeting of integrins using Gd-based paramagnetic liposomes has been shown (Sipkins et al., 1998). In a separate study, TGF-β receptors in the endothelium have been targeted using radioisotope-based particles (Bredow et al., 2000). Studies have also been conducted on imaging gene expression using SPION. For instance, tumor cells expressing an engineered transferrin receptor (ETR) and those not expressing the receptor were implanted onto opposite flanks of a mouse. SPION conjugated to transferrin ligand was introduced into the mouse, and MR imaging was performed *in vivo* after 24 hours (Weissleder et al., 2000). The results showed that the cells expressing the receptor (ETR+) had internalized SPION but not the ETR-negative cells, as determined by the negative contrast.

4.6. OUTLOOK

QDs and nanoparticles have already fulfilled some of their promises as new imaging and detection probes. Through their versatile polymer coatings, nanoparticles have also provided a "building block" to assemble multifunctional nanostructures and nanodevices. Multimodality imaging probes could be created by integrating QDs with paramagnetic or superparamagnetic agents. Indeed, researchers have recently attached QDs to Fe_2O_3 and FePt nanoparticles (Gu et al., 2004; Wang et al., 2004) and even to paramagnetic gadolinium chelates (X. Gao and S. Nie, unpublished data). By correlating the deep imaging capabilities of magnetic resonance imaging (MRI) with ultrasensitive optical imaging, a surgeon could visually identify tiny tumors or other small lesions during an operation and remove the diseased cells and tissue completely. Medical imaging modalities such as MRI and PET can identify diseases noninvasively, but they do not provide a visual guide during surgery. The development of magnetic or radioactive QD probes could solve this problem.

Another desired multifunctional device would be the combination of a QD imaging agent with a therapeutic agent. Not only would this allow tracking of pharmacokinetics, but diseased tissue could be treated and monitored simultaneously in realtime. Surprisingly, QDs may be innately multimodal in this fashion, because they have been shown to have potential activity as photodynamic therapy agents (Samia et al., 2003). These combinations are only a few possible achievements for the future. Practical applications of multifunctional nanodevices will not come without careful research, but the multidisciplinary nature of nanotechnology may expedite these goals by combining great minds of many different fields. The results seen so far with QDs point toward the success of QDs in biological systems and also predict the success of other nanotechnologies for biomedical applications.

REFERENCES

Agrawal, A, Gao, X, Nitin, N, Bao, G, and Nie, S (2003). Quantum Dots and FRET-Nanobeads for Probing Genes, Proteins, and Drug Targets in Single Cells. *ASME International Mechanical Engineering Congress and Exposition*. ASME, Washington, D.C.

Agrawal, A, Tripp, RA, Anderson, LJ, and Nie, S (2005). Real-time detection of virus particles and viral protein expression with two-color nanoparticle probes. *J Virol* **79**:8625–8628.

Akerman, ME, Chan, WC, Laakkonen, P, Bhatia, SN, and Ruoslahti, E (2002). Nanocrystal targeting *in vivo*. *Proc Natl Acad Sci USA* **99**:12617–12621.

Alivisatos, AP (1996). Semiconductor clusters, nanocrystals, and quantum dots. *Science* **271**:933–937.

Arbab, AS, Bashaw, LA, Miller, BR, Jordan, EK, Lewis, BK, Kalish, H, and Frank, JA (2003). Characterization of biophysical and metabolic properties of cells labeled with superparamagnetic iron oxide nanoparticles and transfection agent for cellular MR imaging. *Radiology* **229**:838–846.

Bailey, RE, and Nie, S (2003). Alloyed semiconductor quantum dots: Tuning the optical properties without changing the particle size. *J Am Chem Soc* **125**:7100–7106.

Bhakdi, S, Tranum-Jensen, J, and Sziegoleit, A (1985). Mechanism of membrane damage by streptolysin-O. *Infect Immun* **47**:52–60.

Bhakdi, S, Weller, U, Walev, I, Martin, E, Jonas, D, and Palmer, M (1993). A guide to the use of pore-forming toxins for controlled permeabilization of cell-membranes. *Med Microbiol Immunol* **182**:167–175.

Boal, AK, Ilhan, F, DeRouchey, JE, Thurn-Albrecht, T, Russell, TP, and Rotello, VM (2000). Self-assembly of nanoparticles into structured spherical and network aggregates. *Nature* **404**:746–748.

Bredow, S, Lewin, M, Hofmann, B, Marecos, E, and Weissleder, R (2000). Imaging of tumour neovasculature by targeting the TGF-beta binding receptor endoglin. *Eur J Cancer* **36**:675–681.

Brooks, RA, Moiny, F, and Gillis, P (2001). On T-2-shortening by weakly magnetized particles: The chemical exchange model. *Magn Reson Med* **45**:1014–1020.

Bruchez, M, Jr, Moronne, M, Gin, P, Weiss, S, and Alivisatos, AP (1998). Semiconductor nanocrystals as fluorescent biological labels. *Science* **281**:2013–2016.

Bulte, JWM, and Brooks, RA (1997). Magnetic nanoparticles as contrast agents for MR imaging. In: Häfeli, U, Schütt, W, Teller, J, and Zborowski, M (eds.), *Scientific and Clinical Applications of Magnetic Carriers*. Plenum Press, New York, pp. 527–543.

Bulte, JWM, Douglas, T, Witwer, B, Zhang, SC, Strable, E, Lewis, BK, Zywicke H, Miller, B, van Gelderen, P, Moskowitz, BM, Duncan, ID, and Frank, JA (2001). Magnetodendrimers allow endosomal magnetic labeling and *in vivo* tracking of stem cells. *Nat Biotechnol* **19**:1141–1147.

Chan, WC, and Nie, S (1998). Quantum dot bioconjugates for ultrasensitive nonisotopic detection. *Science* **281**:2016–2018.

Chan, WC, Maxwell, DJ, Gao, X, Bailey, RE, Han, M, and Nie, S (2002). Luminescent quantum dots for multiplexed biological detection and imaging. *Curr Opin Biotechnol* **13**:40–46.

Choquet, D, and Triller, A (2003). The role of receptor diffusion in the organization of the postsynaptic membrane. *Nat Rev Neurosci* **4**:251–265.

Curtis, A, and Wilkinson, C (2001). Nanotechniques and approaches in biotechnology. *Trends Biotechnol* **19**:97–101.

Dahan, M, Levi, S, Luccardini, C, Rostaing, P, Riveau, B, and Triller, A (2003). Diffusion dynamics of glycine receptors revealed by single-quantum dot tracking. *Science* **302**:442–445.

Derfus, AM, Chan, WCW, and Bhatia, SN (2004a). Probing the cytotoxicity of semiconductor quantum dots. *Nano Lett* **4**:11–18.

Derfus, AM, Warren, CWC, and Bhatia, SN (2004b). Intracellular delivery of quantum dots for live cell labeling and organelle tracking. *Adv Mater* **16**:961–966.

Dickson, RB, Willingham, MC, and Pastan, I (1981). Alpha 2-macroglobulin adsorbed to colloidal gold: A new probe in the study of receptor-mediated endocytosis. *J Cell Biol* **89**:29–34.

Dubertret, B, Skourides, P, Norris, DJ, Noireaux, V, Brivanlou, AH, and Libchaber, A (2002). *In vivo* imaging of quantum dots encapsulated in phospholipid micelles. *Science* **298**:1759–1762.

Duda, SH, Laniado, M, Kopp, AF, Gronewaller, E, Aicher, KP, Pavone, P, Jehle, E, and Claussen, CD (1994). Superparamagnetic iron oxide: Detection of focal liver lesions at high-field-strength MR imaging. *J Magn Reson Imaging* **4**:309–314.

Elghanian, R, Storhoff, JJ, Mucic, RC, Letsinger, RL, and Mirkin, CA (1997). Selective colorimetric detection of polynucleotides based on the distance-dependent optical properties of gold nanoparticles. *Science* **277**:1078–1081.

Faulk, WP, and Taylor, GM (1971). An immunocolloid method for the electron microscope. *Immunochemistry* **8**:1081–1083.

Gao, X, Cui, Y, Levenson, RM, Chung, LW, and Nie, S (2004). *In vivo* cancer targeting and imaging with semiconductor quantum dots. *Nat Biotechnol* **22**:969–976.

Gao, X, Yang, L, Petros, JA, Marshall, FF, Simons, JW, and Nie, S (2005). *In vivo* molecular and cellular imaging with quantum dots. *Curr Opin Biotechnol* **16**:63–72.

Gao, XH, and Nie, SM (2003). Doping mesoporous materials with multicolor quantum dots. *J Phys Chem B* **107**:11575–11578.

Grabar, KC, Freeman, RG, Hommer, MB, and Natan, MJ (1995). Preparation and characterization of Au colloid monolayers. *Anal Chem* **67**:735–743.

Gref, R, Minamitake, Y, Peracchia, MT, Trubetskoy, V, Torchilin, V, and Langer, R (1994). Biodegradable long-circulating polymeric nanospheres. *Science* **263**:1600–1603.

Gu, HW, Zheng, RK, Zhang, XX, and Xu, B (2004). Facile one-pot synthesis of bifunctional heterodimers of nanoparticles: A conjugate of quantum dot and magnetic nanoparticles. *J Am Chem Soc* **126**:5664–5665.

Haes, AJ, and Van Duyne, RP (2002). A nanoscale optical biosensor: Sensitivity and selectivity of an approach based on the localized surface plasmon resonance spectroscopy of triangular silver nanoparticles. *J Am Chem Soc* **124**:10596–10604.

Han, M, Gao, X, Su, JZ, and Nie, S (2001). Quantum-dot-tagged microbeads for multiplexed optical coding of biomolecules. *Nat Biotechnol* **19**:631–635.

Henglein, A (1989). Small-particle research—physicochemical properties of extremely small colloidal metal and semiconductor particles. *Chem Rev* **89**:1861–1873.

Hermanson, GT (1996). *Bioconjugate Techniques*. Academic Press (Elsevier Science USA), San Diego.

Ishii, D, Kinbara, K, Ishida, Y, Ishii, N, Okochi, M, Yohda, M, and Aida, T (2003). Chaperonin-mediated stabilization and ATP-triggered release of semiconductor nanoparticles. *Nature* **423**:628–632.

Jaiswal, JK, Mattoussi, H, Mauro, JM, and Simon, SM (2003). Long-term multiple color imaging of live cells using quantum dot bioconjugates. *Nat Biotechnol* **21**:47–51.

Jeanmaire, DL, and Vanduyne, RP (1977). Surface Raman spectroelectrochemistry. 1. Heterocyclic, aromatic, and aliphatic-amines adsorbed on anodized silver electrode. *J Electroanal Chem* **84**:1–20.

REFERENCES

Josephson, L, Tung, CH, Moore, A, and Weissleder, R (1999a). High-efficiency intracellular magnetic labeling with novel superparamagnetic-tat peptide conjugates. *Bioconjug Chem* **10**:186–191.

Josephson, L, Tung, CH, Moore, A, and Weissleder, R (1999b). High-efficiency intracellular magnetic labeling with novel superparamagnetic-tat peptide conjugates. *Bioconjug Chem* **10**:186–191.

Kim, S, Fisher, B, Eisler, HJ, and Bawendi, M (2003). Type-II quantum dots: CdTe/CdSe(core/shell) and CdSe/ZnTe(core/shell) heterostructures. *J Am Chem Soc* **125**:11466–11467.

Kim, S, Lim, YT, Soltesz, EG, De Grand, AM, Lee, J, Nakayama, A, Parker, JA, Mihaljevic, T, Laurence, RG, Dor, DM, Cohn, LH, Bawendi, MG, and Frangioni, JV (2004). Near-infrared fluorescent type II quantum dots for sentinel lymph node mapping. *Nat Biotechnol* **22**: 93–97.

Kirsch, J, Wolters, I, Triller, A, and Betz, H (1993). Gephyrin antisense oligonucleotides prevent glycine receptor clustering in spinal neurons. *Nature* **366**:745–748.

Klarreich, E (2001). Biologists join the dots. *Nature* **413**:450–452.

Kreibig, U, and Genzel, L (1985). Optical-absorption of small metallic particles. *Surf Sci* **156**:678–700.

Larson, DR, Zipfel, WR, Williams, RM, Clark, SW, Bruchez, MP, Wise, FW, and Webb, WW (2003). Water-soluble quantum dots for multiphoton fluorescence imaging *in vivo*. *Science* **300**:1434–1436.

Lewin, M, Carlesso, N, Tung, CH, Tang, XW, Cory, D, Scadden, DT, and Weissleder, R (2000). Tat peptide-derivatized magnetic nanoparticles allow *in vivo* tracking and recovery of progenitor cells. *Nat Biotechnol* **18**:410–414.

Li, M, Schnablegger, H, and Mann, S (1999). Coupled synthesis and self-assembly of nanoparticles to give structures with controlled organization. *Nature* **402**:393–395.

Lidke, DS, Nagy, P, Heintzmann, R, Arndt-Jovin, DJ, Post, JN, Grecco, HE, Jares-Erijman, EA, and Jovin, TM. (2004). Quantum dot ligands provide new insights into erbB/HER receptor-mediated signal transduction. *Nat Biotechnol* **22**:198–203.

Mattoussi, H, Mauro, JM, Goldman, ER, Anderson, GP, Sundar, VC, Mikulec, FV, and Bawendi, MG (2000). Self-assembly of CdSe–ZnS quantum dot bioconjugates using an engineered recombinant protein. *J Am Chem Soc* **122**:12142–12150.

Medintz, IL, Clapp, AR, Mattoussi, H, Goldman, ER, Fisher, B, and Mauro, JM (2003). Self-assembled nanoscale biosensors based on quantum dot FRET donors. *Nat Mater* **2**:630–638.

Meier, J, Vannier, C, Serge, A, Triller, A, and Choquet, D (2001). Fast and reversible trapping of surface glycine receptors by gephyrin. *Nat Neurosci* **4**:253–260.

Melville, D, Paul, F, and Roath, S (1975). Direct magnetic separation of red-cells from whole-blood. *Nature* **255**:706–706.

Mitchell, P (2001). Turning the spotlight on cellular imaging—Advances in imaging are enabling researchers to track more accurately the localization of macromolecules in cells. *Nat Biotechnol* **19**:1013–1017.

Nam, JM, Thaxton, CS, and Mirkin, CA (2003). Nanoparticle-based bio-bar codes for the ultrasensitive detection of proteins. *Science* **301**:1884–1886.

Nie, SM, and Emery, SR (1997). Probing single molecules and single nanoparticles by surface-enhanced Raman scattering. *Science* **275**:1102–1106.

Nitin, N, LaConte, LEW, Zurkiya, O, Hu, X, and Bao, G (2004). Functionalization and peptide-based delivery of magnetic nanoparticles as an intracellular MRI contrast agent. *J Biol Inorg Chem* **9**:706–712.

Palmacci, S, and Josephson, L (1993). Synthesis of polysaccharide covered superparamagnetic oxide colloids. U.S. Patent.

Palmer, M (2001). The family of thiol-activated, cholesterol-binding cytolysins. *Toxicon* **39**:1681–1689.

Palmer, M, Harris, R, Freytag, C, Kehoe, M, Tranum-Jensen, J, and Bhakdi S (1998). Assembly mechanism of the oligomeric streptolysin O pore: The early membrane lesion is lined by a free edge of the lipid membrane and is extended gradually during oligomerization. *EMBO J* **17**:1598–1605.

Perez, JM, Josephson, L, O'Loughlin, T, Hogemann, D, and Weissleder R (2002). Magnetic relaxation switches capable of sensing molecular interactions. *Nat Biotechnol* **20**:816–820.

Perez, JM, Simeone, FJ, Saeki, Y, Josephson, L, and Weissleder, R (2003). Viral-induced self-assembly of magnetic nanoparticles allows the detection of viral particles in biological media. *J Am Chem Soc* **125**:10192–10193.

Rosi, NL, and Mirkin, CA (2005). Nanostructures in biodiagnostics. *Chem Rev* **105**:1547–1562.

Samia, ACS, Chen, XB, and Burda, C (2003). Semiconductor quantum dots for photodynamic therapy. *J Am Chem Soc* **125**:15736–15737.

Savic, R, Luo, LB, Eisenberg, A, and Maysinger, D (2003). Micellar nanocontainers distribute to defined cytoplasmic organelles. *Science* **300**:615–618.

Schmid, G (1992). Large clusters and colloids—metals in the embryonic state. *Chem Rev* **92**:1709–1727.

Shen, T, Weissleder, R, Papisov, M, Bogdanov, A, Jr, and Brady, TJ (1993). Monocrystalline iron oxide nanocompounds (MION): physicochemical properties. *Magn Reson Med* **29**:599–604.

Sipkins, DA, Cheresh, DA, Kazemi, MR, Nevin, LM, Bednarski, MD, and Li, KC (1998). Detection of tumor angiogenesis *in vivo* by alphaVbeta3-targeted magnetic resonance imaging. *Nat Med* **4**:623–626.

Soo, PL, Luo, LB, Maysinger, D, and Eisenberg, A. (2002). Incorporation and release of hydrophobic probes in biocompatible polycaprolactone-block-poly(ethylene oxide) micelles: Implications for drug delivery. *Langmuir* **18**:9996–10004.

Stephens, DJ, and Pepperkok, R (2001). The many ways to cross the plasma membrane. *Proc Natl Acad Sci USA* **98**:4295–4298.

Stimpson, DI, Hoijer, JV, Hsieh, WT, Jou, C, Gordon, J, Theriault, T, Gamble, R, and Baldeschwieler, JD (1995). Real-time detection of DNA hybridization and melting on oligonucleotide arrays by using optical wave guides. *Proc Natl Acad Sci USA* **92**:6379–6383.

Sun, SH, Murray, CB, Weller, D, Folks, L, and Moser, A (2000). Monodisperse FePt nanoparticles and ferromagnetic FePt nanocrystal superlattices. *Science* **287**:1989–1992.

Taton, TA, Mirkin, CA, and Letsinger, RL (2000). Scanometric DNA array detection with nanoparticle probes. *Science* **289**:1757–1760.

Templeton, AC, Wuelfing, MP, and Murray, RW (2000). Monolayer protected cluster molecules. *Acc Chem Res* **33**:27–36.

Triller, A, Cluzeaud, F, Pfeiffer, F, Betz, H, and Korn, H (1985). Distribution of glycine receptors at central synapses: an immunoelectron microscopy study. *J Cell Biol* **101**:683–688.

van de Hulst, HC (1981). Light Scattering by small particles. In: *Light Scattering by Small Particles*. Dover, New York, pp. 397–400.

Vannier, C, and Triller, A (1997). Biology of the postsynaptic glycine receptor. *Int Rev Cytol* **176**:201–244.

Vives, E (2003). Cellular uptake [correction of utake] of the Tat peptide: An endocytosis mechanism following ionic interactions. *J Mol Recognit* **16**:265–271.

Vives, E, Richard, JP, Rispal, C, and Lebleu, B (2003). TAT peptide internalization: seeking the mechanism of entry. *Curr Protein Pept Sci* **4**:125–132.

Wadia, JS, Stan, RV, and Dowdy, SF (2004). Transducible TAT-HA fusogenic peptide enhances escape of TAT-fusion proteins after lipid raft macropinocytosis. *Nat Med* **10**:310–315.

Wang, DS, He, JB, Rosenzweig, N, and Rosenzweig, Z (2004). Superparamagnetic Fe2O3 Beads-CdSe/ZnS quantum dots core-shell nanocomposite particles for cell separation. *Nano Lett* **4**:409–413.

Wehrenberg, BL, Wang, CJ, and Guyot-Sionnest, P (2002). Interband and intraband optical studies of PbSe colloidal quantum dots. *J Phys Chem B* **106**:10634–10640.

Weissleder, R, Moore, A, Mahmood, U, Bhorade, R, Benveniste, H, Chiocca, EA, and Basilion, JP (2000). *In vivo* magnetic resonance imaging of transgene expression. *Nat Med* **6**:351–355.

Wu, XY, Liu, HJ, Liu, JQ, Haley, KN, Treadway, JA, Larson, JP, Ge, NF, Peale, F, and Bruchez, MP (2003). Immunofluorescent labeling of cancer marker Her2 and other cellular targets with semiconductor quantum dots. *Nat Biotechnol* **21**:41–46.

Zhao, M, Kircher, MF, Josephson, L, and Weissleder, R (2002). Differential conjugation of tat peptide to superparamagnetic nanoparticles and its effect on cellular uptake. *Bioconjug Chem* **13**:840–844.

Zhao, M, and Weissleder, R (2004). Intracellular cargo delivery using tat peptide and derivatives. *Med Res Rev* **24**:1–12.

Zhong, X, Feng, Y, Knoll, W, and Han, M (2003a). Alloyed Zn(x)Cd(1-x)S nanocrystals with highly narrow luminescence spectral width. *J Am Chem Soc* **125**:13559–13563.

Zhong, X, Han, M, Dong, Z, White, TJ, and Knoll, W (2003b). Composition-tunable Zn(x)Cd(1-x)Se nanocrystals with high luminescence and stability. *J Am Chem Soc* **125**:8589–8594.

CHAPTER

5

TAILORING NANOPARTICLES FOR THE RECOGNITION OF BIOMACROMOLECULE SURFACES

MRINMOY DE, ROCHELLE R. ARVIZO, AYUSH VERMA, AND VINCENT M. ROTELLO

5.1. INTRODUCTION

Numerous cellular processes rely on biomolecular interactions such as protein–protein interactions, protein–nucleic acid interactions, enzyme activity, and cell surface recognition. These interactions are primarily due to complementary electrostatic, hydrophobic, and polar surface interactions between biomacromolecules. These interactions pave the way for alternative approaches to diagnostic biosensors for rapid monitoring of imbalances and illnesses, as well as therapeutic agents. More than that selective binding of the synthetic receptors at biomacromolecular surfaces provides an alternative approach to the predominantly active-site-based inhibitors, thus generating new classes of enzyme inhibitors.

Recognition of biomacromolecular surfaces using synthetic receptors is based on noncovalent host–guest interaction as observed in small molecule systems. However, the regulation of biomacromolecules remains a far more significant challenge. This challenge is primarily due to two basic requirements for an effective recognition between a biomacromolecule and its receptor. First of all, a large biomacromolecule–receptor contact area is required. For example, relatively large surface areas are required for effective binding of protein surfaces (which are solvent exposed) to successfully inhibit the active site. Insight into this requirement comes from examination of protein–protein interactions, which reveals that a surface area of more than 6 nm^2 per protein are typically involved in such interactions (Conte et al., 1999). The second challenge stems from the complexities of the surfaces involved (Rinaldis et al., 1998) in terms of their multiple electrostatic (Golumbfskie et al., 1999), hydrophobic (Lijnzaad and Argos, 1997), and topological features. A number of "small molecule" systems (Park et al., 2002a) and macromolecular scaffolds have been used to address this challenge for protein surface recognition. "Small molecule" systems include receptors on calixarene and porphyrin scaffolds (Hamuro et al., 1997; Lin et al., 1998; Park et al., 1999; Ernst et al., 2003; Wilson et al., 2003), cyclodextrin dimers (Leung et al., 2000), and transition metal complexes targeted against surface-exposed histidines (Fazal et al., 2001). Macromolecule scaffolds include multivalent libraries of receptors possessing partially constrained backbones (Gordon et al., 1998; Strong and Kiessling, 1999; Gestwicki et al, 2002). Such systems demonstrate a certain level of success in modulation of biomacromolecular function, but the question of protein surface recognition still remains

---

New Frontiers in Ultrasensitive Bioanalysis. Edited by Xiao-Hong Nancy Xu
Copyright © 2007 John Wiley & Sons, Inc.

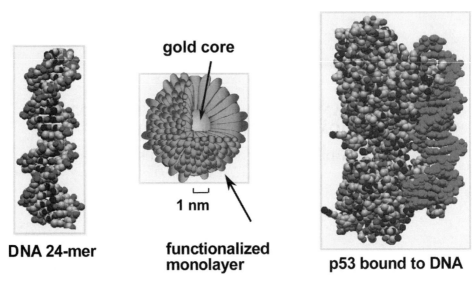

Figure 5.1. Relative sizes of a nanoparticle with a 2-nm core and an octanethiol-functionalized monolayer and possible biological targets.

indistinct. Nanoparticle-based receptors, however, offer a platform for biomacromolecular surface recognition which is unique and distinctive in its own way. The use of core-shell nanoparticle systems, such as monolayer protected clusters (MPCs) and mixed monolayer protected clusters (MMPCs), possesses some distinctive and significant features that make them promising scaffolds for creation of receptors targeted to biomacromolecular surfaces. The first important feature is that the size of the nanoparticle core can be tuned from 1.5 to 8 nm with overall diameters of 2.5–11 nm (Hostetler et al., 1998). This variability of core sizes provides a suitable platform for the interaction of nanoparticles with biomacromolecules on comparable size scales (Figure 5.1).

The second useful property of nanoparticles is that they can be fabricated with a wide range of surface functionality, thus providing a versatile route to creation of surface-specific receptors. This provides a unique tool for achieving efficient and specific recognition of protein surfaces via complementary surface interactions. The third important feature of MMPCs is that a range of metal and semiconductor cores can be generated featuring useful electronic, fluorescence, and magnetic properties that allow for use as probes and/or diagnostic agents. Finally, these systems have been shown to self-template to guest molecules, allowing an increase in the affinity and selectivity upon incubation with the guest molecules (Boal and Rotello, 2000; Verma et al., 2004a). MPCs and MMPCs provide a definite advantage over the conventionally used synthetic receptors, which are limited in their ability to mimic biological interactions due to their inherent rigidity.

Recently, numerous studies have focused on the use of nanoparticles as a "solid-phase" support where they have been used as structural building blocks or as a visualization aid for sensor studies. Examples of such studies include interactions between streptavidin and biotin-labeled particles (Riepl et al., 2002; Haes and Van Duyne, 2002; Raschke et al., 2003) or hybridization of complementary DNA strands conjugated to nanoparticles (Yun et al., 2002; Park et al., 2002b; Cao et al., 2002). However, the focus of this chapter will be the properties and utilization of monolayer and mixed monolayer-protected nanoparticles using the functional organic groups on the nanoparticle surface as multivalent recognition elements targeted at biomacromolecular surfaces. The use of MPCs and MMPCs as synthetic receptors for biomolecular recognition allows for modulation of activities of proteins and nucleic acids, not possible through the traditional use of nanoparticles as support elements. Furthermore, MPCs and MMPCs can allow comparisons of biological complex formation such as protein–protein interactions based on surface complementarity.

5.2. FABRICATION AND PROPERTIES OF MONOLAYER AND MIXED MONOLAYER PROTECTED CLUSTERS

Due to the ease of synthesis and the ability to fabricate surfaces with various functional groups, MPCs and MMPCs provide a useful tool for biomacromolecular surface recognition. Core-shelled nanoparticles contain a cluster of metal atoms forming a truncated octahedron shape, immediately surrounded by a self-assembled monolayer (SAM) or mixed monolayer (SAMM) (Templeton et al., 2000). The monolayer helps prevent aggregation and provides stability to the nanoparticle. Another important function of this monolayer or mixed monolayer is providing possible recognition elements upon the surface of the nanoparticle.

Brust et al. (1994) first reported the synthesis of fabricated MPCs by reducing the metal salt in the presence of capping ligands such as thiols (Figure 5.2a). In this process, mild reducing agents (e.g., $NaBH_4$) were used for the wide range of capping ligand functionality. This procedure was found most suitable for a series of noble metal cores such as Pd, Au, Ag, and Pt. The size of the metal cluster can be tuned in two ways: (1) varying the ratio of the metal salt to capping ligand and (2) controlling the rate of addition of reducing agent. Later it was found that the ligands on the surface of the metal cluster are exchangeable with different functionalized ligands by Murray and co-workers (Templeton et al., 1998a,b). Using this property, MPCs can be further functionalized through the Murray place displacement reaction to obtain MMPCs (Figure 5.2b). This method rapidly generates a wide range of functionalized MMPCs for various applications. In addition to the Murray place displacement reaction to generate functionalities on the surface, direct synthetic approach can be applied to introduce active groups on the surface (Templeton et al., 1998a,b). Fast reductant addition and cooled solutions produce smaller and more monodisperse particles. Also, larger thiol/gold mole ratios give smaller average core-sized particles. A higher abundance

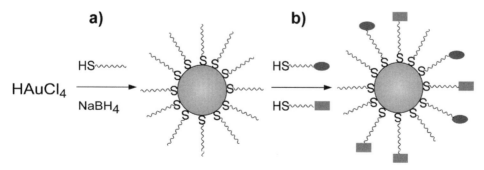

Figure 5.2. Synthesis of gold MPCs using (a) the Brust–Schiffrin reaction and MMPCs via (b) the Murray place-exchange method.

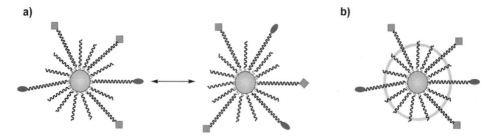

Figure 5.3. (a) Schematic representation of ligand mobility on nanoparticle surface. (b) Incorporation of internal hydrogen bonding functionality restricts the mobility of functionalized ligand.

of small core sizes (≤2 nm) is obtained by quenching the reaction immediately following reduction or by using sterically bulky ligands (Chen and Murray, 1999; Ingram et al., 1997). The number of ligand covered the nanoparticle depends on the core size of the metallic cluster. For example, approximately 100 ligand chains cover a MPC featuring a 2-nm gold core diameter (Templeton et al., 2000), yielding a surface area of ~110 nm^2 (functionalized with undecane thiols as ligands).

Besides the versatility in nanoparticle fabrication, the structural architecture of MPCs and MMPCs contain many unique properties. One such important structural characteristic of these systems is the radial dependence of the monolayer on the MPCs (Templeton et al., 2000). By using this radial dependence property, Rotello and co-workers synthesized amide functionalized nanoparticles featuring amides at various positions in the ligand chains to monitor the hydrogen bonding efficiency (Boal and Rotello, 2000b). The purpose of this study was the control of the position of ligands on the MMPCs' surface.

Since ligands are mobile on the surface of nanoparticle, it is simple to create MMPCs with self-optimizing receptors (Boal and Rotello, 2000a). This mobility of the ligands creates limitations in their application, however. In the absence of a guest molecule, the ligands on the nanoparticle will randomize (Figure 5.3a). One way to control the randomization is by incorporating internal, *intra*monolayer hydrogen-bonding elements in the monolayer (Figure 5.3b). Amide thiols with various chain lengths were used for this purpose because they can form two- and three-dimensional hydrogen-bonding networks. It was observed that hydrogen bonding was weakest when the amide was extremely close to the surface. As the amides were moving away from the gold core, an initial increase and then gradual decrease of hydrogen bonding was also observed. This trend supports the radial dependence of the monolayers on the MPCs. Amides extremely near the surface are too constrained to maximize hydrogen bonding, while at greater distances interactions become increasingly disfavored due to higher degree of freedom.

The mobility of the thiols on the self-assembled monolayer (SAM) surface presents the possibility of creating environmentally responsive systems. This particular property of the thiol ligands maximizes the binding enthalpy by dynamically optimizing the guest molecule association. To demonstrate this property of the nanoparticle, MMPC **1** was synthesized (Figure 5.4b) (Boal and Rotello, 2000a). MMPC **1** was fabricated with the aromatic stacking element pyrene and a hydrogen-bonding moiety, diamidopyridine, in an octanethiol monolayer. To examine the dynamic response of this system, flavin was introduced which can be stabilized by both hydrogen bonding and aromatic substituents. This provides a driving force for rearrangement of the thiols to an optimal binding configuration (Figure 5.4b). The time-course NMR experiment shows that upon addition of flavin in MMPC **1**, the

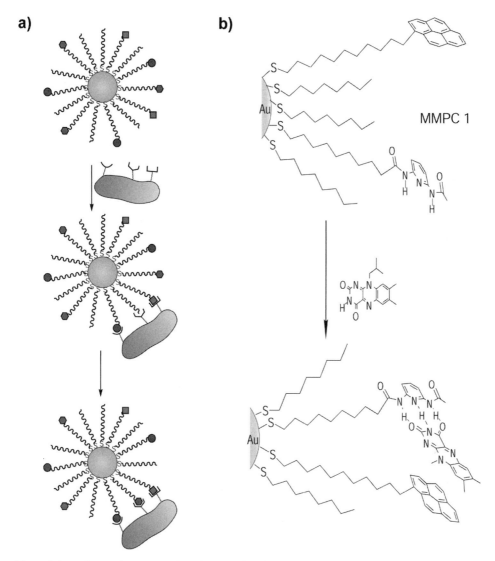

Figure 5.4. (a) Schematic representation of an MMPC optimizing to the surface of a biological target. (b) The templation of MMPC 1 on addition of flavin to allow optimal binding.

corresponding rearrangement of thiol takes place after 73 hours. Further study revealed that ~71% binding constant increases due to this templation process.

The study discussed above demonstrates the big advantage that MMPCs have over conventional rigid receptors. The surface recognition of MMPCs can be extended to systems with complex surface features, such as proteins, nucleic acids, and polysaccharides (Figure 5.4a). Thus, MMPCs functionalized with multiple recognition groups can be optimized against guest molecules targeted with various sites containing high selectivity and binding affinity.

Another important feature as a result of the cluster shape is that the monolayer-protected nanoparticles have faceted surfaces. To demonstrate this property, amide- and ester-functionalized MPCs (MMPC **2** and MMPC **3**) bearing end groups of varying steric

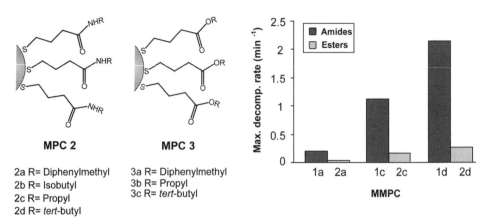

Figure 5.5. Maximum NaCN decomposition rates of the amide- and ester-functionalized MPCs.

bulk and π-stacking ability were synthesized. These MPCs were used to probe into the nature of end groups on hydrogen-bonding efficiency and independently test the protection provided by the monolayer-packing mode against cyanide-induced decomposition (Figure 5.5) (Paulini et al., 2002). The hydrogen-bonding efficiency and the resistance toward NaCN degradation of the core was found to decrease in the order diphenylmethyl > isobutyl > propyl > *tert*-butyl. The π-stacking ability of diphenylmethyl groups provided the best protection, while isobutyl end groups provided better steric packing than propyl groups. *Tert*-butyl groups, however, cause surface crowding and distort the chains. The cyanide-induced decomposition of corresponding ester analogs showed markedly slower decomposition rates compared to amide derivatives. This arises as a consequence of hydrogen-bonding in the amide monolayer which leads to the formation of facet-segregated bundles, exposing the vertices to CN − penetration.

Some chemical and physical properties of MPCs and MMPCs are dependent on the end-functional group of the nanoparticle. The end groups have greater importance toward biomacromolecular recognition. For example, MMPCs with hydrophilic end groups are water-soluble, which is an important prerequisite for most biological studies. Since the nanoparticle possesses a large surface-to-volume ratio, the behavior of the ligand on the surface of a nanoparticle is somewhat different from an unbound single ligand. The nanoparticle surface enhances the strength of the bound ligand, thus allowing for amplified interactions with biomacromolecules. One such good example is carbohydrate–protein interactions. It is well known that the affinity between carbohydrates and protein interactions is low. However, using carbohydrate-functionalized nanoparticles increased the affinity of this interaction due to the presence of multiple groups on the nanoparticles surface. This demonstrates that polyvalent interactions between the ligands and receptors are collectively stronger than corresponding monovalent interactions (Lin et al., 2002). The gold core also affects the properties of fluorophores by quenching the intrinsic fluorescence. The extent of quenching is dependent on the distance of the fluorophore from the metal score (Ipe et al., 2002; Wang et al., 2002a; Gu et al., 2003b).

The ability to create MMPCs with divergent functionalization and other biocompatible properties demonstrates that nanoparticles are suitable scaffolds for biomacromolecular surface recognition.

5.3. DNA SURFACE RECOGNITION

Multifunctional fabricated nanoparticles can be utilized as a receptor for recognition of biomacromolecular surfaces such as proteins, peptides, and DNA. Among all of the macromolecular surfaces, DNA is considered as a relatively simple surface for receptor targeting. In spite of its simplicity, the enormous specificity of the Watson–Crick hydrogen bonding allows for convenient programming of artificial DNA receptors. Another very attractive feature of DNA is the great mechanical rigidity of its short double helices. The helices effectively behave as a rigid rod spacer between two tethered functional molecular components on both ends. Moreover, DNA displays a relatively high physicochemical stability. Finally, nature provides a complete toolbox of highly specific biomolecular reagents, such as endonucleases, ligases, and other DNA-modifying enzymes. DNA can also be considered as a negatively charged flexible strand. Targeting the DNA, cellular activity can be directly regulated. Based on intercalation and major/minor groove binding, small molecules have been utilized to bind to specific DNA sequences (Dervan, 2001; Dervan and Edelson, 2003) to either inhibit (Wang et al., 2002b; Gottesfeld et al., 2002) or promote (Mapp et al., 2000; Coull et al., 2002) DNA transcription. The principles that allow receptor-DNA binding can be extended to nanoparticle systems by incorporating DNA-binding moieties onto the nanoparticle surface. This can be achieved in two ways. First, nanoparticles functionalized with a single strand of DNA is highly selective toward its complementary strand, thus displaying excellent sequence specificity. Second, we can utilize a network of noncovalent interactions such as complementary electrostatic interactions to promote high affinity of nanoparticle-DNA binding. Here we have focused on the utilization of a network of noncovalent interactions to promote the high affinity of nanoparticle-DNA binding.

To verify the possibility of MMPCs modulating DNA activity through noncovalent interactions, Rotello and co-workers used cationic MMPC **4** to bind DNA (McIntosh et al., 2001). MMPC **4** has a gold core of ∼2 nm, whose surface is encapsulated by positively charged trimethylammonium ligands, which can bind to the negatively charged phosphate backbone of 37mer duplex DNA via electrostatic complementarity (Figure 5.6a). The binding between the nanoparticle and DNA was examined by the use of UV centrifugation assay. This requires the DNA to change to a "bound" conformation in order to precipitate

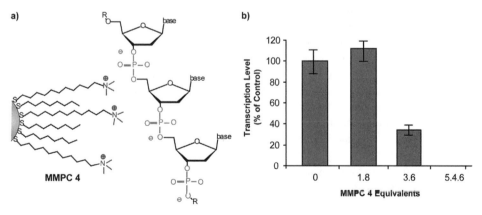

Figure 5.6. (a) Structure of MMPC 4 scaffold and the DNA backbone. (b) Percent transcription level increases with increase in the amount of MMPC 4.

from the solution. The study demonstrated that nanoparticles bind DNA with a 4:1 stoichiometry (nanoparticle to DNA). In order to determine the strength of binding between the DNA and nanoparticles, the ability of the nanoparticles to inhibit DNA transcription *in vitro* was measured (Figure 5.6b). In this experiment, Rotello and co-workers demonstrated that upon incubation with DNA, MMPC **4** effectively inhibited DNA transcription by T7 RNA polymerase. The DNA:polymerase complex is estimated to have a K_d of approximately 5 nM (Kuzmine and Martin, 2001), indicating that either the altered conformation of the nanoparticle-bound DNA interrupts the recognition process or MMPC **4** binds with higher affinity than the T7 RNA polymerase. Previously, extended aggregates of nanoparticles in solid phase have been assembled using DNA templates (Sauthier et al., 2002; Warner and Hutchison, 2003; Iacopino et al., 2003). However, discrete DNA-MMPC clusters of 20 nm in diameter were obtained in solution, as characterized by dynamic light scattering (DLS), thus displaying the presence of nonaggregated structures.

From the above experiment, the interaction between MMPCs and DNA opens the possibility for various biological applications like gene delivery into cells. To study the feasibility of this idea, MMPCs were synthesized with varing amounts of quaternary ammonium functionalized groups (Sandhu et al., 2002). The MMPCs were briefly incubated with DNA plasmid encoding for galactosidase and then were introduced into human embryonic kidney cells. Excess MMPCs were used for optimal transfection because an overall positive charge of DNA–nanoparticle complex is required for cellular uptake (Wolfert et al., 1996; Truong-Le et al., 1999; Kneuer et al., 2000). Further studies explored the extent of the overall positive charge, and hydrophobicity of the MMPC can control the rate of transfection. The effect of varying cationic coverage was investigated by constructing different MMPCs using differing ratios of octanethiol to ammonium thiol. The hydrophobic effect was studied by using MMPCs with undecanethiol (MMPC **5**) and tetradecanethiol (MMPC **6**) instated of octanethiol (Figure 5.7a). The experiment shows that the most efficient nanoparticle-mediated internalization of the plasmid was observed with a ∼68% coverage of the cationic charge (Figure 5.7b). This suggests the importance of amphiphilic particles for interaction with the cell membrane for subsequent release from the endosomal vesicle. Significantly, the most efficient MMPC was ∼8-fold more effective than 60-kDa polyethylenimine (PEI), a widely used transfection agent, which proves that MMPCs can be used as potential transfection agent. This result also suggests that either the charge/particle ratio or the hydrophobicity of the MMPCs is an important determinant of transfection efficiency. The next step was to determine the effect of hydrophobicity for DNA transfection. This was studied with MMPC **5** and MMPC**6** with increasing lengths (Bielinska et al., 1999) of unfunctionalized alkane thiols (Figure 5.7a). It was observed that increasing the chain length of the alkane thiol resulted in an increase in plasmid transfection to ∼85% with a 14-carbon alkane thiol chain (Figure 5.7c). However, a further increase in alkane chain length decreased the solubility of the nanoparticle–DNA complex, thus hindering transfection. These studies demonstrate that MMPCs can be successfully employed as transfection vectors, displaying the utility of nanoparticles in modulation of an important biological process.

Since polyethylenimine is widely used as a transfection agent, conjugating this ligand onto gold nanoparticles may provide effective transfection into cells. Based on this idea, Klibanov and co-workers covalently attached branched 2-kDa polyethylenimine (PEI2) to gold nanoparticles to synthesis MPC **7** (Figure 5.8a) to investigate the delivery efficiency into monkey kidney cells *in vitro* (Thomas and Klibanov, 2003). The outcome from this study was that conjugating PEI2 to the nanoparticles increased its effective molecular weight, consequently enhancing DNA binding and condensation and resulting in improved transfection. Their studies revealed that the transfection efficiency varied with the

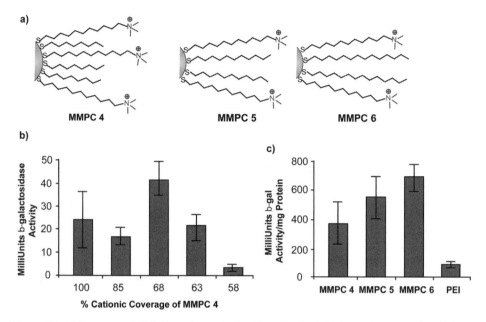

Figure 5.7. (a) Structures of MMPCs used to study the effect of hydrophobicity on the transfection efficiency. (b) Optimal transfection observed with ~68% functionalized MMPC 4 displaying the importance of amphiphilic nanoparticles for delivery. (c) Greater hydrophobic character of MMPCs results in a higher transfection.

PEI-to-gold molar ratio in the conjugates, with the best conjugate being 12 times more potent than the unmodified polycation. Furthermore, examining the addition of the N-dodecyl-PEI2 to the conjugate during complex formation revealed that the efficiency of the delivery could be doubled (Figure 5.8b). Consistent with our studies, the study indicated that conjugating dodecyl-PEI2 to the nanoparticles leads to self-assembly due

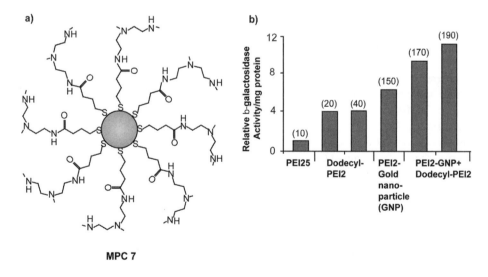

Figure 5.8. (a) MPC 7 scaffold featuring branched 2 kDa polyethylenimine (PEI2) conjugated to a gold core. (b) Addition of dodecyl-PEI2 to MPC 7 increases the transfection efficiency.

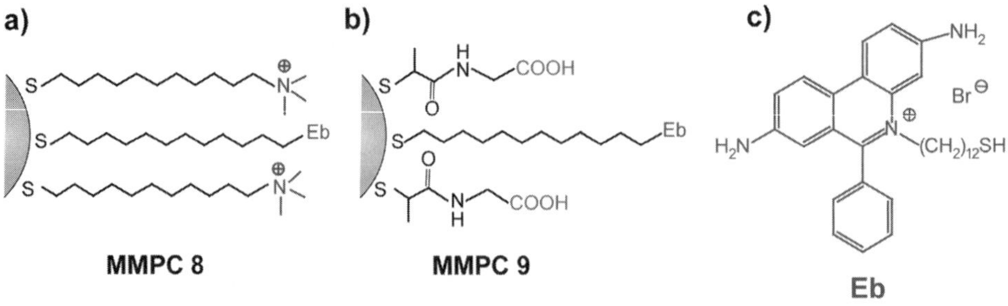

Figure 5.9. Structures of mixed monolayers of (a) tiopronin and ethidium thiolate and (b) trimethylammonium and ethidium thiolate. (c) Ethidium bromide structure.

to hydrophobic interaction between the dodecyl substitution. Moreover, the hydrophobicity of the particle increases the interaction of the polyplexes with plasma membrane and with endosomal–lysosomal membrane components, thus influencing cellular uptake and endosomal escape. Importantly, although unmodified PEI2 transfects just 4% of the cells, the PEI2–gold nanoparticle complex transfects 25% and further addition of dodecyl-PEI2 shows transfection into 50% of the cells, as assessed by histochemical staining by X-Gal. The intracellular trafficking of the polyplexes was monitored by transmission electron microscopy (TEM), which revealed the entry of the complexes in the nucleus less than 1 hour after transfection.

In the MMPC-mediated transfection studies, DNA binds to the nanoparticles via complementary electrostatic interactions; however, an alternate route to the design of nanoparticles featuring complementary elements can be achieved by introduction of DNA base-pair intercalating moieties into the nanoparticle monolayer. Murray and co-workers have used ethidium bromide (Eb) (Figure 5.9c) as a means of binding cationic (Figure 5.9a) and anionic (Figure 5.9b) gold nanoparticles to DNA (Wang et al., 2002a). In their study, each nanoparticle contains only one or two ethidium thiolate ligands. The binding of the nanoparticles was monitored by increase in EtBr fluorescence on binding to DNA. Binding of the cationic trimethylammonium functionalized nanoparticle, MMPC **8**, to the DNA was efficient and rapid. However, the binding of the tiopronin carboxylate functionalized nanoparticle (MMPC **9**) did not occur until NaCl concentrations were greater than 0.1M. The slower binding of tiopronin/ethidium MMPC allowed analysis of two competing binding interactions: (1) the binding of EtBr with DNA and (2) pairing of cationic EtBr with the anionic tiopronin. This dual mode in binding raises other interesting possibilities in the design of MMPC systems.

The interaction of a DNA base pair attached to nanoparticles was also used in optical biosensor devices as reported by T. Melvin and co-workers (Dyadyusha et al. 2005), They observed that the emission of CdSe quantum dots attached with 5′-end of a DNA sequence was competently quenched by gold nanoparticles functionalized with the complementary DNA strand. The emission was completely recovered by the addition of external 10 times complementary DNA, which is described schematically in Figure 5.10.

Nanoparticles encoded with DNA molecules were used for ultrasensitive detection of proteins by using the complementary DNA strand interaction. In an experiment, Mirkin and co-workers demonstrated this concept by analysis of a prostate-specific antigen (PSA) (Nam et al., 2000). In the experiment, PSA antibodies were conjugated to a Fe_2O_3 microparticle to bind PSA, as shown in Figure 5.11a. Then a gold NP functionalized with PSA antibodies

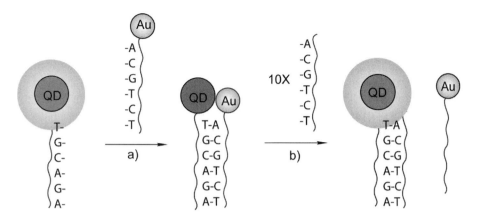

Figure 5.10. Schematic representation showing the following: (a) Addition of one equivalent of the Au-DNA to the QD-DNA quenched the emission. (b) Addition of the ten equivalent of the unlabeled complementary oligonucleotide to the complex displaced the Au-DNA and established the emission.

and DNA that is unique to the protein target was added to bind with PSA and immobilized by the magnetic particle (Figure 5.11b). Magnetic separation of this assembly followed by thermal dissociation of the DNA duplex provided the free DNA strand (i.e., bar-code DNA) and the single-stranded-oligonucleotide-functionalized gold NPs. The isolated bar-code DNA was added in a polymerase chain reaction (PCR) for amplification and chip-based DNA detection (Figure 5.11c). Since gold nanoparticles can bind a large number

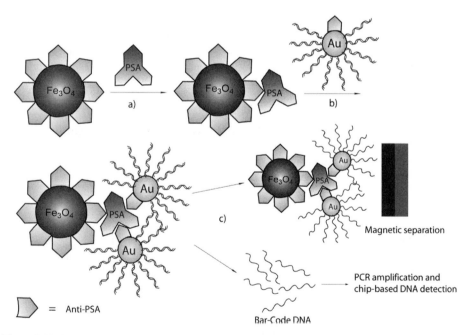

Figure 5.11. Schematic representation of amplified immunosensing of prostate-specific antigen (PSA) with gold NPs that are encoded with DNA. (a) Target protein PSA captured by MMP probes. (b) Sandwich captured target protein with AuNP probes. (c) Magnetic separation of MMP probes and bar-code DNA dehybridization.

of oligonucleotides, considerable amplification was observed and PSA was detected at a 30 attomolar concentration.

Recently, di-functional DNA–gold nanoparticles containing two different DNA sequences for the sensitive detection of protein antigens were used by Niemeyer and coworkers (Hazarika et al., 2005) In this system, one of the two sequences is attached with immobilized antibodies and the other is used for signal amplification by means of the DNA-directed assembly of multiple layers of NPs. It is anticipated that sensitive proteins with very low concentration can be detected by using this scaffold at low cost in the very near future.

5.4. SURFACE RECOGNITION OF PROTEIN AND PEPTIDES

The above studies demonstrate the affinity of MMPCs to bind and regulate DNA. This opens the possibility for further investigations between the interaction of nanoparticles and more complex systems such as proteins and peptides. The traditional approach for synthetic receptors is targeting the proteins active site and regulating its activity via active site inhibition (Toogood, 2002; Gadek and Nicholas, 2003). But this approach is sometimes subjugated for modulation of proteins without a well-defined active site such as proteins involved in signal transduction (Chrunyk et al., 2000) and dimerization (Zhang et al., 1991; Cochran, 2000). Moreover, such receptors could provide a potent tool to control protein–protein and protein–nucleic acid interactions that are central to cellular processes. In this respect, as mentioned earlier, MMPCs are very useful tools for examining nanoparticle–protein interaction due to their diversity in functionalization and the ability to maximize interaction with a guest over time through templation resulting in increased binding affinity.

Since it is a well-studied protein, the Rotello group uses chymotrypsin (ChT) to study recognition of protein surfaces using MPCs and MMPCs. The ring of cationic residues around ChTs active site (Figure 5.12a) (Capasso et al., 1997) is a suitable target for negatively charged receptors, thus providing a useful model for protein–protein interfaces. In addition, ChT features a well-defined catalytic activity (Blow, 1976) CD, (Schechter et al., 1995), and fluorescence markers (Desie et al., 1986). The carboxylate-functionalized gold MMPC **10** (Figure 5.12b) (2-nm core diameter and 6-nm overall diameter) is used for the negatively charged receptors. The negatively charged MMPC is expected to bind electrostatically to the active site of ChT. This binding should sterically block the active site to substrate access, resulting in complete enzyme inhibition.

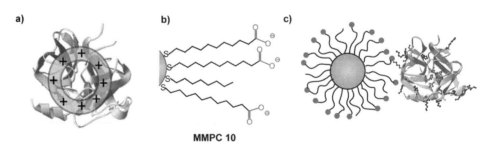

Figure 5.12. (a) Chymotrypsin (ChT): the active site is surrounded by cationic residues. (b) Structure of MMPC 10. (c) Relative sizes of MMPC 10 (2-nm core diameter) and ChT.

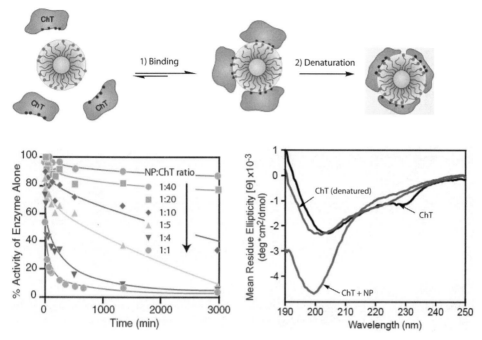

Figure 5.13. Nanoparticle-mediated enzymetic inhibition of ChT. (a) ChT binds to MPC 15 and denatures on the nanoparticle surface. (b) Inhibition of activity of ChT in various nanoparticle concentrations. (c) Circular dichroism shows ChT denaturation occurs over time at nanoparticle surface. See insert for color representation of this figure.

It was observed that upon incubation with MMPC **10**, the ChT was completely inhibited, with the extent of inhibition being dependent on time and concentration of the anionic nanoparticles. It was also observed that the protein was denatured slowly after binding with MMPC **10**, which was determined through circular dichroism (CD). These results imply that the inhibition displayed is a two-step process. The initial step is fast and reversible due to complementary electrostatic binding followed by a slower irreversible process, resulting in ChT denaturation on the nanoparticle surface (Figure 5.13).

From the activity assay study the complete inhibition is observed at NP:ChT 1:4. Kinetic analysis revealed that the inhibition was very effective with a K_d of 10.4 ± 1.3 nM. The binding ratio of the nanoparticle with ChT was found to be 1:5, which indicated a complete saturation of the MMPC **10** surface with the protein, given their relative surface areas. This study also revealed a certain level of selectivity, because elastase, galactosidase, and cellular retinoic acid-binding protein displayed no significant interaction with MMPC **10**. Additionally, positively charged MMPC **4** displayed no inhibition of ChT activity.

Since the interaction between MMPCs and ChT is noncovalent, it is possible the unique nature of the MMPC scaffold can be used for restoring enzymatic activity via electrostatically mediated release and refolding of the protein. For this purpose derivatives of trimethylammonium-functionalized surfactants (Figure 5.14a) were used to explore this possibility. These surfactants were added to the preincubated MMPC-ChT complex (Fischer et al., 2003). For surfactant **11** (Figure 5.14a) up to 50% of the native enzyme activity was recovered just after addition. Dynamic light scattering (DLS), fluorescence, and fluorescence anisotropy experiments were used to confirm the release of ChT from the nanoparticle

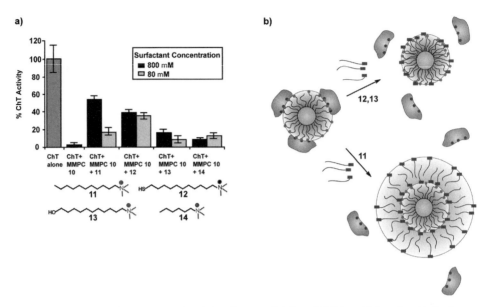

Figure 5.14. (a) Restoration of enzymatic activity of nanoparticle-bound ChT by addition of positively charges surfactants. (b) Schematic representation of monolayer modification on surfactant addition resulting in disassociation of ChT from MMPC.

surface and to determine the structure of the desorbed protein. Using the different surfactants and based on DLS data of MMPC–protein assemblies, two different mechanisms were postulated to explain the enzymatic recovery upon addition of surfactants (Figure 5.14b). In the first mechanism, the surfactants **12** and **13** directly modify the monolayer by intercalation and/or chain displacement, resulting in attenuation of monolayer charge mediating protein release and subsequent restoration of activity. However, the second mechanism indicated that the alkane surfactant **11** forms a bilayer around the nanoparticle, resulting in protein release. But the effect depends on the chain length of alkane surfactant because the addition of smaller length surfactant **14** did not result in significant reactivation.

The above studies show that MMPCs can be used as an effective tool for recognition of proteins and inhibition of enzyme activity; however, binding with protein proceeds to denaturation. In practical application it is important to retain the native enzyme structure upon binding. Moreover, templation of nanoparticle to a protein surface relies on retention of native protein conformation.

To determine the receptors that permit inhibition but prevent denaturation of proteins, thiolalkyl and thioalkylated oligo(ethylene glycol) (OEG) ligands with chain-end functionality—namely, (i) OEG terminated with hydroxyl group (Figure 5.15b), (ii) carboxylate-terminated thiolalkyl ligand (Figure 5.15a), and (iii) carboxylate terminated OEG (Figure 5.15c)—were used to fabricate water-soluble CdSe nanoparticle scaffolds (Hong et al., 2004). It was all ready reported that ethylene glycol units resist nonspecific interactions with biomacromolecules (Kane et al., 2003). Thus, inserting an ethylene glycol spacer between the functional end group and the alkane monolayer is expected to diminish nonspecific interactions of the MMPC-bound protein.

CdSe nanoparticles can be used as fluorescent tags for bioimaging (Chan and Nie, 1998; Bruchez et al., 1998), allowing for a better understanding of the interaction between

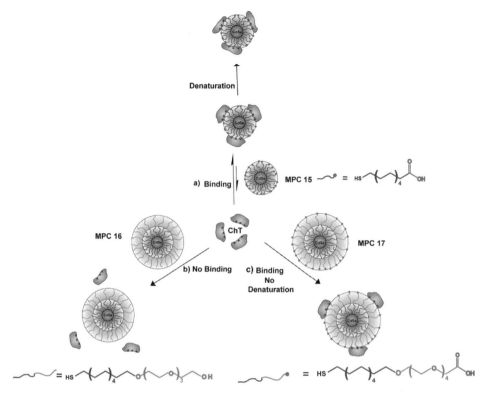

Figure 5.15. Three levels of control over ChT structure and function by CdSe-based MPCs. (a) ChT binds to MPC 15 and denatures on the nanoparticle surface. (b) No binding of ChT to MPC 16 surface is observed. (c) ChT binds to MPC 17 but retains native conformation.

CdSe nanoparticles and proteins. Upon incubation with ChT, three levels of control of enzyme activity and structure were observed. No interaction was observed with MPC **16**, which contains a terminal hydroxyl end group. Nanoparticles containing the carboxylate-terminated thiolalkyl ligand (MPC **15**) bound and denatured ChT, which is consistent with our earlier studies using gold MMPCs with the same functionality. However, while the enzyme bound to nanoparticles displaying carboxylate terminated OEG thiols MPC **17**) showed substantial loss of enzymatic activity, no significant loss in the native structure of the bound enzyme was seen as investigated through circular dichroism (CD) and fluorescence experiments. The binding in the latter case arose primarily from complementary electrostatic interactions of the enzyme and nanoparticle, which were confirmed through ionic strength studies. This study demonstrates that MMPCs can be used to modulate protein activity and structure, which can form the basis of a number of pragmatic biological applications. Based on same concept, photocleavable monolayers on gold NPs have also been designed to switch the inhibition of enzymes by UV light (Fischer et al., 2004).

The above concept of complementary electrostatic interaction was also used for positively charged nanoparticles with negatively charged proteins. Positively charged nanoparticle such as MMPC **18** and MMPC **19** bind with negatively charged protein, β-galactosidase (β-Gal), and inhibits its activity (Verma et al., 2004). When glutathione (GSH) was added to the MMPC **18**-β-Gal complex at intracellular concentrations, the activity of β-Gal was

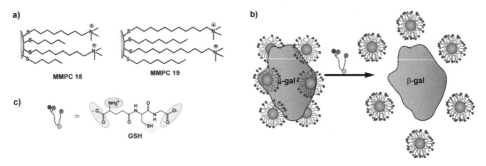

Figure 5.16. (a) Structures of MMPC scaffolds. (b) Schematic representation of GSH-mediated disruption of MMPC-â-gal binding. (c) Structure of reduced glutathione. See insert for color representation of this figure.

completely reversed (Figure 5.16b). However, the MMPC **19** monolayer provides effective protection against GSH-mediated reactivation of the bound enzyme. This result suggests that the delivery of proteins by NPs and their release *in vivo* mediated by GSH is feasible, which has great potential in future therapeutics study.

In addition to the above-mentioned surface complementary interaction systems, specific biomacromolecular interactions have been employed to construct nanoparticles with protein specificity. Zheng and Huang introduced either a biotin group or glutathione onto the surface of gold NPs protected by tri(ethylene glycol) thiols (Zheng and Huang, 2004). Their gel electrophoresis results showed that the NPs can bind specifically to streptavidin and glutathione-*S*-transferase, respectively, eliminating the nonspecific binding with other proteins (Figure 5.17).

Another route to specifically bind proteins has been accomplished by using transition metal complex-functionalized MMPCs. In order to show this, Xu et al. (2004) fabricated

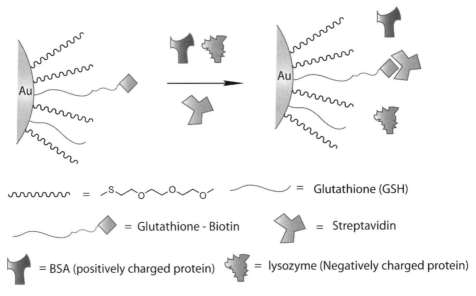

Figure 5.17. Schematic representation of the biotin-functionalized MMPC, which specifically interacts with a atreptavidin molecule without any nonspecific binding.

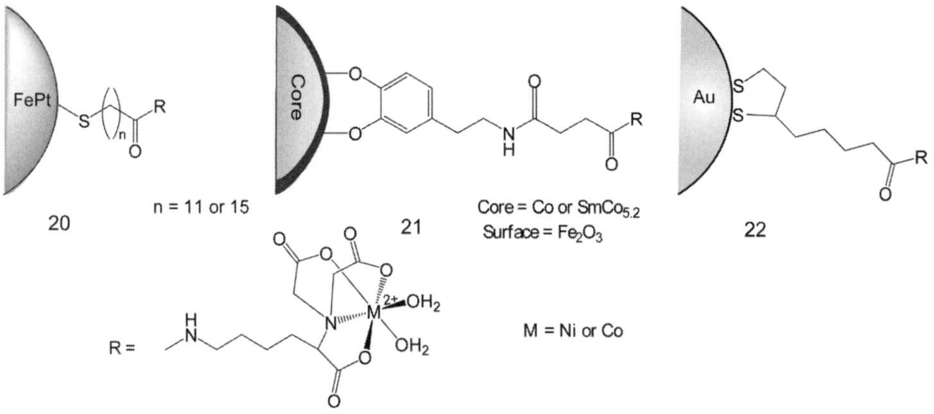

Figure 5.18. Structure of nanoparticles targeting histidine-tagged proteins.

FePt magnetic MMPC **20**, which possesses nickel terminated nitrilotriacetic acid (NTA) (Figure 5.18). Through metal chelation, these NPs show high affinity and specificity toward histidine-tagged proteins (proteins with six consecutive histidines residues). This can be used to manipulate the histidine-tagged recombinant proteins and bind other biological substrates at low concentrations. Superior to commercial magnetic microbeads, these NPs have a great protein binding capacity due to their high surface-to-volume-to ratio. Furthermore, the magnetic core of the NPs allows proteins to be purified by external magnets. Generally, such proteins are separated via metal-chelate affinity chromatography (MCAC), which makes use of NTA-attached resin to immobilize nickel ions (Ni^{2+}). The above study reveals the ability of the magnetic nanoparticles to obtain pure proteins directly from lysed cell mixtures through magnetic separation within 10 min. The work also indicates the superiority of the magnetic nanoparticles to MCAC columns because they do not exhibit nonspecific binding. By employing the same concept, Xu et al. (2004) immobilized a nickel–NTA complex on the iron oxide shell of magnetic NPs **21** with dopamine as an anchor and used them to target the histidine-tagged proteins. In a more recent publication, Abad et al. (2005) constructed NTA chelated cobalt complex-capped gold NP **22** and investigated its interaction with histidine-tagged proteins. Again, their study demonstrated that the NTA-Co(II) platform is most suited for the specific immobilization of histidine-tagged proteins. Such proteins are usually separated via metal-chelate affinity chromatography (MCAC), which employs NTA-attached resin to immobilize nickel ions (Ni^{2+}). The study demonstrates the ability of the magnetic nanoparticles to obtain pure proteins directly from lysed cell mixtures through magnetic separation within 10 min. The work also indicates the superiority of the nanoparticles to MCAC columns because they do not exhibit nonspecific binding.

In nature, sometimes α-helices are responsible for protein–protein interactions. This type of interaction was also targeted by the use of artificial scaffolds. With this in mind, Rotello and co-workers used MMPC **4** to target a tetraaspartate peptide (peptide **23**) featuring the aspartate residues in alternating $i, i + 3$ and $i, i + 4$ positions (Figure 5.19a) for recognition and stabilization of α-helices (Verma et al., 2004). The circular dichroism (CD) studies show that the addition of MMPC **4** to the peptide solution increases the helicity of the peptide from ∼4% to ∼60%. Further studies show that the helicity was increased with

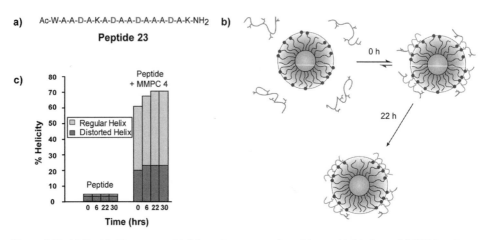

Figure 5.19. (a) Peptide 23 sequence. (b) Schematic representation of the peptide binding to MMPC 4 surfaces. (c) Increase in helicity over time on incubation of the peptide with the nanoparticles demonstrating receptor templation.

time (~20%) and gave more stabilized peptide helix (Figure 5.19b). This work exhibits the ability of MMPCs to template to large biomolecular surface areas which can be useful for generation of protein-surface-specific nanoparticle receptors through strategies such as monolayer cross-linking.

As nanoparticles can be fabricated with multiple ligands, they can also be used to enhance low-affinity interactions such as carbohydrate–protein interactions. This type of interaction by using nanoparticle was reported by Wu et al. (Ipe et al., 2002; Wang et al., 2002b; Gu et al., 2003). The authors have extended their investigation to the binding of mannose-conjugated gold NPs to mannose-specific adhesion with type 1 pili in *E. coli*. Type 1 pili are filamentous proteinaceous appendages that extend from the surface of many gram negative organisms and are composed of FimA, FimF, FimG, and FimH proteins. Among these proteins, FimA presents more than 98% and FimH is highly specific to bind with D-mannose (Harris et al., 2001; Krogfelt et al., 1990). To demonstrate specific binding of *m*-AuNP to FimH, the other used two *E. coli* strains ORN178 and ORN208. The ORN178 strain has type 1 pili, whereas the ORN208 strain is deficient of the FimH. The TEM results showed that *m*-AuNP selectively bound the pili of the ORN178 strain but not those of the ORN208 strain, demonstrating specific binding of D-mannose-AuNP to FimH. The nanoparticles were found to be stable in various media, high ionic strength, and pH values ranging from 1.5 to 12, indicating high stability and applicability in various biological conditions.

The study on carbohydrate–protein interactions was continued with lectins and lactose-coated nanoparticles. Ethylene-glycol-containing lactose-conjugated gold nanoparticles have used to bind to agglutinin, a bivalent lectin by Kataoka and co-workers (Otsuka et al., 2001). Since β-D-galactose shows specific binding with lectin, when ethylene-glycol-containing lactose-conjugated gold nanoparticles were exposed to lectin, selected aggregation leading to distinct changes in the absorption spectrum was exhibited. The aggregation is fully reversible by addition of excess galactose because galactose binds strongly with lectin (Figure 5.20). Importantly, the system can be utilized for quantitative assay since the degree of aggregation is proportional to the lectin concentration. Protein recognition using carbohydrates has also been examined successfully by mannose-, glucose-, and

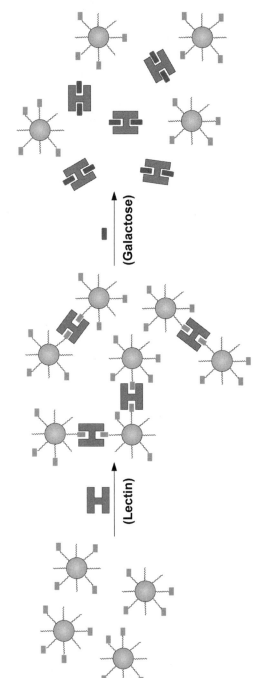

Figure 5.20. Lactose-conjugated gold nanoparticles aggregate on addition of lectin. This can be reversed upon addition of excess galactose.

galactose-encapsulated gold nanoparticles (Lin et al., 2003). In another study by Gervay-Hague and co-workers, gold nanoparticles featuring galactosyl and glucosyl headgroups were used to display their relative ability to displace HIV-associated glycoprotein gp120 from the cellular receptor GalCer, which is a molecular event involved in HIV recognition of mucosal membranes (Nolting et al., 2003). This shows the potential utility of polyvalent ligand arrays on nanoplatforms. Collectively, these studies demonstrate the applicability of carbohydrate gold nanoparticles as effective inhibitors of protein–carbohydrate interactions that could be extended to *in vivo* biological systems.

Based on antibody–antigen recognition, the self-assembly of nanoparticles is also reported. Mann and co-workers demonstrated antibody–antigen coupling as a strategy for the directed self-assembly of metallic nanoparticles into extended 3D networks that can exhibit higher-order structures such as wires and filaments. The IgE or IgG antibodies have specificities to dinitrophenyl (DNP) and biotin, respectively. They first attached the IgE or IgG antibodies to individual NPs. Afterward, the addition of bivalent antigens with appropriate double-headed functionalities caused the aggregation of NPs to 3D networks, as evidenced by TEM examination (Figure 5.21).

Along with gold nanoparticles, magnetic nanoparticles are also used in recognizing biological surfaces with great efficiency compared to their magnetic bead counterparts with

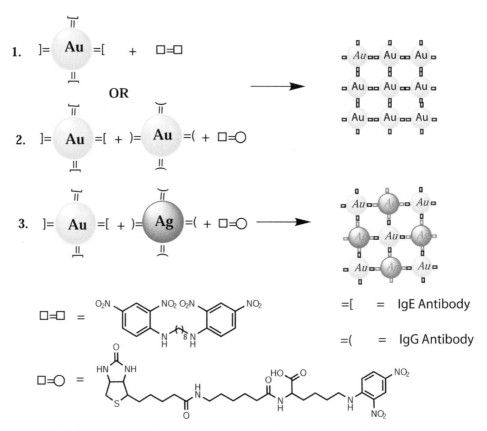

Figure 5.21. Schematic representation showing self-assembly of metallic (routes 1 and 2), and bimetallic (route 3) macroscopic materials using antibody–antigen cross-linking.

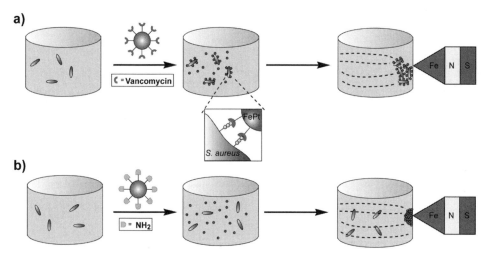

Figure 5.22. Schematic illustration of (a) capturing bacteria with vancomycin-functionalized magnetic nanoparticles through recognition of terminal peptides on the bacterial cell wall and (b) a control experiment with amine-functionalized magnetic nanoparticles displaying no interaction with the bacteria.

some advantages. The most important advantage of magnetic nanoparticles over beads is higher surface-to-volume ratio. Generally, beads are 1–5 μm in size as compared to <10 nm for the magnetic nanoparticles. An other advantage is size; the smaller size of nanoparticles allows faster movement and easier entry into cells, making the magnetic nanoparticles more suitable for *in vivo* applications. Additionally, the magnetic property helps separation and purification of nanoparticles as well as nanoparticle–biomolecule complexes. One such example is exhibited by Xu and co-workers (Gu et al., 2003a). FePt nanoparticles functionalized with vancomycin (Van) (a broad-spectrum antibiotic) were synthesized with a 3- to 4-nm diameter. Van can bind specifically to a terminal peptide D-Ala–D-Ala on the cell wall of a gram-positive bacterium via hydrogen bonding. The FePt–Van nanoparticle captured vancomycin-resistant enterococci (VME) and other gram-positive bacteria at concentrations of ∼10 cfu (colony-forming units)/mL within an hour and separated by magnetic field from the solution (Figure 5.22a). The FePt–NH$_2$ nanoparticles were not captured by either of the bacteria, indicating the specificity of biomolecular interaction (Figure 5.22b). The author also reported another application using the pragmatic property of FePt magnetic nanoparticles by synthesizing nitrilotriacetic acid (NTA)-modified magnetic nanoparticles (see page 107).

Binding of nanoparticles and proteins is also used in generating nanoparticle-based protein sensors. For instance, the aggregation of lactose-conjugated gold nanoparticles with bivalent lectins induced the visible color change from pinkish-red to purple (Otsuka et al., 2001). This property can be used as a chromogenic sensor for lectins. In another study, Kim and co-workers used fluorescent resonance energy transfer (FRET) between quantum dots (QDs) and gold nanoparticles to determine the inhibition assay in biomolecules in presence of nanoparticle (Oh et al., 2005). They described that, in the absence of avidin, gold nanoparticles functionalized with biotin bind with streptavidin on the QDs and quench the fluorescence, but after the external addition of avidin they inhibit the streptavidin–biotin interactions and releases gold nanoparticles from QDs and releasing fluorescence (Figure 5.23).

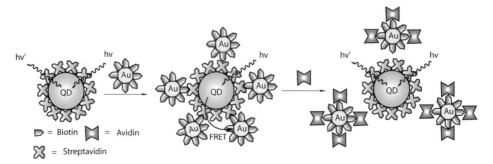

Figure 5.23. Schematic representation of quenching and release of fluorescence of streptavidin-QDs by biotin-AuNPs.

Another report on effective fluorescent resonance energy transfer (FRET) was made by Nagasaki et al. (2004) by using biotin-PEG/polyamine CdS quantum dot. They used the specific interaction of Au-nanoparticles with Texas Red-labeled streptavidin. It was observed that the extent of the energy transfer was proportional to the concentration of the dye-labeled protein. Therefore, this system can be utilized as a highly sensitive bioanalytical system.

The optical properties of metallic NPs were also used in protein detection. Willner and co-workers reported the catalytic enlargement of aptamer-functionalized gold nanoparticles to amplify the optical detection of aptamer–thrombin complexes in solution and on surface (Pavlov et al., 2004). According to their procedure, thrombin was first bound to the aptamer, which was covalently attached to a glass substrate. The aptamer-functionalized gold NPs then bound to the other side of the thrombin molecule. The gold nanoparticles attached with thrombin were enlarged in the presence of $HAuCl_4$, cetyl trimethylammonium bromide (CTAB) as surfactant, and NADH (Figure 5.24).

The concentration of thrombin was measured by UV absorbance, and it was observed that absorbance increases with the increase of thrombin concentration. Due to the enlargement of NPs, a new coupled plasmon absorbance peak at 650 nm at higher thrombin concentration was observed.

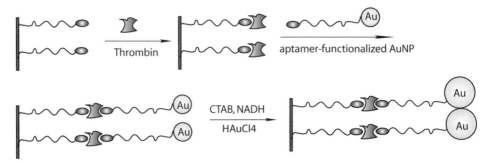

Figure 5.24. Amplified detection of thrombin on surfaces by the catalytic enlargement of thrombin aptamer-functionalized gold NPs.

SUMMARY AND OUTLOOK

This chapter summarizes various applications of metal nano-clusters in biomolecular recognition. The success of this recognition arises from the fact that nanoparticles and biomacromolecules are virtually similar in size. The tunable core size and various core materials can be used in different systems for biomacromolecular recognition. These nanoparticles not only demonstrate selective binding with biomacromolecules but also allow thermodynamically stable, kinetically inert, and stoichiometrically well-defined bio-recognition. Other than the tunable size advantage, the surface of nanoparticle can be fabricated with wide range of functionality. More importantly, the nanoparticle surface can be templated to the guest surface, given that there exists an environmentally responsive receptor for biomolecules. Considering these properties along with other features, it is well established that a nanoparticle can be used as an important synthetic receptor for recognition and control over structure and function of different biomolecules such as DNA, proteins, and peptides. Additionally, the ability of biomacromolecular surface recognition of MPCs and MMPCs can be extended to development of novel hybrid materials composed of biomolecules and nanoparticles featuring novel functional properties for generating advance materials which might have lot of application in sensing, catalysis, signal transduction, transport, or other applications in biomedical science and tissue engineering. Therefore, the development of nanoparticle–biomolecular interaction not only advances the fundamental research but also helps to bring about progress in supramolecular sciences at the interface of biotechnology and materials research. Also, the extensive present research and much more attention in this field promises plenty of exciting results from future investigation.

REFERENCES

Abad, JM, Mertens, SFL, Pita, M, Fernandez, VM, Schiffrin, DJ (2005). Stereochemical recognition of doubly functional aminotransferase in 2-deoxystreptamine biosynthesis. *J Am Chem Soc* **127**:5689–5694.

Blow, MD (1976). Structure and mechanism of chymotrupsin. *Acc Chem Res* **9**:145–152.

Boal, AK, and Rotello, VM (2000a). Fabrication and self-optimization of multivalent receptors on nanoparticle scaffolds. *J Am Chem Soc* **122**:734–735.

Boal, AK, and Rotello, VM (2000b). *Intra-* and *inter*monolayer hydrogen bonding in amide-functionalized alkanethiol self-assembled monolayers on gold nanoparticles. *Langmuir* **16**:9527–9532.

Bielinska, AU, Chen, CL, Johnson, J, and Baker, JR (1999). DNA Complexing with polyamidoamine dendrimers: Implication for transfection. *Bioconjug Chem* **10**:843–850.

Bruchez, M, Moronne, M, Gin, P, Weiss, S, and Alivisatos, AP (1998). Semiconductor nanocrystals as fluorescent biological labels. *Science* **281**:2013–2016.

Brust, M, Walker, M, Bethell, D, Schiffrin, DJ, and Whyman, R (1994). Synthesis of thiol-derivatized gold nanoparticles in a 2-phase liquid-liquid system. *J Chem Soc Chem Commun* 801–802.

Cao, YWC, Jin, RC, and Mirkin, CA (2002). Nanoparticles with Raman spectroscopic fingerprints for DNA and RNA detection. *Science*, **297**:1536–1540.

Capasso, C, Rizzi, M, Menegatti, E, Ascenzi, P, and Bolognesi, M (1997). Crystal structure of the bovine alpha-chymotrypsin: Kunitz inhibitor complex. An example of multiple protein:protein recognition sites. M, *J Mol Recognit* **10**:26–35.

Chan, WCW, and Nie, SM (1998). Quantum dot bioconjugates for ultrasensitive nonisotopic detection. *Science* **281**:2016–2018.

Chen, S, and Murray, RW (1999). Arenethiolate monolayer-protected gold clusters. *Langmuir* **15**:682–689.

Chrunyk, BA, Rosner, MH, Cong, Y, McColl, AS, Otterness, IG, and Daumy, GO (2000). Inhibiting protein–protein interactions: A model for antagonist design. *Biochemistry* **39**:7092–7099.

Cochran, AG (2000). Antagonists of protein–protein interactions. *Chem Biol* **7**:R85–R94.

Conte, LLo, Chothia, C, and Janin, J (1999). The atomic structure of protein–protein recognition sites. *J Mol Biol* **285**:2177–2198.

Coull, JJ, He, GC, Melander, C, Rucker, VC, Dervan, PB, Margolis, DM (2002). Targeted derepression of the human immunodeficiency virus type 1 long terminal repeat by pyrrole-imidazole polyamides. *J Virol* **76**:12349–12354.

Dervan, PB (2001). Molecular recognition of DNA by small molecules. *Bioorg Med Chem* **9**:2215–2235.

Dervan, PB, and Edelson, BS (2003). Recognition of the DNA minor groove with pyrrole-imidazole polyamides. *Curr Opin Struct Biol* **13**:284.

Desie, G, Boens, N, and De, Schryver, FC (1986). Intramolecular distances determined by energy transfer. Dependence on orientational freedom of donor and acceptor. *Biochemistry* **25**:8301–8308.

Dyadyusha, L, Yin, H, Jaiswal, S, Brown, TJ. Baumberg, J, Booy, FP, and Melvin, T. (2005). Quenching of CdSe quantum dot emission, a new approach for biosensing, *Chem Commun* (25), 3201–3203.

Ernst, JT, Becerril, J, Park, HS, Yin, H, and Hamilton, AD (2003). Design and application of an α-helix-mimetic scaffold based on an oligoamide-foldamer strategy: Antagonism of the Bak BH3/Bcl-xL complex. *Angew Chem-Int Edit* **42**:535–539.

Fazal, MA, Roy, BC, Sun, SG, Mallik, S, and Rodgers, KR (2001). Surface recognition of a protein using designed transition metal complexes. *J Am Chem Soc* **123**:6283–6290.

Fischer, NO, Verma, A, Goodman, CM, Simard, JM, and Rotello, VM (2003). Reversible "irreversible" inhibition of chymotrypsin using nanoparticle receptors. *J Am Chem Soc* **125**:13387–13391.

Fischer, NO, Paulini, R, Drechsler, U, and Rotello, VM (2004). Light-induced inhibition of chymotrypsin using photocleavable monolayers on gold nanoparticles. *Chem Commun* 2866–2867.

Gadek, TR, and Nicholas, JB (2003). Small molecule antagonists of proteins. *Biochem Pharmacol* **65**:1–8.

Gestwicki, JE, Cairo, CW, Strong, LE, Oetjen, KA, and Kiessling, LL (2002). Influencing receptor-ligand binding mechanisms with multivalent ligand architecture. *J Am Chem Soc* **124**:14922–14933.

Golumbfskie, AJ, Pande, VS, and Chakraborty, AK (1999). Simulation of biomimetic recognition between polymers and surfaces. *Proc Natl Acad Sci USA* **96**:11707–11712.

Gordon, EJ, Sanders, WJ, and Kiessling, LL (1998). Synthetic ligands point to cell surface strategies. *Nature*, **392**:30–31.

Gottesfeld, JM, Belitsky, JM, Melander, C, Dervan, PB, and Luger, K (2002). Blocking transcription through a nucleosome with synthetic DNA ligands. *J Mol Biol* **321**:249–263.

Gu, HW, Ho, PL, Tsang, KWT, Wangm, L, and Xu, B (2003a). Using biofunctional magnetic nanoparticles to capture vancomycin-resistant enterococci and other gram-positive bacteria at ultralow concentration. *J Am Chem Soc* **125**:15702–15703.

Gu, T, Ye, T, Simon, JD, Whitesell, JK, and Fox, MA (2003b). Subpicosecond transient dynamics in gold nanoparticles encapsulated by a fluorophore-terminated monolayer. *J Phys Chem B* **107**:1765–1771.

Haes, AJ, and Van, Duyne, RP (2002). A nanoscale optical biosensor: Sensitivity and selectivity of an approach based on the localized surface plasmon resonance spectroscopy of triangular silver nanoparticles. *J Am Chem Soc* **124**:10596–10604.

Hamuro, Y, Calama, MC, Park, HS, and Hamilton, AD (1997). A calixarene with four peptide loops: an antibody mimic for recognition of protein surfaces. *Angew Chem-Int Edit* **36**:2680–2683.

Harris, SL, Spears, PA, Havell, EA, Hamrick, TS, Horton, JR, and Orndorff, PE (2001). Characterization of *Escherichia coli* type 1 pilus mutants with altered binding specificities. *J Bacteriol* **183**:4099–4102.

Hazarika, P, Ceyhan, B, and Niemeyer, CM (2005). Sensitive detection of proteins using difunctional DNA-gold nanoparticles. *Small* **1**:1–4.

Hong, R, Fischer, NO, Verma, A, Goodman, CM, Emrick, T, and Rotello, VM (2004). Control of protein structure and function through surface recognition by tailored nanoparticle scaffolds. *J Am Chem Soc* **126**:739–743.

Hostetler, MJ, Wingate, JE, Zhong, CJ, Harris, JE, Vachet, RW, Clark, MR, Londono, JD, Green, SJ, Stokes, JJ, Wignall, GD, Glish, GL, Porter, MD, Evans, ND, and Murray, RW (1998). Alkanethiolate gold cluster molecules with core diameters from 1.5 to 5.2 nm: Core and monolayer properties as a function of core size. *Langmuir* **14**:17–30.

Iacopino, D, Ongaro, A, Nagle, L, Eritja, R, and Fitzmaurice, D (2003). Imaging the DNA and nanoparticle components of a self-assembled nanoscale architecture. *Nanotechnology* **14**:447–452.

Ingram, RS, Hostetler, MJ, and Murray, RW (1997). Poly-hetero-α-functionalized alkanethiolate-stabilized gold cluster compounds. *J Am Chem Soc* **119**:9175–9178.

Ipe, BI, Thomas, KG, Barazzouk, S, Hotchandani, S, and Kamat, PV (2002). Photoinduced charge separation in a fluorophore-gold nanoassembly. *J Phys Chem B* **106**:18–21.

Kane, RS, Deschatelets, P, and Whitesides, GM (2003). Kosmotropes form the basis of protein-resistant surfaces. *Langmuir* **19**:2388–2391.

Kneuer, C, Sameti, M, Bakowsky, U, Schiestel, T, Schirra, H, and Schmidt, H, and Lehr, CM (2000). A nonviral DNA Delivery system based on surface modified silica-nanoparticles can efficiently transfect cells *in vitro*. *Bioconjug Chem* **11**:926–932.

Krogfelt, KA, Bergmans, H, and Klemm, P (1990). Direct evidence that the FimH protein is the mannose specific adhesin of *Escherichia coli* type 1 fimbriae. *Infect Immun* **58**:1995–1998.

Kuzmine, I, and Martin, CT (2001). Pre-steady state kinetics of initiation of transcription by T7 RNA polymerase—A new kinetic model. *J Mol Biol* **305**:559–566.

Leung, DK, Yang, ZW, Breslow, R (2000). Selective disruption of protein aggregation by cyclodextrin dimmers. *Proc Natl Acad Sci USA* **97**:5050–5053.

Lijnzaad, P, and Argos, P (1997). Hydrophobic patches on protein subunit interfaces: Characteristics and prediction. *Proteins* **28**:333–343.

Lin, CC, Yeh, YC, Yang, CY, Chen, CL, Chen, GF, Chen, CC, and Wu, YC (2002). Selective binding of mannose-encapsulated gold nanoparticles to type 1 pili in *Escherichia coli*. *J Am Chem Soc* **124**:3508–3509.

Lin, CC, Yeh, YC, Yang, CY, Chen, GF, Chen, YC, Wu, YC, Chen, CC (2003). Quantitative analysis of multivalent interactions of carbohydrate-encapsulated gold nanoparticles with concanavalin A. *Chem Commun* 2920–2921.

Lin, Q, Park, HS, Hamuro, Y, Lee, CS, and Hamilton, AD (1998). Protein surface recognition by synthetic agents: Design and structural requirements of a family of artificial receptors that bind to cytochrome c. *Biopolymers* **47**:285–297.

Mapp, AK, Ansari, AZ, Ptashne, M, and Dervan, PB (2000). Activation of gene expression by small molecule transcription. *Proc Natl Acad Sci USA* **97**:3930–3935.

McIntosh, CM, Esposito, EA, Boal, AK, Simard, JM, Martin, CT, and Rotello, VM (2001). Inhibition of DNA transcription using cationic mixed monolayer protected gold clusters. *J Am Chem Soc* **123**:7626–7629.

Nagasaki, Y, Ishii, T, Sunaga, Y, Watanabe, Y, Otsuka, H, and Kataoka, K (2004). Novel molecular recognition via fluorescent resonance energy transfer using a biotin-PEG/polyamine stabilized CdS quantum dot. *Langmuir* **20**:6396–6400.

Nam, J-M, Thaxton, CS, and Mirkin, CA (2000). Nanoparticle-based bio-bar codes for the ultrasensitive detection of proteins. *Science* **301**:1884–1886.

Nolting, B, Yu, JJ, Liu, GY, Cho, SJ, Kauzlarich, S, and Gervay-Hague, J (2003). Synthesis of gold glyconanoparticles and biological evaluation of recombinant Gp120 Interactions. *Langmuir* **19**:6465–6473.

Oh, E, Hong, M-Y, Lee, D, Nam, S-H, Yoon, HC, and Kim, H-S (2005). Inhibition assay of biomolecules based on fluorescence resonance energy transfer (FRET) between quantum dots and gold nanoparticles. *J Am Chem Soc* **127**:3270–3271.

Otsuka, H, Akiyama, Y, Nagasaki, Y, and Kataoka, K (2001). Quantitative and reversible lectin-induced association of gold nanoparticles modified with α-lactosyl-ω-mercapto-poly(ethylene glycol). *J Am Chem Soc* **123**:8226–8230.

Park, HS, Lin, Q, and Hamilton, AD (1999). Protein surface recognition by synthetic receptors: A route to novel submicromolar inhibitors for α-chymotrypsin. *J Am Chem Soc* **121**:8–13.

Park, HS, Lin, Q, and Hamilton, AD (2002a). Supramolecular chemistry and self-assembly special feature: Modulation of protein–protein interactions by synthetic receptors: Design of molecules that disrupt serine protease–proteinaceous inhibitor interaction. *Proc Natl Acad Sci USA* **99**:5105–5109.

Park, SJ, Taton, TA, Mirkin, CA (2002b). Array-based electrical detection of DNA with nanoparticle probes. *Science* **295**:1503–1506.

Paulini, R, Frankamp, BL, and Rotello, VM (2002). Effects of branched ligands on the structure and stability of monolayers on gold nanoparticles. *Langmuir* **18**:2368–2373.

Pavlov, V, Xiao, Y, Shlyahovsky, B, and Willner, I (2004). Aptamer-functionalized Au nanoparticles for the amplified opictal detection of thrombin. *J Am Chem Soc* **126**:11768–11769.

Raschke, G, Kowarik, S, Franzl, T, Sonnichsen, C, Klar, TA, Feldmann, J, Nichtl, A, and Kurzinger, K (2003). Biomolecular recognition based on single gold nanoparticle light scattering. *Nano Lett* **3**:935–938.

Riepl, M, Enander, K, Liedberg, B, Schaferling, M, Kruschina, M, and Ortigao, F (2002). Functionalized surfaces of mixed alkanethiols on gold as a platform for oligonucleotide microarrays. *Langmuir* **18**:7016–7023.

Rinaldis, M, Ausiello, G, Cesareni, G, and Helmer-Citterich, M (1998). Three-dimensional profiles: A new tool to identify protein surface similarities. *J Mol Biol* **284**:1211–1221.

Sandhu, KK, McIntosh, CM, Simard, JM, Smith, SW, and Rotello, VM (2002). Gold nanoparticle-mediated Transfection of mammalian cells. *Bioconjug Chem* **13**:3–6.

Sauthier, ML, Carroll, RL, Gorman, CB, and Franzen, S (2002). Nanoparticle layers assembled through DNA hybridization: Characterization and optimization. *Langmuir*, **18**:1825–1830.

Schechter, NM, Eng, GY, Selwood, T, and McCaslin, DR (1995). Structural changes associated with the spontaneous inactivation of the serine proteinase human tryptase. *Biochemistry* **34**:10628–10638.

Strong, LE, and Kiessling, LL (1999). A general synthetic route to defined, biologically active multivalent arrays *J Am Chem Soc* **121**:6193–6196.

Templeton, AC, Hostetler, MJ, Kraft, CT, Murray, RW (1998a). Reactivity of monolayer-protected gold cluster molecules: Steric effects. *J Am Chem Soc* **120**:1906–1911.

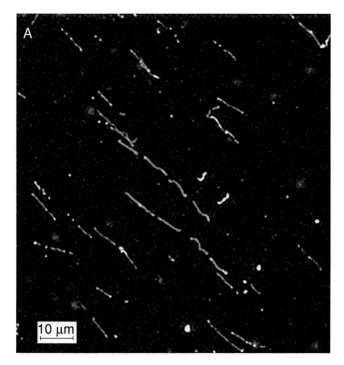

Figure 1.8A. See page 16 for text discussion of this figure.

Figure 1.9A. See page 17 for text discussion of this figure.

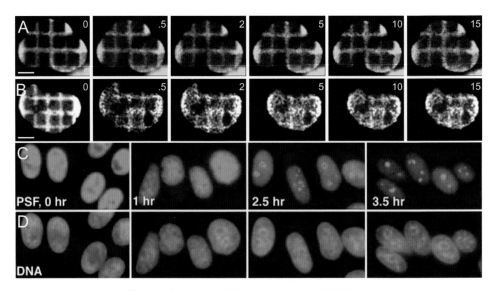

Figure 2.1. See page 30 for text discussion of this figure.

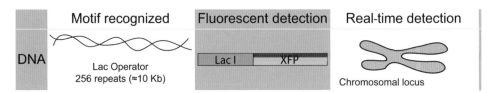

Figure 2.2. See page 31 for text discussion of this figure.

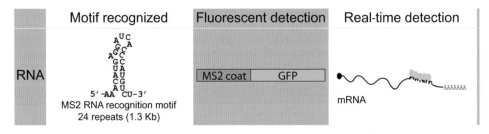

Figure 2.3. See page 33 for text discussion of this figure.

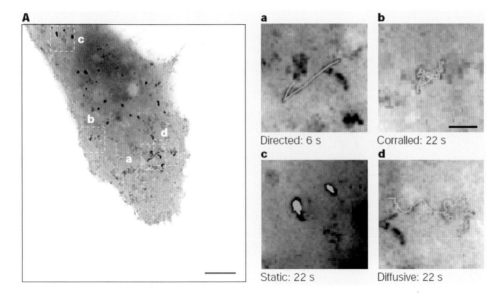

Figure 2.4. See page 34 for text discussion of this figure.

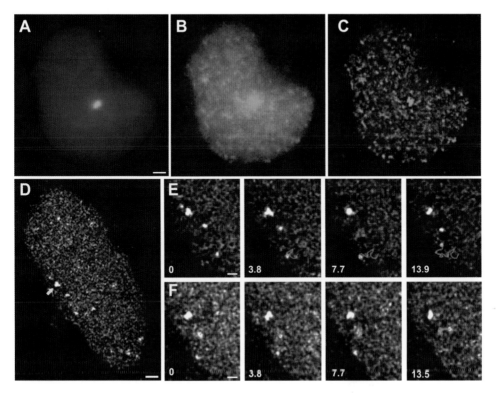

Figure 2.5. See page 35 for text discussion of this figure.

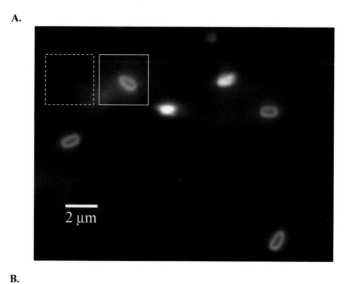
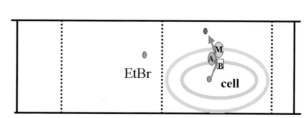

Figure 3.4. See page 52 for text discussion of this figure.

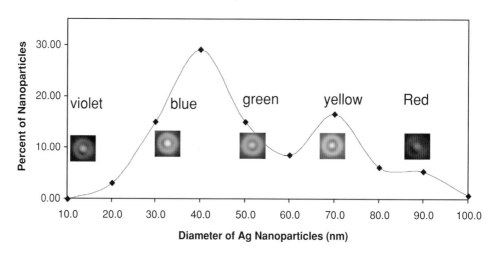

Figure 3.7. See page 56 for text discussion of this figure.

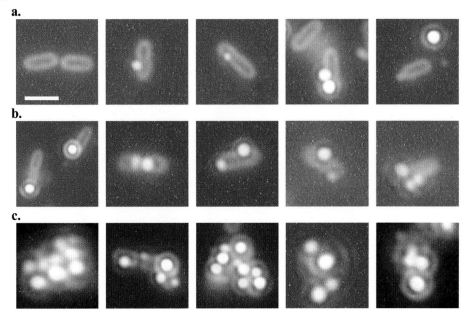

Figure 3.8. See page 59 for text discussion of this figure.

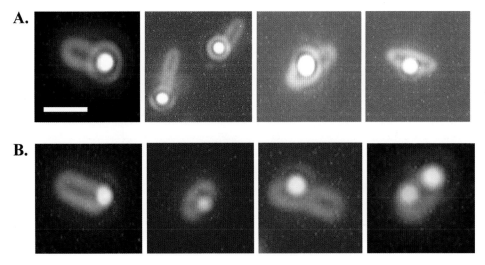

Figure 3.9. See page 60 for text discussion of this figure.

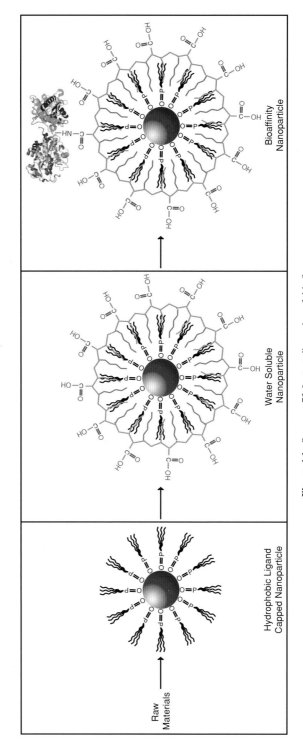

Figure 4.1. See page 73 for text discussion of this figure.

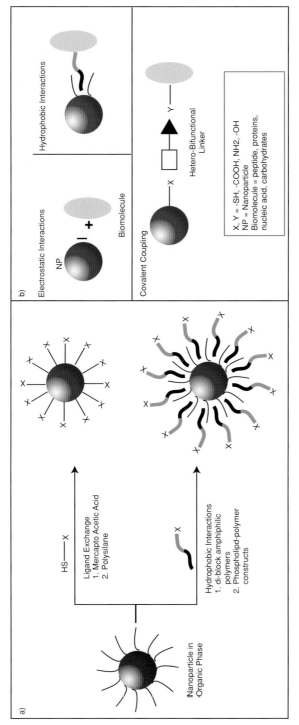

Figure 4.2. See page 75 for text discussion of this figure.

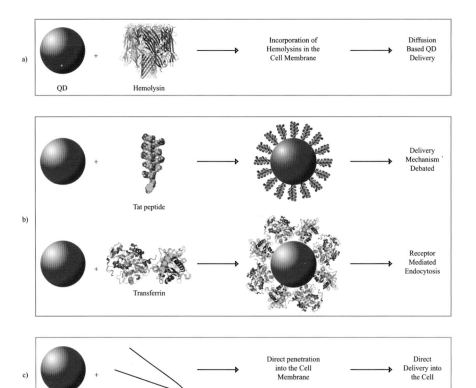

Figure 4.3. See page 76 for text discussion of this figure.

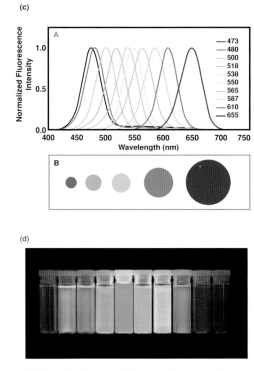

Figures 4.4C and D. See page 77 for text discussion of these figures.

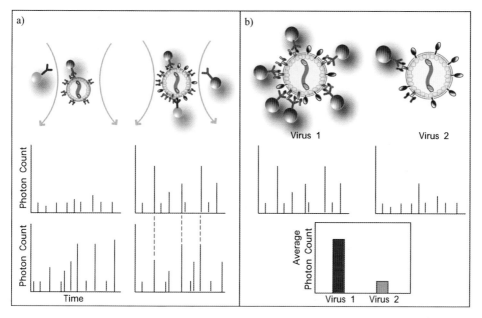

Figure 4.5. See page 78 for text discussion of this figure.

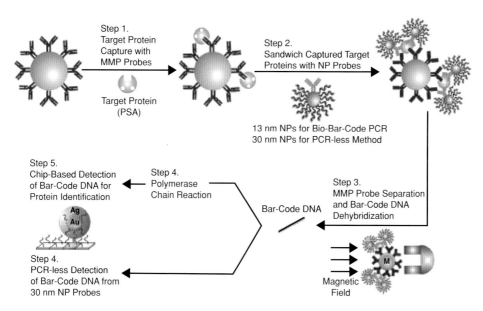

Figure 4.6. See page 80 for text discussion of this figure.

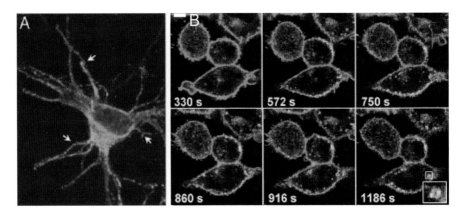

Figure 4.7. See page 81 for text discussion of this figure.

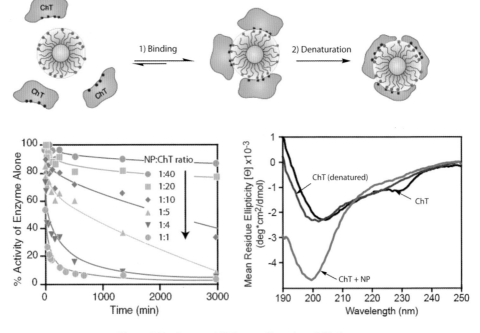

Figure 5.13. See page 103 for text discussion of this figure.

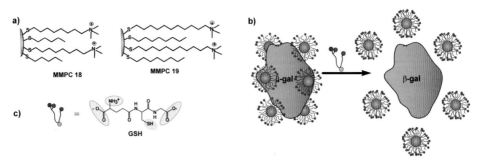

Figure 5.16. See page 106 for text discussion of this figure.

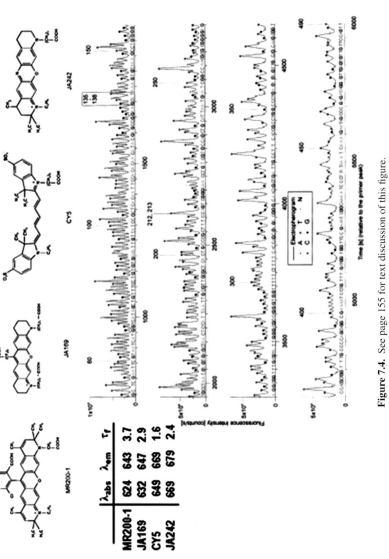

Figure 7.4. See page 155 for text discussion of this figure.

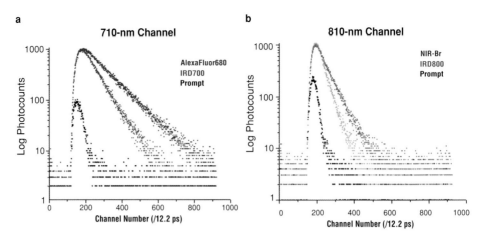

Figure 7.8. See page 160 for text discussion of this figure.

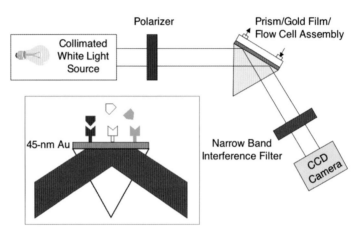

Figure 8.1. See page 171 for text discussion of this figure.

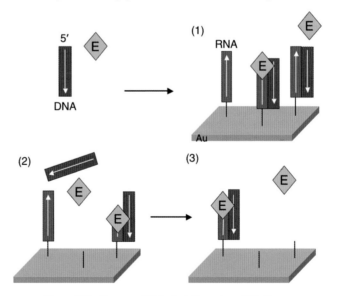

Figure 8.11. See page 183 for text discussion of this figure.

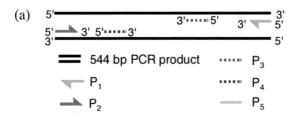
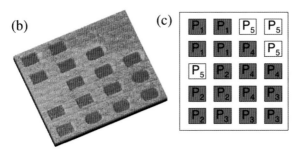

Figure 8.15. See page 186 for text discussion of this figure.

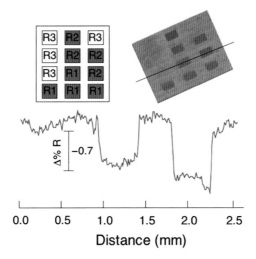

Figure 8.16. See page 187 for text discussion of this figure.

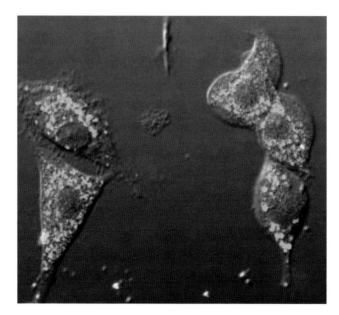

Figure 9.3. See page 204 for text discussion of this figure.

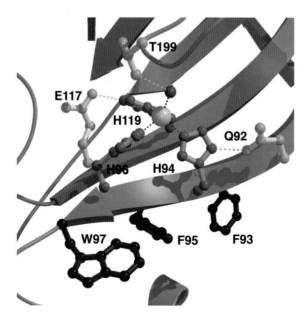

Figure 9.4. See page 206 for text discussion of this figure.

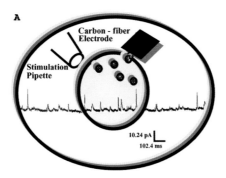

Figures 10.1A. See page 216 for text discussion of this figure.

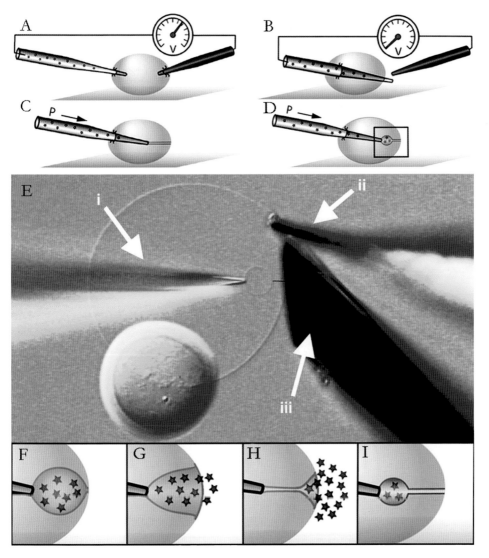

Figure 10.3. See page 224 for text discussion of this figure.

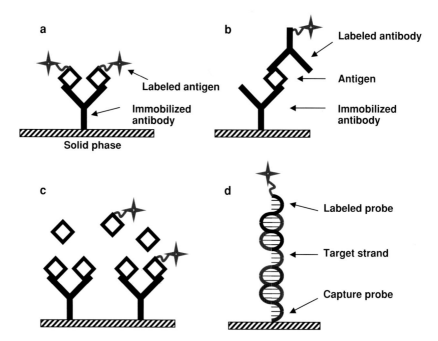

Figure 11.1. See page 239 for text discussion of this figure.

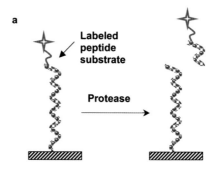

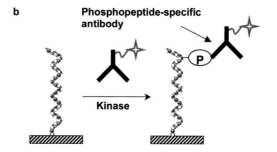

Figure 11.2. See page 240 for text discussion of this figure.

Templeton, AC, Hostetler, MJ, Warmoth, EK, Chen, SW, Hartshorn, CM, Krishnamurthy, VM, Forbes, MDE, and Murray, RW (1998b). Gateway reactions to diverse, polyfunctional monolayer-protected gold clusters. *J Am Chem Soc* **120**:4845–4849.

Templeton, AC, Wuelfing, MP, and Murray, RW (2000). Monolayer protected cluster molecules. *Acc Chem Res* **33**:27–36.

Thomas, M, and Klibanov, AM (2003). Conjugation to gold nanoparticles enhances polyethylenimines transfer of plasmid DNA into mammalian cells. *Proc Natl Acad Sci USA* **100**:9138–9143.

Toogood, PL (2002). Inhibition of protein–protein association by small molecules: Approaches and progress. *J Med Chem* **45**:1543–1558.

Truong-Le, VL, Walsh, SM, Schweibert, E, Mao, HQ, Guggino, WB, August, JT, and Leong, KW (1999). Gene transfer by DNA-gelatin nanospheres. *Arch Biochem Biophys* **361**:47–56.

Verma, A, Nakade, H, Simard, JM, Rotello, VM (2004). Recognition and stabilization of peptide α-helices using templatable nanoparticle receptors. *J Am Chem Soc* **126**:10806–10807.

Verma, A, Simard, JM, Worrall, JWE, and Rotello, VM (2004). Tunable reactivation of nanoparticle-inhibited β-galactosidase by glutathione at intracellular concentrations. *J Am Chem Soc* **126**:13987–13991.

Wang, GL, Zhang, J, and Murray, RW (2002a). DNA binding of an ethidium intercalator attached to a monolayer-protected gold cluster. *Anal Chem* **74**:4320–4327.

Wang, TX, Zhang, DQ, Xu, W, Yang, JL, Han, R, and Zhu, DB (2002b). Preparation, characterization, and photophysical properties of alkanethiols with pyrene units-capped gold nanoparticles: Unusual fluorescence enhancement for the aged solutions of these gold nanoparticles. *Langmuir* **18**:1840–1848.

Warner, MG, and Hutchison, JE (2003). Formation of linear and branched nanoassemblies of gold nanoparticles by electrostatic assembly in solution on DNA scaffolds. *Nat Mater* **2**:272–276.

Wilson, AJ, Groves, K, Jain, RK, Park, HS, and Hamilton, AD (2003). Directed denaturation: Room temperature and stoichiometric unfolding of cytochrome *c* by a metalloporphyrin dimer. *J Am Chem Soc* **125**:4420–4421.

Wolfert, MA, Schacht, EH, Toncheva, V, Ulbrich, K, Nazarova, O, and Seymour, LW (1996). Characterization of vectors for gene therapy formed by self-assembly of DNA with synthetic block co-polymers. *Hum Gene Ther* **7**:2123–2133.

Xu, C, Xu, K, Gu, H, Zhong, X, Guo, Z, Zheng, R, Zhang, X, and Xu, B.(2004). Nitrilotriacetic acid-modified magnetic nanoparticles as a general agent to bind histidine-tagged proteins. *J Am Chem Soc* **126**:3392–3393.

Yun, CS, Khitrov, GA, Vergona, DE, Reich, NO, and Strouse, GF (2002). Enzymatic manipulation of DNA-nanomaterial constructs. *J Am Chem Soc* **124**:7644–7645.

Zhang, ZY, Poorman, RA, Maggiora, LL, Heinrikson, RL, and Kezdy, FJ (1991). Dissociative inhibition of dimeric enzymes. Kinetic characterization of the inhibition of HIV-1 protease by its COOH-terminal tetrapeptide. *J Biol Chem* **266**:15591–15594.

Zheng, M, and Huang, X. (2004). Nanoparticles comprising a mixed monolayer for specific bindings with biomolecules. *J Am Chem Soc* **126**:12047–12054.

CHAPTER

6

NANOSCALE CHEMICAL ANALYSIS OF INDIVIDUAL SUBCELLULAR COMPARTMENTS

GINA S. FIORINI AND DANIEL T. CHIU

6.1. INTRODUCTION

6.1.1. Subcellular Compartments—Motivation and Challenges of their Analysis

Most of the information currently provided in cellular biology texts on subcellular compartments centers on the overall structure, function, and general molecular makeup of specific organelles. The information is typically based upon research performed using biochemical assays of bulk organelle samples, which provides generalized or averaged data, or microscopy, optical- and electron-based imaging, which can provide information specific to a single organelle (Lodish et al., 2000). Fluorescence microscopy takes advantage of site-specific fluorescent dyes, labeled antibodies, DNA probes, or fluorescent protein (e.g., GFP) transfection to selectively image different types of cellular components or the location of a type of molecule (e.g., a particular protein), providing dynamic information. Because cellular contents can autofluoresce and interfere with fluorescence microscopy, the use of scanning confocal or two-photon microscopy techniques can be used to improve the sensitivity and resolution of the final images; however, optical resolution has a theoretical maximum of about 0.2 µm in the visible wavelength. In contrast to optical microscopy, the use of electron microscopy can record static images of cellular structures in the nanometer range ($\sim$0.1-nm resolution for transmission electron microscopy and 10 nm for scanning electron microscopy, depending on the sample). Despite the general power of microscopy in providing high-resolution visual images, minimal chemical information is accumulated about the subcellular compartments that are being visualized.

There is a lack of techniques for obtaining chemical information from nanoscale subcellular structures due to the minute amount of samples that are available for analysis. The sample volume scales as the third power with the diameter of the subcellular compartment, and a typical single mammalian cell with a diameter of $\sim$10 µm has a volume of $\sim$$10^{-12}$ L, which is 1000 times the volume for a large subcellular structure of $\sim$1-µm diameter and a staggering 10 million times the volume for a small vesicle with a diameter of $\sim$50 nm. A typical single organelle may range in diameter from tens of nanometers to a couple micrometers, with a corresponding volume of $\sim$$6 \times 10^{-20}$ L (e.g., for a 50-nm synaptic vesicle) to $\sim$$8 \times 10^{-15}$ L (e.g., for a 2-µm mitochondrion, although individual mitochondria vary

---

New Frontiers in Ultrasensitive Bioanalysis. Edited by Xiao-Hong Nancy Xu
Copyright © 2007 John Wiley & Sons, Inc.

widely in size). Within a volume of 6×10^{-20} L, even at a high concentration of 100 mM, the number of molecules present is only ~3600 (Chiu, 2003). The small sample volume, limited number of molecules, and the complex mixtures of molecules contained within each subcellular compartment require an approach that is both highly sensitive and capable of isolating, identifying, and quantifying each of the components of the mixture.

As discussed elsewhere in this book, single-cell analysis techniques are available; however, the sensitivity required for manipulating and detecting the small nanometer-sized samples of subcellular compartments must be dramatically increased beyond that of current single-cell analysis techniques. Despite this demanding challenge, there is a possible advantage: With this drastic decrease in volume, there is the corresponding reduction in the chemical complexity of the sample (Chiu, 2003). Rather than having perhaps hundreds of thousands of small molecules and proteins present in a single cell, a small subcellular compartment might contain only a fraction of these molecules. Provided that we can achieve adequate sensitivity, this reduction in complexity makes the comprehensive chemical characterization of single subcellular structures more manageable in comparison with the chemical analysis of single cells.

6.1.2. Relevant Research from the Past and Present

Although no analytical methods currently exist for the total microanalysis of subcellular compartments, several techniques have been demonstrated for the analysis of secretions from individual vesicles of a single cell. The use of microelectrodes for electrochemical detection of vesicle secretions is common (Travis and Wightman, 1998). For this type of analysis, microelectrodes are positioned close to a single cell at selected locations, thus minimizing dilution of the analytes by diffusion and providing spatial information at high sensitivity. The technique is capable of detecting zeptomoles of analytes (Hochstetler et al., 2000), but is limited to the detection of electroactive species only and is not optimal for the analysis of complex mixtures. An advantage of the use of microelectrodes for electrochemical detection is that the method is compatible with *in vivo* measurements, which some high-resolution microscopy-based methods lack (e.g., transmission electron microscopy).

For studying complex mixtures of cellular secretions, a method to separate each component of the mixture for subsequent detection and analysis is necessary. The most common separation technique used is capillary electrophoresis (CE). With inner diameters of fused silica capillaries on the order of tens of micrometers and down to 400 nm, CE is capable of handling small samples of cellular secretions and cellular cytoplasm (Lillard et al., 1996; Gordon and Shear, 2001; Liu et al., 1999; Page et al., 2002; Olefirowicz and Ewing, 1990; Woods et al., 2005; Wu et al., 2004); however, with the extremely low analysis volumes of single vesicles, there is a significant degree of diffusion within the capillary that leads to an overall sensitivity that is too low for the real-time analysis of single vesicle secretions.

The analysis of the contents of single vesicles has been demonstrated by the research teams of Zare (Chiu et al., 1998; Lillard et al., 1998) and Sweedler (Rubakhin et al., 2000). Both experiments were performed on the fairly large (~1 μm diameter, ~10^{-15} L volume) vesicles found in the atrial gland of *Aplysia californica*. Zare's team used capillary electrophoresis with laser-induced fluorescence detection (CE-LIF). A single vesicle was optically trapped and translated into the inlet of the capillary and allowed to react with a fluorogenic dye prior to CE-LIF. The experiment was able to detect the amino acids in the vesicle. A major disadvantage of this approach is the dilution of the sample inside the much larger capillary volume, preventing the analysis of samples much smaller than *Aplysia* vesicles.

In contrast, Sweedler and co-workers detected peptides using matrix-assisted laser desorption/ionization time-of-flight mass spectrometry (MALDI-TOF). For this technique, single vesicles were placed on a sample plate and covered in MALDI matrix prior to MALDI-TOF analysis. The low-molecular-weight ions from the MALDI matrix obscure low-molecular-weight analytes, thus leading to the analysis of the peptides in the vesicles rather than the amino acids. An advantage of this approach is that it does not require the tagging of the analytes as in CE-LIF. However, MALDI-TOF currently exhibits a generally reduced sensitivity as compared to CE-LIF, which limits the application of MALDI-TOF to fairly large or concentrated vesicle samples.

In addition to the analysis of vesicles, researchers have also been working on the analysis of other subcellular compartments. For example, Arriaga and co-workers have been focusing on studying mitochondria (Duffy et al., 2002). Their work uses CE-LIF to study intact mitochondria isolated from cells. Mitochondria have different surface charge based upon their outer membrane composition, enabling individual mitochondria to be separated and detected using CE-LIF. Arriaga and co-workers have extended their efforts to measuring the cardiolipin content of mitochondria as well (Fuller et al., 2002). In general, the work described in this section illustrates the desire of researchers to probe ever smaller samples, moving from single-cell releasates to individual vesicles and organelles and emphasizing the need for a powerful universal method to analyze subcellular organelles.

6.1.3. Overview of Nanosurgery/Droplet Platform

To extend our research efforts to better understand the localization and function of complex cellular activities within the cell, we need a set of analytical techniques capable of providing a complete chemical inventory—from large protein molecules to small metabolites—contained within nanometer-sized subcellular compartments. A technological platform that offers comprehensive biochemical information on individual subcellular compartments with submicrometer spatial resolution should bridge the visual information accessible with imaging methods and the biochemical information obtained from bulk assays that are currently available.

To address this need, we are working to develop a technology platform for carrying out the profiling of single subcellular compartments. Our strategy is based on the following: (1) Visually select a subcellular compartment (organelle) using high-resolution microscopy. (2) Remove the organelle from the cell using laser-based single-cell nanosurgery. (3) Encapsulate the captured organelle into a droplet (femtoliter volume). (4) Create another droplet that contains reactive dyes for chemically labeling the organelle contents. (5) Fuse the two droplets to carry out the labeling reaction. (6) Release the labeled contents of the organelle from the reaction droplet and perform a separation of the contents using a selected microscale separation technique (e.g., CE). (7) Detect, identify, and quantitate the separated contents of the organelle by optically based single-molecule detection. Figure 6.1 is a flow chart that summarizes this strategy. The following sections in this chapter discuss what has been accomplished to date, as well as future challenges.

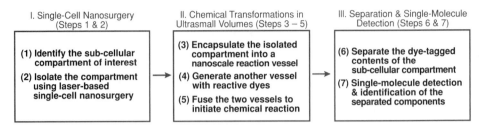

Figure 6.1. Overview of our technology platform for subcellular analysis.

6.2. EXTRACTION OF SUBCELLULAR COMPARTMENTS FROM SINGLE CELLS

6.2.1. Background on Laser Manipulation

Phase I of the subcellular organelle analysis platform (Figure 6.1) encompasses the technique of single-cell nanosurgery, which includes the selection of a subcellular compartment and the subsequent laser-based procedure to capture the organelle and remove it from the cell. Two different types of laser manipulation are used during nanosurgery: (1) optical trapping and (2) laser ablation. Optical trapping is used to hold onto the selected organelle, while laser ablation creates a small hole in the cell membrane, through which the organelle is transported using the optical trap. These laser manipulations are not limited to use in nanosurgeries; laser manipulations are also used in subsequent steps of the subcellular analysis platform.

Optical trapping, also known as laser tweezers, has been employed for the manipulation of various cells, vesicles, and biomolecules for applications in biomedical fields, biopolymers, and, of special interest in this chapter, bioanalysis (Kuyper and Chiu, 2002). Optical trapping is made possible by the fact that photons can impart momentum despite having no mass. A tightly focused laser beam, which passes through a high numerical aperture (N.A. > 1.3) objective, creates a special distribution of light intensity that exerts a force on particles located within the generated photon flux region (Figure 6.2). In-depth theoretical descriptions of optical trapping forces have been discussed previously (Kuyper and Chiu, 2002; Svoboda and Block, 1994; Ashkin, 1997; Smith et al., 1999).

Single-beam gradient force traps are one of the most common optical trap designs used (Ashkin et al., 1986; Kuyper and Chiu, 2002). Near-infrared (~ 800–1200 nm) lasers are commonly used, minimizing unwanted damage as biological samples do not absorb light in this range. The laser beam is operated in the TEM_{00} mode to provide a Gaussian intensity profile in the x and y directions, and it is sent through a telescope and overfills the back aperture of the objective to provide control and sharpness to the z direction of the intensity gradient. In general, particles with a higher refractive index than the surrounding medium are successfully trapped with this configuration.

In contrast to optical trapping where damage to the sample is minimized, laser surgery, or laser ablation, relies on the capability of lasers to physically alter or damage the sample and is commonly employed in well-known biomedical procedures (Niemz, 2004; Berns et al., 1991; Srinivasan, 1986; Vogel and Venugopalan, 2003). More applicable to this chapter is the use of laser surgery in microscale applications, such as laser capture microdissection, *in vitro* fertilization, and procedures on single cells (optoporation, injection, or lysis). Laser ablation can occur under more than one mechanism (Srinivasan, 1986; Vogel and Venugopalan, 2003). Photoablation involves the absorption of light by molecules in the

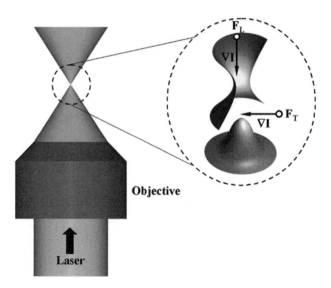

Figure 6.2. Three-dimensional drawing of an expanded view of the laser focus of a single-beam gradient trap. The Gaussian laser beam is sent through a high numerical aperture objective to produce two intensity gradients (∇I) at the focus. F_T represents the transverse force exerted on the particle from the intensity gradient produced by the TEM$_{00}$ mode of the laser. F_L is the longitudinal force on the particle due to the intensity gradient generated by the tight focusing of the objective.

sample and subsequent chemical bond breakage or heating of the sample. However, for the specific laser surgery described in this section, the pulsed ultraviolet (UV) laser causes plasma-induced ablation and/or photodisruption (Shelby et al., 2005). More specifically, plasma-induced ablation is the removal of material by the plasma formed where the laser intereacts with the sample, while photodisruption is caused by the shock waves and cavitation bubbles that occur after plasma formation.

6.2.2. Single-Cell Nanosurgery

Figure 6.3 shows an optical setup used for single-cell surgery, including a pulsed UV laser (3-ns pulse width, 337-nm N_2 laser) for ablation of the cell membrane and a Nd:YAG laser (continuous wave (CW), 1064 nm) for optical trapping of the selected organelle (Shelby et al., 2005). An Ar^+ laser is used to illuminate the fluorescently labeled organelles within the cell, enabling visual identification of a specific target organelle. Under our experimental conditions, the power of the N_2 laser delivered to the cell membrane needs to be above a threshold value of $\sim$0.5 μJ and below 1–2 μJ at which cells are irreversibly damaged. Approximately 0.7 μJ is the optimum pulse energy to use to open the membrane enough to remove an organelle ($\sim$1–3 μm). Cells can be slightly swelled so that organelles experience Brownian motion and are easy to trap and translate out of the cell using the optical trap. Mitochondria and lysosomes have been successfully removed using nanosurgery.

We determined that the location of the N_2 laser focus significantly determines the success of the nanosurgery. If the focus of the laser is on the glass coverslip that the cell rests on, the cell is quickly lysed when the glass surface is ablated. If the focus is just outside of the membrane, in the extracellular environment, then photodisruption is the main mechanism,

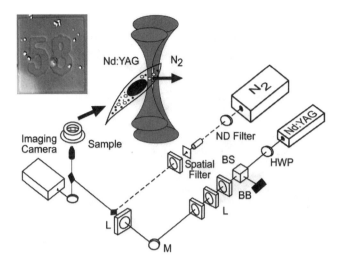

Figure 6.3. Schematic of optical setup. The outputs of a pulsed N_2 laser (337 nm; 3-ns pulse width) and a CW Nd:YAG laser (1064 nm) were aligned collinearly and sent into the microscope, where the two beams were directed into the objective (N.A. = 1.3) by a dichroic and focused onto the cell by the objective. The microscope was equipped with Nomarski optics, but the components were omitted in the schematic for clarity. The pulsed N_2 laser (~0.7 μJ) was used to ablate a small hole (~1–3 μm) in the cell membrane, and the YAG trapping laser (200 mW) was employed to extract a subcellular organelle (arrow). To track individual cells post-surgery, we cultured cells in PDMS wells formed by sealing PDMS to photoetched coverslips (upper left inset). HWP, half-wave plate; BS, polarizing beam splitter; BB, beam blocker; ND, neutral density filter; M, mirror; L, lens. [Reproduced with permission from *Photochem Photobiol* **81**: 994–1001 (2005). Copyright 2005, American Society for Photobiology.]

with cavitation bubbles causing delocalized damage to the cell. The optimal placement of the laser focus is on the cell membrane. With a focal volume of about 2 μm and a precision for locating the focus within 2 μm of the membrane surface, laser ablation of the membrane is a fairly consistent procedure. Figure 6.4 contains images of nanosurgery on three different cell lines: neuroblastoma cells fused with glialoma cells (NG108) (top panels), Chinese hamster ovary cells (CHO) (middle panels), and murine embryonic stem cells (ES-D3) (bottom panels). From left to right, the images show the cell before surgery, immediately after the N_2 laser pulse, and with the trapped organelle outside of the cell membrane. The optical trap as used in this procedure can photobleach the labeled organelles during the procedure, which hinders visualization of the trapped organelle for extended periods. This issue needs to be resolved for future work.

6.2.3. Cell Viability Following Nanosurgery

We were able to successfully remove organelles from all three cell lines that were tested; however, the fate of the cells following nanosurgery varied with cell line (Shelby et al., 2005). CHO cells survived nanosurgery with higher viability rates than the other two cell lines (NG108 and ES-D3 stem cells). Within 1 h of surgery, 85% of CHO cells survived, and the rate slightly dropped to approximately 75% 4 h following surgery. NG 108 and ES-D3 cells showed significantly lower viability rates than the CHO cells. NG108 cells had a viability rate of 40% at +1 h and 30% at +4 h. ES-D3 cells were adversely affected by the surgery with 15% viability at +1 h and close to 0% beyond 1 h following surgery. In general, decreases in viability rates occurred during the first hour following surgery and

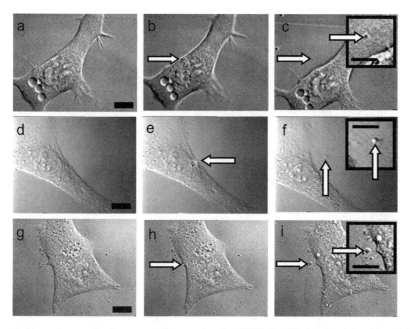

Figure 6.4. Organelle extraction from successful surgeries. An NG108-15 cell before (a), during (b), and after (c) surgery; the arrows point to the organelle that was being extracted. A similar sequence showing a CHO cell before (d), during (e), and after (f) surgery. A small cavitation bubble is visible in (e); (f) shows the organelle (arrow) that was completely removed from the cell. A ES-D3 stem cell before (g), during (h), and after (i) surgery. Scale bars in parts a, d, and g = 10 μm. Inset scale bars = 10 μm. [Reproduced with permission from *Photochem Photobiol* **81**: 994–1001 (2005). Copyright 2005, American Society for Photobiology.]

then leveled off beyond the first hour, indicating that cellular repair likely occurs during the first hour following surgery. Differences in repair mechanisms among cell lines may account for the overall differences in viability following nanosurgery, because some cell types are more resilient than others.

The primary cause of cell death in nanosurgery is the use of the N_2 laser pulse to ablate the cellular membrane. The control experiments detailed in Figure 6.5A–C illustrate that the most significant drops in cell viability rates for all three cell lines coincided with the introduction of the N_2 laser to the procedure. Cell survival is similar for cells exposed to the entire surgery procedure, for cells exposed to both the N_2 laser and the optical trap laser without extraction of an organelle, and for cells only exposed to a N_2 laser pulse.

6.3. NANOSCALE MANIPULATIONS OF SUBCELLULAR COMPARTMENTS AND CHEMICAL REACTIONS OF THEIR CONTENTS

6.3.1. Effects of Diffusion and the Need for Small Reaction Volumes

To manipulate chemically (e.g., dye tagging the metabolites) the contents of a nanometer-sized subcellular compartment, the main challenge is to overcome dilution by spatially confining the chemical reaction to an ultrasmall volume. Figure 6.6 is a simple estimation that shows the importance of spatially localizing the reactants and the rapid nature of diffusion at the nanometer and micrometer length scale.

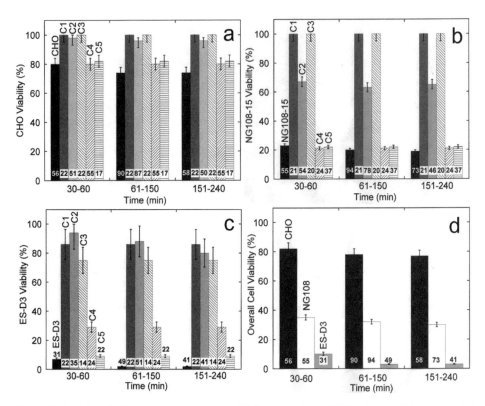

Figure 6.5. Viability of cells after organelle extraction. The bar graphs show viability of each cell line over a period of four hours. (a) The raw data for the complete surgical procedure (% viability) for CHO cells are represented by the solid black bars. C1 control is cells with no laser exposure in native media only. C2 control is cells swelled with 25% pure water with no laser exposure. C3 control is swelled cells with IR laser exposure only. C4 control is swelled cells with UV laser exposure only. C5 control is swelled cells exposed to both IR and UV laser irradiation but no organelle removal. (b, c) These are similar controls for the NG108-15 and ES-D3 cells, respectively. (d) The percent viability for each cell line was obtained by normalizing the survival rate of cells that had undergone surgery with the media plus swelling control (C2). The number of cells used for each experiment is given at the bottom of each bar, and error bars represent the higher of either the experimental error or standard error of the binomial distribution. [Reproduced with permission from *Photochem Photobiol* **81**: 994–1001 (2005). Copyright 2005, American Society for Photobiology.]

Diameter	Volume	Concentration
50 nm	6.5×10^{-20} L	0.1 M
	1 ms $\downarrow D = 5 \times 10^{-10}$ m^2/s	
2 μm	4.2×10^{-15} L	1.5 μM
	1 s $\downarrow$	
64 μm	1.4×10^{-10} L	45 pM

Figure 6.6. A simple estimation illustrating the nonlinear effects of diffusion, which makes dilution an extremely rapid process at the nano- and micrometer regime. For example, small molecules (with a diffusion coefficient of 5×10^{-6} cm^2/s) present at 100 mM in a 50-nm-diameter vesicle will be diluted to a concentration of 45 pM (10^{-12} M) in approximately 1 s.

6.3.2. Droplet Formation

To overcome the detrimental effects of dilution, the molecules to be reacted must be confined within a small reaction volume (Chiu et al., 1999). In the past years, we have explored suitable nanoscopic reaction vessels to conduct and control chemical transformations in femtoliter (10^{-15} L) volumes. Such reaction vessels must satisfy a number of criteria. For example, individual vessels must be formed easily on demand with precise control over their dimensions, the isolated organelle must be encapsulated within the vessel during its formation and stably remain in the vessel, the vessel must be physically and chemically stable, each vessel must be manipulated and transported easily (either optically, fluidically, or electrically) in a chip-based system, the reaction vessels must be fused together without leakage so their respective contents can be combined, and the vessel must be compatible with conditions used for most biologically relevant reactions (e.g., derivatization). After much characterization, we found the most suitable reaction vessels to be aqueous droplets present in an immiscible medium (He et al., 2004; He et al., 2005a, 2005b).

Droplets can be easily produced in micro and nano fluidic channels in a variety of different manners (Tice et al., 2003; Tan et al., 2004). Here we discuss three methods: (1) shear force in a T-channel design, (2) using a channel constriction, and (3) electrogeneration (He et al., 2005a, 2005b). All three methods employ the use of hydrophobic poly(dimethylsiloxane) (PDMS) channels so that the generated aqueous droplets remain in the oil phase and do not associate with the channel walls. The first method is shown in Figures 6.7A and 6.7B. The oil phase (e.g., soy oil or silicone oil) flows through two of the branches of the T channel, and the aqueous phase flows in the third branch (perpendicular to the other two branches) (He et al., 2005a). Droplets are formed at the interface of the aqueous and oil phases as the oil exerts a shear force on the interface, creating an aqueous drop in an immiscible oil phase. With appropriate settings for the flow rates of the two phases, a continuous stream of droplets is easily formed using this method. The size of the droplets formed with this method is approximated by equating the Laplace pressure with the shear stress. Increasing the velocity of the oil phase and decreasing the channel dimensions will increase the shear stress and thus decrease the droplet radius. However, only the local T-branch channels were narrowed, while the remainder of the design utilizes fairly large channels, minimizing the pressure drop associated with both of the phases and allowing for easier control of the fluid phases for the generation of droplets.

The second method of droplet generation is to pass the aqueous phase through a channel constriction (Figures 6.7G and 6.7H) (He et al., 2005a). Essentially, the aqueous phase is forced through an orifice into a larger field of the oil phase, similar to techniques used to create emulsifications. We used channel constrictions with a minimum width of 15 μm as compared to 1000 μm in width for the reservoir and exit channels. The entire channel system is first filled with the oil phase, and then the aqueous phase is introduced. Because the channels were treated to be hydrophobic, the oil phase is always in contact with the channel walls; also, as the aqueous phase is injected into the oil, discrete water plugs are formed. The water plugs take on the shape of droplets after they enter the larger oil chamber. High flow rates are not necessary with this method, which helps subsequent analysis steps as slow moving droplets are easier to manipulate and track.

The third method of droplet generation is electrogeneration, in which high-voltage pulses are used to generate the droplets (He et al., 2005b). This technique is based upon the formation of a conical interface (Taylor cone) where the applied electric stress is balanced against the surface tension. We found that at high voltages the cone elongates and forms a jet

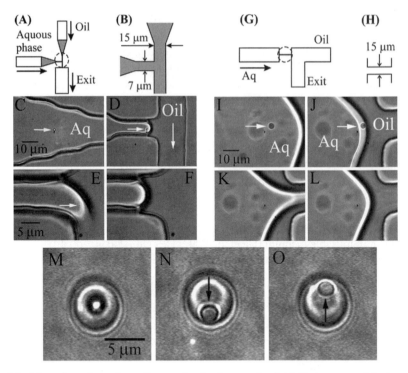

Figure 6.7. Schematics and sequences of images showing the two microfluidic designs we used for the encapsulation of single micro- and nanoparticles into droplets. (A and B) Schematic of a T channel, in which the aqueous droplet is sheared off at the junction by the flowing oil phase; the circled area in part A is expanded in part B. Shaded areas in parts A and B are the shallow (10 μm) channels and the rest are deep (50 μm) channels. (C–F) Images showing the optical trapping and transport of a selected bead to the T-junction (parts C and D). Once the bead was parked and maintained at the interface of the two fluids, a pressure pulse was applied to the aqueous phase to shear off a single droplet (parts E and F). The bead was entrapped within the droplet during this formation process, and thus it was removed from the T junction (F). (G, H) Schematic of a second design we used for forming droplets, in which the aqueous phase is pushed through a channel constriction into an immiscible phase. Part H is the expanded drawing of the circled area in part G. To introduce a bead into the droplet, we again optically trapped and transported the bead to the interface (I and J) so that the bead was enclosed within the droplet as the droplet was being generated (K and L). Once a bead was encapsulated within the droplet (M), it can be freely move within the droplet by optical trapping, but it cannot be forced across the interface and be removed from the droplet (N and O). The immiscible phase was soybean oil. Scale bar in part C applies to part D, seale bar in part E applies to part F, scale bar in part I applies to parts J–L, and scale bar in part M applies to parts N and O. [Reproduced with permission from *Anal Chem* **77**:1539–1544 (2005). Copyright 2005, American Chemical Society.]

where Raleigh instability occurs to produce droplets approximately 1.8 times the original jet diameter. Droplet dimensions varied between 3 and 25 μm with the channel dimensions shown in Figure 6.8, translating to a range of approximately 14 fL to 8 pL. The droplets can be easily formed on demand with this method, making it very useful for the subcellular analysis platform where single droplets are needed and not a continuous stream. Figure 6.8 is a series of images taken from a video of the electrogeneration of a single droplet at an electrical pulse of 800 V for 10 ms. After the droplet is ejected, the interface returns to its original shape and the droplet quickly attains a spherical shape. We found that by applying slight external pressure to the aqueous phase, the interface becomes highly curved and jet formation was more reproducible and easier to achieve. Because cone formation is

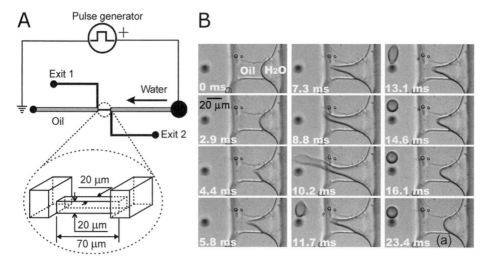

Figure 6.8. (A) Schematic of the experimental setup. The microfluidic channels were fabricated in PDMS. Each of the main channels (light shading) had a length of 1.1 cm. The heights of the main channels and the exit channels (dark shading) were 100 μm, and the widths were 1000 μm for the main channels and 500 μm for the exit channels. The circular area shows the details of the small channel constriction (orifice), with the right entrance being where the immiscible interface was maintained prior to initiating the pulse; the immiscible phase was silicone oil (density of 1.008 g/mL and dynamic viscosity of 20 mPa·s at 25°C) mixed with 3% (w/w) Span 85. Voltage pulses were applied using a custom-built high-voltage pulse generator with additional circuits to rapidly discharge voltage during the trailing edge of the pulse. The images were acquired with an inverted microscope (Nikon TE 300, Nikon, Tokyo, Japan) and a high-speed camera (CPL-MS10K, Canadian Photonic Labs, Minnedosa, Manitoba, Canada). (B) Sequence of images showing the generation of a single droplet upon application of a voltage pulse (800 V, 10 ms). [Reproduced with permission from *Appl Phys Lett* **87**:031916 (2005). Copyright 2005, American Institute of Physics.]

a balancing act between the electrical stress and the surface tension, this observation was hypothesized to be due to the fact that the applied pressure counteracts the surface tension that maintains the interface, thus allowing the electric pulse energy to be applied more directly to forming a jet and producing a droplet.

6.3.3. Encapsulation of Single Cells and Subcellular Compartments in Droplets

Prior to chemically labeling the contents of a single isolated organelle, the organelle must be encapsulated into a droplet where it can later be lysed so its contents are released into the droplet. Encapsulation is enabled by combining the optical control offered by laser trapping with the versatility of droplet generation in a microfluidic system. We have demonstrated the encapsulation of polymer beads, cells, and organelles into droplets using both the T-channel and constriction-based droplet formation methods (He et al., 2005a). In both cases the optical trap is used to position the selected particle at the interface of the two immiscible fluids, then slight pressure is applied to the aqueous and immiscible phase to shear or pinch off the single droplet. The main challenge to the procedure is to controllably generate a single droplet on demand once the particle has been positioned at the interface. For this requirement, the channel constriction method has a definite advantage over the T-channel design; from this perspective the electrogeneration method may prove to be the most compatible method for encapsulating particles, cells, and organelles into single droplets.

Figure 6.7 shows the encapsulation of beads in droplets. Figures 6.9A–D show the encapsulation of a single B lymphocyte cell in a droplet, whereas Figures 6.9E–H show the encapsulation of a single mitochondrion in a droplet. In the scope of the subcellular analysis platform, once an organelle is encapsulated in a droplet, the next stages are to lyse the organelle to release the contents of the organelle into the droplet for subsequent chemical reactions. Figure 6.9H nicely illustrates that the volume of the droplet can be kept similar to the volume of a single mitochondrion, reducing the detrimental effects of diffusion and dilution that would occur if the organelle were lysed in an open microchannel. We were able to use an ~5-ns pulse from a frequency-tripled YAG (335 nm) to photolyze a cell encapsulated in a droplet. Laser photolysis occurs on such a quick timescale that the cell does not really have time to respond to the damage, thus enabling a quick "snapshot" of the cell's state at that time. An enzymatic assay was performed in this manner: A droplet containing a fluorogenic substrate (fluorescein di-β-D-galactopyranoside for conversion by intracellular β-galactosidase) was used to encapsulate a cell, and once the cell was lysed, the enzymatic reaction began and the fluorogenic substrate was converted to a fluorescent species enabling detection and monitoring of the reaction (He et al., 2005a).

6.3.4. Concentration of Analytes within Droplets

We found that one remarkable and important characteristic of droplet nanoreactors is their ability to concentrate reactants via the slow dissolution of water molecules into the surrounding immiscible phase (He et al., 2004). Because most biological molecules of interest (e.g., proteins, peptides, metabolites, and amino acids) are significantly larger than water molecules and are often charged, they do not dissolve in the surrounding organic phase as the water droplet, in which they are contained, shrinks. Once we have concentrated molecules of interest to the desired level, we can optically trap and move the droplet into another organic solution in which droplet shrinkage does not occur. This capability is especially pertinent in the analysis of the smallest subcellular compartment (e.g., a 50-nm vesicle) because even if the reaction can be performed in a 1-μm-diameter droplet, it would still correspond to a 8000-fold dilution. We can now overcome such dilution problems by re-concentrating the molecules of interest within the nanodroplet.

To quantify this concentration effect, Figure 6.10 plots the increase in fluorescence intensity as the water droplet containing a fluorescent dye, Alexa-488, decreases in volume in a continuous phase of soybean oil (He et al., 2004b). The ~2.3-fold increase in the fluorescence intensity matches perfectly the ~2.3 fold decrease in volume, which indicates that Alexa-488 was concentrated gradually inside the aqueous droplet as anticipated and did not diffuse into the organic phase.

One point to note is that this phenomenon is prominent mostly in the micrometer length scale, because the rate of change in concentration scales as the fifth power of the surface-area-to-volume ratio of the droplet. We have systematically studied the dynamics of droplet shrinkage as well as the dependence of this shrinkage behavior on different immiscible phases, which ranges from no shrinkage (e.g., for perfluoro solvents) to moderately rapid shrinkage (~μm/min) such as the ones shown in Figure 6.10. By tuning the composition of the organic phase and by placing individual droplets in different organic environments created within microfluidic systems, the rate and extent of the droplet shrinkage can be exquisitely controlled.

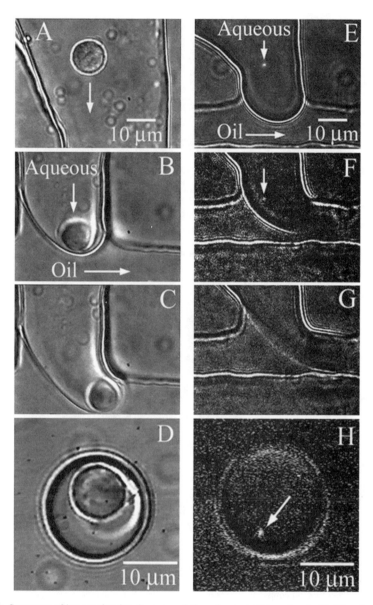

Figure 6.9. Sequences of images showing the encapsulation of a single B lymphocyte (A–D) and a single mitochondrion (E–H) into an aqueous droplet in silicone oil. Optical trapping was used to transport and position the cell close to the water/oil interface (A–C). An entrapped cell in a droplet is shown in part D. We visualized under fluorescence and optically manipulated a single mitochondrion stained with Mitotracker Green FM at the interface of the two fluids (E and F). Upon application of a pressure pulse to the microchannels (F and G), the mitochondrion was carried away by the flow as the droplet was sheared off. (H) A mitochondrion encapsulated in an aqueous droplet downstream from the T junction. The scale bar in part A applies to parts B and C, and the one in part E applies to parts F and G. [Reproduced with permission from *Anal Chem* **77**:1539–1544 (2005). Copyright 2005, American Chemical Society.]

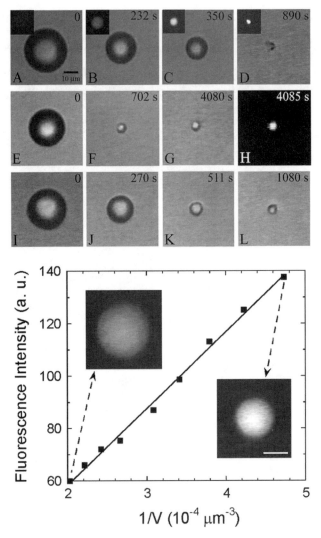

Figure 6.10. (Top panel) Sequence of micrographs showing the concentration of different types of nanoparticles and solutes within free-floating aqueous microdroplets in soybean oil. (A–D) Fluorescent polystyrene nanoparticles (27 nm in diameter) were concentrated inside an aqueous droplet; the insets are the corresponding fluorescence images showing increases in fluorescence intensity as the concentration of beads increased. (E–H) Shrinkage of an aqueous droplet containing dye-labeled carbonic andydrase; the microdroplet shrank gradually over the first ~12 min (E and F), after which the size of the droplet remained fairly constant (F and G). Part H shows the fluorescence image of the concentrated dye-tagged protein. (I–L) Concentration of sodium chloride in an aqueous droplet. The droplet shrank over the first ~9 min (I–K), after which the shape and size of the droplet remained fairly constant for the remaining 18 min (L). The scale bar applies to all panels (A–L). (Bottom panel) A plot showing the measured fluorescence intensity of an aqueous droplet, which contained the fluorescent dye Alexa-488, increased linearly with the reciprocal of the volume of the droplet. The solid line is a linear fit. The insets show the corresponding fluorescence images of the droplet at the indicated time points. The scale bar represents 10 μm and applies to both insets. [Reproduced with permission from *Anal Chem* **76**:1222–1227 (2004). Copyright 2004, American Chemical Society.]

6.3.5. Transport of Droplets

Once the droplet containing the subcellular organelle of choice has been formed, the droplet will need to be transported to other regions of the microfluidic device for further manipulation or analyses. Transport can be accomplished through optical trapping or electrowetting, among other potential methods. Just as optical trapping can be used for extracting organelles and positioning them at the fluid interface for encapsulation into a droplet, trapping is also well-suited to transport the droplet to a desired location in the microfluidic device. Aqueous droplets have a lower refractive index than the surrounding oil phase and cannot be trapped by the standard single-beam gradient force trap. A modified trap that is capable of trapping low refractive index species is needed to make this concept feasible.

In addition to optical methods, another promising technique for transporting droplets on chip is electrowetting (Kuo et al., 2003). The phenomenon of electrowetting, which changes the wetting properties of a surface upon application of an electric field, can be used to shuttle droplets efficiently and rapidly in an immiscible medium despite the fact that the aqueous droplet being moved is not in contact with the surface. Figure 6.11 shows this electrowetting-induced movement of a droplet in oil, in which the immiscible medium acts as a lubrication layer that significantly increases the velocity of droplet motion and the response of the system. This technique permits the transport of discrete droplets over long distances along a predefined path (e.g., centimeters) from one region of the microchip to another with good directional control and at high speed (cm/sec).

6.3.6. Fusion of Droplets to Enable Chemical Reactions

Once produced, individual droplets can be selected and fused, thereby mixing the contents of the different droplets and allowing chemical reactions to occur (Chiu et al., 1999; Chiu, 2003). Figure 6.12 shows the fusion of two aqueous droplets in oil, which took place over several seconds (Chiu, 2003). Since a single-beam gradient optical trap cannot trap aqueous droplets with an index of refraction lower than the surrounding medium, the repulsive force from the intensity gradient is used instead to displace the two aqueous droplets and force them together to fuse. Using a repulsive force rather than trapping leads to a tedious and low yield process. A different variety of optical trapping that is capable of directly trapping and manipulating aqueous droplets could significantly improve the fusion of droplets. In general, the speed and ease with which droplets are fused depend on several parameters, such as the size of the droplets and the amount of force by which the droplets are pushed together.

6.3.7. Fluidic Devices

To develop and apply these analytical techniques for subcellular analysis, they must be integrated seamlessly into a platform that is both robust and easy to use. We need to design and fabricate the interface among these techniques onto a single fluidic chip, in which single-cell nanosurgery is performed on-chip followed by in-channel droplet generation and nanoscale chemical reaction, with each component of the reacted mixture subsequently separated by CE in microchannels and then detected by single-molecule techniques. To accomplish this feat, we need methods of microfabrication that are capable of producing both complex structures (e.g., three-dimensional networks of microchannels) and small

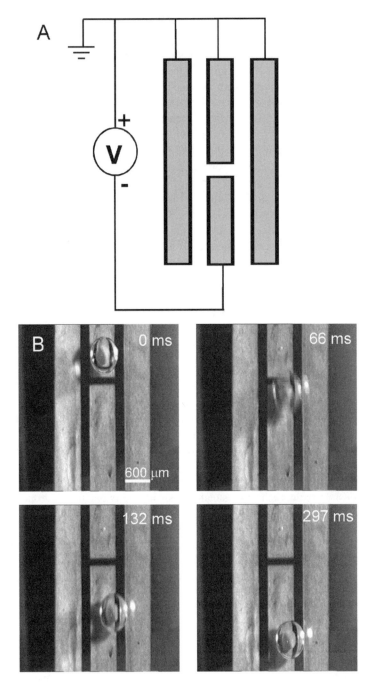

Figure 6.11. (A) Schematic of the electrode design used to demonstrate the gliding of KCl droplet in olive oil. (B) Successive video frames showing the gliding of the KCl droplet with the application of –700 V to the bottom electrode. [Reproduced with permission from *Langmuir* **19**, 250–255 (2003). Copyright 2003, American Chemical Society.]

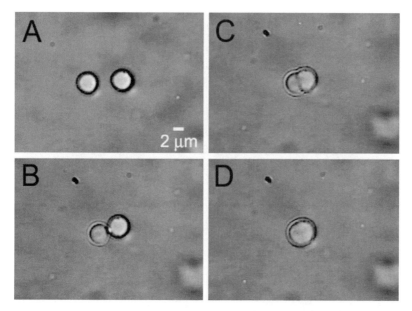

Figure 6.12. (A–D) Sequence of images showing the fusion of two aqueous droplets in oil. The droplets measure ~4 μm in diameter, which corresponds to a volume of ~2 × 10^{-13} L. [Reproduced with permission from *Trends Anal Chem* **22**:528–536 (2003). Copyright 2003, Elsevier.]

features (e.g., down to ~1 μm). But most importantly, these methods must be rapid and economical. If each step of the fabrication procedure is tedious and expensive as is the case for most traditional techniques of silicon micromachining, then it becomes impractical and difficult to build complexities (e.g., complex geometry and integrated components) onto the platform. It would also be tedious to iterate the designs and to test the different approaches for integration. To address this issue, rapid prototyping techniques that are both easy to implement and economical for fabricating disposable devices are an excellent choice (Fiorini and Chiu, 2005).

Rapid prototyping is most common with PDMS (McDonald and Whitesides, 2002); however, rapid prototyping procedures are also available with glass (Rodriguez et al., 2003) and thermoset polyester (TPE) (Fiorini et al., 2004). Each of these substrates has their own set of advantages and drawbacks. Glass is well-known for its robustness, reusability, and stability of surface conditions. Despite the fact that glass microfluidic devices can be rapid prototyped, glass is still not as easy to work with as polymers and the complexity of the microfluidic system is still limited to planar one-level systems and low aspect-ratio structures. Rapid prototyping in PDMS is the benchmark for ease of fabrication of microfluidic devices. Multilevel systems are easily fabricated in PDMS, and access ports (fluid reservoirs and connections to fluidic tubing) are simple to create (Anderson et al., 2000). PDMS does have unstable surface characteristics to deal with, and it is incompatible with many organic solvents (Lee et al., 2003). If immiscible phases other than the soybean oil or perfluoro solvents currently used for droplet formation are desired, PDMS would not be the best choice of substrate. TPE is rapid prototyped in a manner similar to PDMS, yet it provides more stable surface characteristics and is compatible with organic solvents (Fiorini et al., 2004). TPE may be used more often when these traits are required.

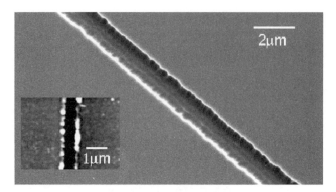

Figure 6.13. SEM of an etched channel in glass using the patterned Ni layer as an etch mask; inset is an AFM image of the corresponding Ni mask on glass prior to HF etching. [Reproduced with permission from *Langmuir* **20:**1833–1837 (2004). Copyright 2004, American Chemical Society.]

For small channel features on the order of a few microns or submicrons, the use of transparency masks as used in traditional rapid prototyping are not possible; high-definition masks are used instead. Electron-beam lithography can create these high-resolution masks, but the cost of these masks can be prohibitive. To overcome the inability of existing rapid prototyping techniques for fabricating small microchannels, we have developed an approach that utilizes direct laser writing on electroless and electrolytically deposited metal masks for the fabrication of microchannels having submicrometer to several micrometers in width (Lorenz et al., 2004). This capability is useful in the fabrication of channels that are one to a few micrometers in width, and we can go from design to device typically in one to two days using this technique. The resist-patterned master created from the masks can be used for the replication of PDMS or TPE microchannels. Alternatively, we can use this method to fabricate glass microchannels directly (Figure 6.13). This ability to make small channels will be useful for both generating small aqueous droplets that are one to a few micrometers in diameter and for single-molecule detection after CE separation.

6.4. ANALYSIS OF THE CONTENTS OF SUBCELLULAR COMPARTMENTS

6.4.1. Separation of the Contents of Subcellular Compartments

After chemical transformation, the contents contained in the aqueous droplet in the immiscible phase can be emptied into the separation channel that is filled with a buffer solution. Although the number of molecules contained within a subcellular compartment is small, the chemical complexity of such samples can be significant. This complexity necessitates a method that can efficiently separate the different components of the sample for subsequent detection and identification. The most suitable technique to carry out this step is CE, owing to its inherent ability to handle very small volumes, its high separation efficiency, and its compatibility with high-sensitivity detection techniques [e.g., laser-induced fluorescence (LIF)]. CE-LIF has been used previously for the chemical analysis of a single atrial gland vesicle ($\sim$1 µm in diameter) from the gastropod mollusk, *Aplysia californica* (Chiu et al.,

1998). By further reducing the cross section of the channel such that it is matched with the dimension of the laser excitation focus, it should be possible to achieve detection limits of a single molecule or a few molecules, which is required for the analysis of subcellular organelles.

6.4.2. Single-Molecule Detection of the Separated Contents

Once the analytes are tagged with good fluorophores and separated, their detection is relatively straightforward, because the field of single-molecule detection has advanced considerably in the last few years (Kuyper et al., 2005). The detection of single fluorescent molecules can be achieved with an excellent signal-to-noise ratio (SNR) and has become more facile and routine with improved detectors and optics. To record the signal, avalanche photodiodes are commonly used owing to remarkable quantum efficiency (>70%) and low dark noise (<25 counts/sec), which, along with low background, give impressive SNRs as compared to other optical methods (e.g., epifluorescence) (Kuyper et al., 2005). Figure 6.14A shows the detection volume of a single-molecule confocal fluorescence microscope that we have constructed, which is defined latitudinally by the laser focus and longitudinally by the pinhole present at the image plane (Nie et al., 1995). This far-field method for single-molecule detection shows excellent SNR at ∼1-ms photon integration time, which is illustrated in Figure 6.14B. One decisive factor to achieve such high SNR is to minimize the laser probe volume by using a diffraction limited laser focus, which is ∼0.5 μm in diameter and 2 μm in height, depending on the degree of spherical aberration present in the optical system (Figure 6.14A). This small detection volume ensures low background noise that may be caused by scattering and the presence of impurities. To detect and count every molecule that is present and separated within the capillary, therefore, the size of the channel at the detection region must match the size of the laser probe volume (Chiu, 2003).

To achieve this goal, several technology developments that we have described must converge, including the ability to fabricate such small channels, perform CE in these channels, and carry out single-molecule measurements. It is important, therefore, to note that while the feasibility of portions of our subcellular analysis platform have been demonstrated as described above, much work still remains to fully realize this platform.

6.5. CONCLUSION

Genomics and proteomics techniques, such as two-dimensional gel electrophoresis/MALDI MS or DNA microarrays, can identify and determine the abundance of specific protein and DNA molecules present in different types of cells; yet, these techniques utilize a collection of bulk samples (e.g., many cells from a particular tissue sample) and can only provide averaged data. Information on the localization of a particular molecule within the subcellular structures or the variation among these structures is not provided by contemporary genomics or proteomics technologies, and the heterogeneities that are critical to the functions of biological systems remain unnoticed. A strategy that is capable of cataloging the biomolecules contained within nanometer-sized organelles will be essential to investigating the biological complexities of cells and their subcellular compartments that are currently unanswered.

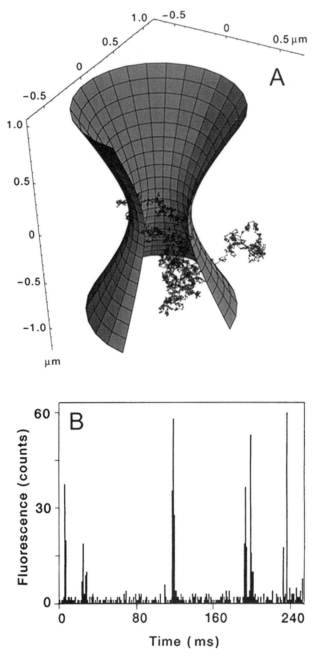

Figure 6.14. (A) Simulation showing the probe volume of a single-molecule confocal fluorescence microscope. The wiggly lines are the simulated diffusive trajectory of a single molecule. (B) Photon trace showing the detection of single Rhodamine 6G molecules diffusing in solution, in which each spike signals the presence of a single dye molecule within the probe volume. [Reproduced with permission from *Anal Chem* **67**:2849–2857 (1995). Copyright 1995, American Chemical Society.]

REFERENCES

Anderson, JR, Chiu, DT, Jackman, RJ, Cherniavskaya, O, McDonald, JC, Wu, H, Whitesides, SH, and Whitesides, GM (2000). Fabrication of topologically complex three-dimensional microfluidic systems in PDMS by rapid prototyping. *Anal Chem* **72:**3158–3164.

Ashkin, A (1997). Optical trapping and manipulation of neutral particles using lasers. *Proc Natl Acad Sci USA* **94:**4853–4860.

Ashkin, A, Dziedzic, JM, Bjorkholm, JE, and Chu, S (1986). Observation of a single-beam gradient force optical trap for dielectric particles. *Opt Lett* **11:**288–290.

Berns, MW, Wright, WH, and Steubing, RW (1991). Laser microbeam as a tool in cell biology. *Int Rev Cyt* **129:**1–44.

Chiu, DT (2003). Micro- and nano-scale chemical analysis of individual sub-cellular compartments. *Trends Anal Chem* **22:**528–536.

Chiu, DT, Lillard, SJ, Scheller, RH, Zare, RN, Rodriguez-Cruz, SE, Williams, ER, Orwar, O, Sandberg, M, and Lundqvist, JA (1998). Probing single secretory vesicles with capillary electrophoresis. *Science* **279:**1190–1193.

Chiu, DT, Wilson, CF, Ryttsen, F, Stromberg, A, Farre, C, Karlsson, A, Nordholm, S, Gaggar, A, Modi, BP, Moscho, A, Garza-Lopez, RA, Orwar, O, and Zare, RN (1999). Chemical transformations in individual ultrasmall biomimetic containers. *Science* **283:**1892–1895.

Duffy, CF, Fuller, KM, Malvey, MW, O'Kennedy, R, and Arriaga, EA (2002). Determination of electrophoretic mobility distributions through the analysis of individual mitochondrial events by capillary electrophoresis with laser-induced fluorescence detection. *Anal Chem* **74:**171–176.

Fiorini, GS, and Chiu, DT (2005). Disposable microfluidic devices: Fabrication, function, and application. *Biotechniques* **38:**429–446.

Fiorini, GS, Lorenz, RM, Kuo, JS, and Chiu, DT (2004). Rapid prototyping of thermoset polyester microfluidic devices. *Anal Chem* **76:**4697–4704.

Fuller, KM, Duffy, CF, and Arriaga, EA (2002). Determination of the cardiolipin content of individual mitochondria by capillary electrophoresis with laser-induced fluorescence detection. *Electrophoresis* **23:**1571–1576.

Gordon, MJ, and Shear, JB (2001). Electrophoretic characterization of dynamic biochemical microenvironments. *J Am Chem Soc* **123:**1790–1791.

He, MY, Sun, CH, and Chiu, DT (2004). Concentrating solutes and nanoparticles within individual aqueous microdroplets. *Anal Chem* **76:**1222–1227.

He, MY, Edgar, JS, Jeffries, GDM, Lorenz, RM, Shelby, JP, and Chiu, DT (2005a). Selective encapsulation of single cells and subcellular organelles into picoliter- and femtoliter-volume droplets. *Anal Chem* **77:**1539–1544.

He, MY, Kuo, JS, and Chiu, DT (2005b): Electro-generation of single femtoliter- and picoliter-volume aqueous droplets in microfluidic systems. *Appl Phys Lett* **87:**031916.

Hochstetler, SE, Puopolo, M, Gustincich, S, Raviola, E, and Wightman, RM (2000). Real-time amperometric measurements of zeptomole quantities of dopamine released from neurons. *Anal Chem* **72:**489–496.

Kuo, JS, Spicar-Mihalic, P, Rodriguez, I, and Chiu, DT (2003). Electrowetting-induced droplet movement in an immiscible medium. *Langmuir* **19:**250–255.

Kuyper, CL, and Chiu, DT (2002). Optical trapping: A versatile technique for biomanipulation. *Appl Spectrosc* **56:**300A–312A.

Kuyper, CL, Jeffries, GDM, Lorenz, RM, and Chiu, DT (2005). Single-molecule detection and manipulation in nanotechnology and biology. In: CSSR. Kumar, J. Hormes, and Leuschner, C (eds.). *Nanofabrication Towards Biomedical Applications*. Wiley-VCH, Weinheim.

Lee, JN, Park, C, and Whitesides, GM (2003). Solvent compatibility of poly(dimethylsiloxane)-based microfluidic devices. *Anal Chem* **75**:6544–6554.

Lillard, SJ, Yeung, ES, and McCloskey, MA (1996). Monitoring exocytosis and release from individual mast cells by capillary electrophoresis with laser-induced native fluorescence detection. *Anal Chem* **68**:2897–2904.

Lillard, SJ, Chiu, DT, Scheller, RH, Zare, RN, Rodriguez-Cruz, SE, Williams, ER, Orwar, O, Sandberg, M, and Lundqvist, JA (1998). Separation and characterization of amines from individual atrial gland vesicles of *Aplysia californica*. *Anal Chem* **70**:3517–3524.

Liu, YM, Moroz, T, and Sweedler, JV (1999). Monitoring cellular release with dynamic channel electrophoresis. *Anal Chem* **71**:28–33.

Lodish, H, Berk, A, Zipursky, SL, Matsudaira, P, Baltimore, D, and Darnell, J (2000). *Molecular Cell Biology*, 4th edition. W. H. Freeman and Company, New York.

Lorenz, RM, Kuyper, CL, Allen, PB, Lee, LP, and Chiu, DT (2004). Direct laser writing on electrolessly deposited thin metal films for applications in micro- and nanofluidics. *Langmuir* **20**:1833–1837.

McDonald, JC, and Whitesides, GM (2002). Poly(dimethylsiloxane) as a material for fabricating microfluidic devices. *Acc Chem Res* **35**:491–499.

Nie, SM, Chiu, DT, and Zare, RN (1995): Real-time detection of single-molecules in solution by confocal fluorescence microscopy. *Anal Chem* **67**:2849–2857.

Niemz, M (2004) *Laser-Tissue Interactions: Fundamentals and Applications*. Springer, Heidelberg.

Olefirowicz, TM, and Ewing, AG (1990). Dopamine concentration in the cytoplasmic compartment of single neurons determined by capillary electrophoresis. *J Neurosci Meth* **34**:11–15.

Page, JS, Rubakhin, SS, and Sweedler, JV (2002). Single-neuron analysis using CE combined with MALDI MS and radionuclide detection. *Anal Chem* **74**:497–503.

Rodriguez, I, Spicar-Mihalic, P, Kuyper, CL, Fiorini, GS, and Chiu, DT (2003): Rapid prototyping of glass microchannels. *Anal Chim Acta* **496**:205–215.

Rubakhin, SS, Garden, RW, Fuller, RR, and Sweedler, JV (2000). Measuring the peptides in individual organelles with mass spectrometry. *Nat Biotechnol* **18**:172–175.

Shelby, JP, Edgar, JS, and Chiu, DT (2005). Monitoring cell survival after extraction of a single subcellular organelle using optical trapping and pulsed-nitrogen laser ablation. *Photochem Photobiol* **81**: 994–1001.

Smith, SP, Bhalotra, SR, Brody, AL, Brown, BL, Boyda, EK, and Prentiss, M (1999). Inexpensive optical tweezers for undergraduate laboratories. *Am J Phys* **67**:26–35.

Srinivasan, R (1986). Ablation of polymers and biological tissue by ultraviolet-lasers. *Science* **234**:559–565.

Svoboda, K, and Block, SM (1994). Biological applications of optical forces. *Annu Rev Biophys Biomol Struct* **23**:247–285.

Tan, YC, Fisher, JS, Lee, AI, Cristini, V, and Lee, AP (2004). Design of microfluidic channel geometries for the control of droplet volume, chemical concentration, and sorting. *Lab Chip* **4**:292–298.

Tice, JD, Song, H, Lyon, AD, and Ismagilov, RF (2003). Formation of droplets and mixing in multi-phase microfluidics at low values of the Reynolds and the capillary numbers. *Langmuir* **19**:9127–9133.

Travis, ER, and Wightman, RM (1998). Spatio-temporal resolution of exocytosis from individual cells. *Annu Rev Biophys Biomol Struct* **27**:77–103.

Vogel, A, and Venugopalan, V (2003). Mechanisms of pulsed laser ablation of biological tissues (Vol. 103, pp. 577, 2003). *Chem Rev* **103**:2079–2079.

Woods, LA, Powell, PR, Paxon, TL, and Ewing, AG (2005). Analysis of mammalian cell cytoplasm with electrophoresis in nanometer inner diameter capillaries. *Electroanalysis* **17**:1192–1197.

Wu, HK, Wheeler, A, and Zare, RN, (2004). Chemical cytometry on a picoliter-scale integrated microfluidic chip. *Proc Natl Acad Sci USA* **101**:12809–12813.

CHAPTER

7

ULTRA SENSITIVE TIME-RESOLVED NEAR-IR FLUORESCENCE FOR MULTIPLEXED BIOANALYSIS

LI ZHU AND STEVEN A. SOPER

7.1. BACKGROUND AND RELEVANCE

The completion of the Human Genome Project (HGP) in 2003 is believed to be a significant milestone in the scientific world and represents one of the greatest achievements of humanity. It has provided a wealth of information, including the number and average size of human genes, the fraction of the genome that codes for proteins, and the degree of sequence similarity, both among humans and compared with other organisms (Collins et al., 2003; Lander et al., 2001). The availability of this information greatly facilitates the identification and isolation of genes that contribute to many human diseases, provides probes that can be used in genetic testing, diagnosis of diseases, and drug development, and offers important information about many basic cellular processes as well.

The Human Genome Project, with goals of identifying all of the approximately 36,000 genes and determining the primary structure of the entire human genome comprised of its 3 billion base pairs, was greatly spawned by new technologies evolving over the past 15 years. Tremendous improvements have occurred in every field related to DNA sequencing, including developments in electrophoretic separation methods, instrumentation for fluorescence detection, DNA purification and cloning methods, and computational methods for the analysis of sequences. The successful completion of the Human Genome Project, however, does not signify an end to further pursuing novel techniques for providing sequence information or improving the throughput and reducing the cost. To the contrary, the number of known sequences increases the need for even more sequence data for verification (comparative genomics) and diagnostic purposes (Hardison, 2003; Onyango, 2004). This chapter focuses primarily on advances in using fluorescence detection for DNA sequencing, including the development of near-IR fluorescent labeling dyes and time-resolved fluorescence for multiplexing. Other basics associated with DNA sequencing will be described briefly.

7.2. WHAT IS DNA SEQUENCING?

Located inside the nucleus of each cell of any living organism, deoxyribonucleic acid (DNA) carries genetic instructions, which consist of a master code that directs the building of all cellular structures and functions. This code is written in a language using only four

New Frontiers in Ultrasensitive Bioanalysis. Edited by Xiao-Hong Nancy Xu
Copyright © 2007 John Wiley & Sons, Inc.

genetic letters, which represent four different nucleotide bases: adenine (A), cytosine (C), guanine (G), and thymine (T). DNA is a well-known macromolecule that consists of many nucleotides as the basic building blocks. These building blocks form a long continuous chain, which is tightly coiled together as a double helix. These double strands are organized into chromosomes and compressed by associated proteins. For humans, a total of 3 billion nucleotides are ordered and arranged within 23 pairs of chromosomes. Each chromosome contains many genes, pieces of DNA that contain information for directing protein synthesis in the cell. They are the basic functional and physical units of heredity and comprise about 2% of the human genome; the remainder consists of noncoding regions, whose functions may include providing chromosomal structural integrity and regulating where, when, and in what quantity proteins are made. Simply put, the order of the three billion A, C, G, and T bases arranged along the chromosomal DNA spells out who we are and how our cells operate. The goal of DNA sequencing is to elucidate the primary structure of this biopolymer by reading the order of the nucleotide units comprising the polymer.

While there are many different strategies for obtaining sequence information, the primary steps typically involve the following: (1) isolation and clonal amplification of large DNA inserts (~50 kbp); shearing these inserts into manageable fragments (~2 kbp); (3) subcloning the short fragments into carrier vectors; (4) Sanger chain-termination reactions of the short DNA fragments; (5) gel electrophoretic size-sorting of the Sanger products; (6) reading the order of the four nucleotide bases using fluorescence detection; and (7) assembling the sequence of the entire genome from the Sanger reads.

7.3. FLUORESCENCE DETECTION FOR DNA SEQUENCING

Following the electrophoretic separation of DNA sequencing fragments, they need to be detected and identified. Since the first demonstrations of the use of fluorescence tags to label DNA oligonucleotides for sequencing (Ansorge et al., 1986, 1987; Prober et al., 1987; Smith et al., 1986), fluorescence detection has been accepted as the predominant detection protocol for DNA sequencing. It has allowed DNA sequencing to be performed in an "automated" fashion with base-calling performed during the gel separation. In this method, Sanger sequencing fragments terminated with different bases are fluorescently labeled. Following gel electrophoresis, the individual components are identified based on unique properties of the fluorescent dyes used to tag each of them. Fluorescence detection possesses high sensitivity, intrinsic simplicity, and the ability to perform on-line detection, which makes it very appealing for high-throughput DNA sequencing applications. More importantly, multiplexing capabilities, which allow multiple tracks of information to be processed in a single lane, allow higher-throughput capabilities. In addition, the recent focus on miniaturizing the separation platforms onto planar chips for sequencing has placed severe demands on detection due to the smaller sample injection volumes. Fluorescence-based sequencing instrumentation and strategies as well as a variety of fluorescent dyes used for gel-based DNA sequencing strategies will be described in the following sections.

7.3.1. Dye-Primer/Dye-Terminator Chemistry in DNA Sequencing

Fluorescence-based DNA sequencing can generally be divided into two categories: dye-primer or dye-terminator sequencing. This categorization is based on the position of where the fluorescent dyes are attached to the substrates used for sequencing, either the oligonucleotide primer or the dideoxynucleotides (i.e., terminators).

In dye primer sequencing, four fluorescent dyes are attached to each sequencing primer and four separate sequencing reactions are carried out. Each reaction contains one dye-labeled primer, four dNTPs, and a particular ddNTP. Dye-primer chemistry is widely used in various sequencing applications due partly to the fact that dye-labeled primers are typically less expensive than dye-labeled terminators. Also, in most sequencing, small pieces of DNA (1–2 kbp in length) are cloned into M13 vectors for propagation. The M13 vectors have a known sequence and can serve as ideal priming sites. However, dye-labeled primers do present some disadvantages. For example, the sequencing reactions must be run in four separate tubes during polymerization and then pooled prior to the gel electrophoresis. In addition, unextended primer can result in a large electrophoretic peak with high fluorescence intensity at the beginning of the trace, which often masks the bases close to the primer-annealing site and makes them difficult to be called.

In dye-terminator sequencing, the fluorescent dyes are attached to the dideoxynucleotides (ddNTPs). Sanger sequencing reactions are run using unlabeled primers, four dNTPs and four fluorescently labeled ddNTPs, with each ddNTP labeled with a unique reporter (dye). The primary advantage of using dye-labeled terminators is that the sequencing reactions can be performed in a single reaction tube, which reduces reagent consumption and minimizes sample transfer steps. Dye-labeled terminators also can be appealing in certain applications—for example, when using primer-walking strategies (Giesecke et al., 1992; Kieleczawa et al., 1992; Voss et al., 1993). In primer-walking, the sequence of the DNA template is initiated from a common priming site, using a primer that is complementary to that site. After reading the sequence at that site, the template is subjected to another round of sequencing, with the priming site occurring at the end of the first read. In this way, a long DNA can be sequenced by walking down the template. Dye terminators are particularly attractive in primer-walking strategies since primer sequence changes frequently and the use of unlabeled primers simplifies the synthetic preparation of these reagents. Additionally, in many cases, dye terminators improve the quality of sequencing data since the excess terminators can be easily removed prior to electrophoresis and, as such, give clean gel reads free from intense primer peaks. However, many polymerase enzymes are very sensitive to the type of dye attached to the ddNTP, which can produce uneven peak heights (broad distribution of fluorescence intensities) during electrophoresis. Uneven peak heights are due to differences in incorporation efficiency of the dye-modified ddNTPs by the particular polymerase enzyme. In dye-primer chemistry, this disparity is absent due to the large displacement of the dye from the polymerization site (Lee et al., 1992).

The instrumentation that can be used for dye-terminator reads can also be used for dye-primer sequencing as well. The difference rests in terms of the sample preparation protocols and the software corrections required for the sequencing data, such as different mobility-correction factors. Using dye-terminator chemistry, a purification step is necessary following DNA polymerization to remove excess dye-labeled terminators because they are negatively charged and can mask the sequencing data embedded within the trace. This can be efficiently accomplished using size exclusion columns and/or a cold ethanol precipitation step.

7.3.2. Fluorescent Dyes for DNA Labeling and Sequencing

A dye set for DNA sequencing is typically comprised of four dyes, with each one specific to a particular nucleotide base. An ideal dye set should possess the following properties: (1) Each dye in the set is spectroscopically distinct so as to be distinguishable. In most cases, the discrimination is based on differences in the emissive properties of the dye. The dyes should also be carefully chosen to minimize spectral leakage (cross-talk) among different detection channels. Other physical properties associated with the dyes, such as the fluorescence lifetime, can also be used to identify them. (2) The dye set should possess good photophysical properties such as high extinction coefficients and large quantum yields. These properties are necessary to ensure high sensitivity and detectability. (3) The dye set should show favorable chemical stability at high temperatures. This is because typical Sanger sequencing reactions utilize multiple temperature cycles that include a 95°C denaturing step. Therefore, the dyes are required to have good thermal stability for extended periods of time. (4) Dyes within a set should produce minimal mobility shifts during the gel separation. The post-run mobility corrections can be very complex and involved. The mobility shift is dependent not only upon the dye and linker structure, but also upon the separation platform used. For example, dyes that show uniform mobility shifts in slab gel electrophoresis may not show the same effect in capillary gel electrophoresis. Uniform mobilities of all dyes are desired to alleviate the need for extensive post-electrophoresis data corrections. (5) An ideal dye set should require minimum numbers of excitation sources (lasers) and detection channels and possess highly efficient excitation and detection. These properties dramatically simplify the instrumentation and the cost of the sequencer. (6) The dyes must not disrupt the activity of the polymerase enzyme. This is particularly an issue in dye-terminator chemistry, since the proximity of the dyes to the polymerase enzyme can dramatically influence its ability to be incorporated into the DNA chain.

Traditional dye sets used in many automated fluorescence-based DNA sequencers consist of a set of four dyes that have absorption and emission spectra in the visible region of the electromagnetic spectrum (400–650 nm). This is due primarily to the readily available excitation sources (such as Ar, He–Ne, and Kr ion lasers) and the high photon detection efficiencies of detectors like photomultiplier tubes (PMTs) in this region. Also, fluorophores that can be detected in the visible are readily available from various commercial sources. The initial set of four fluorescent dyes used for sequencing were introduced by Smith et al. (1986). These were fluorescein, 4-chloro-7-nitrobenzo-2-1-diazole (NBD), tetramethyl-rhodamine, and Texas Red. The dyes were covalently attached to a sequencing primer. A set of interference filters, centered at 520 nm, 550 nm, 580 nm and 610 nm, were used to spectrally resolve the four dyes based on their emission maxima. The use of these four dyes allowed fluorescence four-color sequencing in a single gel lane that also provided real-time base calling and, thus, increasing data throughput.

Following this work, alternative dye sets were synthesized and linked to primers for sequencing (Ansorge et al., 1987; Carson et al., 1993; Karger et al., 1991; Swerdlow and Gesteland, 1990). Among them, the most commonly used dye set included carboxyfluorescein (FAM), carboxy-4',5'-dichloro-2',7'-dimetoxyfluorescien (JOE), carboxytetramethylrhodamine (TAMRA), and carboxy-X-rhodamine (ROX). FAM and JOE are Applied Biosystems trade names for fluorescein-based dyes that can be excited by the 488-nm line from an argon ion laser. TAMRA and ROX are rhodamine-based dyes that can be excited by the 514.5-nm line from the argon ion laser.

While many successful sequencing experiments have been performed with the visible dyes, they are susceptible to some limitations. The primary limitation is the broad emission profiles associated with many of these dyes, which makes it difficult to discriminate among the four colors. The inability to correctly match each detected photon to its source creates an inherent background (cross-talk) in each detection channel, which limits the accuracy of the base-call during sequencing. A limited number of excitation sources in the visible region of the spectrum, the wide, overlapping emission spectra of the visible dyes, a high intrinsic background, and different electrophoretic mobilities of fluorescent dyes are all complicating factors that must be accounted for during DNA sequence reconstruction.

7.3.3. Near-IR Fluorescent Dyes

Performing fluorescence experiments in the near-IR region (650–1000 nm) has some unique advantages (Williams and Soper, 1995): (1) Background interference from impurity molecules in the sample matrix can be substantially reduced in the near-IR since very few biological molecules possess intrinsic fluorescence (autofluorescence) in this region. On the other hand, many biological compounds have autofluorescence in the visible region, which is difficult to reduce using optical filters without compromising the sensitivity of the measurement. As such, detection limits in the near-IR can be significantly improved compared to the visible. (2) Enhanced sensitivity can also be achieved through significant reduction of scattering effects. The amplitude of Raman or Rayleigh scattering is inversely proportional to the fourth power of the excitation wavelength (λ^{-4}) (Gilson and Hendra, 1970). As the wavelength of the excitation light increases, the efficiency of the Raman scattering decreases dramatically, thus decreasing the background signal. (3) With the availability of laser diodes and photodiode transducers, the instrumentation required for near-IR detection can be rather simple and inexpensive. These solid-state lasers and detectors can be run for extended periods of time while supplying ample power and requiring minimal maintenance or operator expertise. Semiconductor laser diodes are now available from 630 nm to longer wavelengths and can be operated in either a continuous or pulsed mode, allowing time-resolved fluorescence measurements as well. Moreover, most photodetectors, such as single-photon avalanche diodes (SPADs), show high single-photon detection efficiencies in the near-IR. Additionally, with the recent major push toward miniaturization of separation platforms, some aiming at portable systems, the compact size and low cost of these semiconductors operated in the near-IR provide the ability to build complete miniaturized systems. Table 7.1 summarizes the comparison of a typical laser diode to an air-cooled argon-ion laser (Middendorf et al., 1993).

All the above merits associated with near-IR fluorescence detection make it a very attractive alternative to visible fluorescence for DNA sequencing. In sequencing, the separation and detection occur within a highly scattering medium, the gel matrix. The reduced amplitude of scattering and the intrinsically lower backgrounds associated with near-IR fluorescence compared to the visible region allow ultrasensitive detection to be achieved. Several groups of researchers have explored the use of near-IR dyes for sequencing.

Table 7.1. An Operational Comparison of a Far-Red Laser Diode and Argon Laser as Excitation Sources for Fluorescence (Middendorf et al., 1993)

	Laser Diode	Argon-Ion Laser
Wavelength (nm)	785 nm	488 nm
Life span	>100,000 h	~3000 h
Optical power output	20 mW	20 mW
Power consumption	0.150 W	1800 W
Replacement cost	<$150	>$5000

The near-IR fluorophores that have been typically used belong to the cyanine-class of dyes (Patonay and Antoine, 1991). The most frequently employed are the dicarbocyanines or the tricarbocyanines. The carbocyanines consist of heteroaromatic structures linked by a polymethine chain containing conjugated carbon–carbon double bonds. The absorption and emission maxima can be altered by changing the length of this polymethine chain or changing the heteroatom within the heteroaromatic fragments. For example, dicarbocyanines show absorption maxima near 630 nm, while the tricarbocyanines show absorption maxima near 780 nm. These dyes have large extinction coefficients and can be augmented with various functional groups to increase their solubility in water (Mujumdar et al., 1989; Strekowski et al., 1996). However, these dyes have poor photochemical stabilities in aqueous media and relatively low fluorescence quantum yields, which result primarily from high rates of internal conversion (Soper and Mattingly, 1994).

Near-IR fluorescence has been demonstrated in sequencing applications using slab gel electrophoresis, where the detection sensitivity has been reported to be ~2000 molecules (Middendorf et al., 1992). Willams and Soper (1995) demonstrated superior detection limits using a tricarbocyanine dye for sequencing by CGE. A direct comparison of detection limits between visible and near-IR excitation (488 nm and 780 nm) was made by electrophoresing sequencing primers, one labeled with FAM and the other labeled with a near-IR dye. It was determined that the detection limits for these systems were 3.4×10^{-20} moles for 780-nm excitation and 1.5×10^{-18} moles for 488-nm excitation. The improvement in the limit of detection for the near-IR system resulted primarily from the significantly lower background observed in the near-IR, in spite of the fact that the fluorescence quantum yield of the near-IR dye was far lower than that of the FAM dye.

Strekowski and co-workers reported the synthesis of several heptamethine cyanine derivatives that contained an isothiocyanate labeling group appropriate for primary amine containing targets (Lipowska et al., 1993; Strekowski et al., 1992 a, b). In their subsequent work, they reported the use of these heptamethine derivatives as labels for DNA sequencing primers (Shealy et al., 1995). Flanagan et al. (1998) developed a series of heavy-atom-modified tricarbocyanine dyes, which possessed succinimidyl esters or isothiocyanate groups for labeling DNA primers that contained primary amine groups (Table 7.2). The interesting feature of these dyes was that they possessed an intramolecular heavy atom to perturb the fluorescence lifetime of the base chromophore. The incorporation of the heavy atom, however, did not alter the absorption and emission profiles of the dyes. Therefore, the dyes possess similar absorption and emission wavelengths but distinct lifetimes, the values of which depend upon the identity of the heavy atom modification. This group of dyes has been targeted for use in DNA sequencing based on lifetime discrimination, in which a

Table 7.2. Chemical Structures and Photophysical Properties of Heavy-Atom Modified Tricarbocyanine Dyes, Which Contain an Isothiocyanate Labeling Group (Flanagan et al., 1998)

Sulfonated Heavy-Atom Modified Near-IR dyes X = I, Br, Cl, or F	Dye	Absorption, λ_{max}	Emission, λ_{max}	ε ($M^{-1}cm^{-1}$)	τ_f (ps)	ϕ_f
	I	766	796	216,000	947	0.15
	Br	768	798	254,000	912	0.14
	Cl	768	797	239,000	880	0.14
	F	768	796	221,000	843	0.14

single excitation source can excite all the dyes, and a single detection channel will process the fluorescence signals. An additional advantage of these dyes is that the primers labeled with these dyes show uniform electrophoretic mobilities; therefore, post-electrophoresis corrections would not be necessary.

An attractive alternative to the cyanine-based near-IR fluorophores are the phthalocyanines (Pc) or naphthalocyanine (NPcs) family of compounds. The Pc's and NPc's possess a conjugated ring structure, which is linked together by four aromatic-dicarbonitrile fragments. The absorption and emission maxima of these dyes can easily be altered by appending different functional groups to the dye macrocycle (Kobayashi et al., 2003; Wrobel and Boguta, 2002). For example, the annulation of benzene rings onto the Pc core will produce the NPc dyes, which have absorbance maxima red-shifted by 50–100 nm compared to the Pc dyes (Bradbrook and Linstead, 1936; Kovshev and Luk'yanets, 1972). The basic structures of Pc and NPc dyes are shown in Figure 7.1. Although the Pc and NPc dyes are very hydrophobic and essentially insoluble in water, functionalization of the periphery of the dye with charged groups (SO_3^-, CO_2^-, PO_3^{2-}, etc.) or attachment of a polar functionality to the metal center (PEG-OSi) will improve water solubility (Sharman et al., 1996). Like the tricarbocyanines, Pc and NPc dyes can be functionalized for conjugation to biomolecules. The attractive feature associated with the Pc dyes are their superior quantum yields and favorable photochemical stabilities compared to the tricarbocyanines (Casay et al., 1994; McCubbin and Phillips, 1986; Wheeler et al., 1984). In addition, Pc and NPc dyes typically have longer fluorescence lifetimes compared to the carbocyanines, which can be adjusted by coordinating different metals to the core of the macromolecule (Shen et al., 1989). Table 7.3 shows a comparison of a few photophysical parameters of Pc and NPc dyes to several other classes of near-IR dyes. Unfortunately, the synthetic route for preparation of these dyes typically produces symmetrically substituted structures. Preparing asymmetrical analogues with functional groups for facile conjugation to target molecules has been somewhat difficult, since purifying the Pc isomers can be problematic. Hammer et al. (2002) have prepared and isolated asymmetric metallo-phthalocyanine dyes containing a reactive isothiocyanate functional group for covalent labeling of oligonucleotide primers, which has

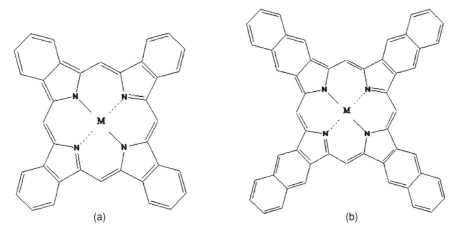

Figure 7.1. Base structure of (a) phthalocyanine (Pc) family of near-IR fluorescence compounds and (b) naphthalocyanine (NPc) dyes. The M represents the metal center and can be, for example, Ni, Ga, Pt, Pd, Cu, Al, Si, and Zn. The identity of the metal center affects both the fluorescence and spectroscopic properties of the metal Pc and NPc dyes.

Table 7.3. Comparison of the Photophysical Properties Associated with Several Classes of Near-IR Dyes

Parameter	Phthalocyanine	Naphthalocyanine	Tricarbocyanines
Absorption maximum	676 nm[a]	769 nm[b]	764 nm[d]
Emission maximum	684 nm[a]	777 nm[b]	73 nm[d]
Quantum yield (ϕ_f)	0.37[a]	0.25[b]	0.05[d]
Lifetime (τ_f)	5.0 ns[a]	2.5 ns[c]	0.76 ns[d]
Photochemical stability (ϕ_d)	1.7×10^{-6c}	3.0×10^{-3c}	7.0×10^{-3e}
Photon yield /molecule ((ϕ_f/ϕ_d))	2.2×10^5	83	7.1

[a] Tetra-sulfonated Al-Pc; aqueous (Ambroz et al. (1991).
[b] Tetra-sulfonated Al-naphthalocyanine; aqueous; (Casay et al. (1994).
[c] Measured in aqueous buffer pH 9.3 (McCubbin and Phillips, (1986).
[d] Measured in aqueous buffer pH 8.3 (Flanagan et al., (1998).
[e] Unpublished data, Soper et al.

shown potential for use in near-IR-based DNA analysis. The same group is proposing the route of making a series Pc and NPc dyes with different metal centers for lifetime-based DNA sequencing.

The ability to tailor the spectroscopic properties with subtle changes in dye structure makes the carbocyanine, Pc, and NPc dyes excellent candidates for bioanalytical applications. Given the interest in sequencing, accompanied by the vigorous growth in solid-state technology, one can expect continued efforts in development of new near-IR dyes for bioanalytical applications, such as DNA sequencing.

Efforts have also been directed toward developing the deep-red cyanine-based fluorescent dyes for DNA sequencing applications as well. These dyes have high extinction coefficients and are available as phosphoramidites and NHS esters for labeling oligonucleotides (primers) or dideoxy terminators. For example, Cy5 and Cy5.5 dye-labeled terminators have been used in single-color sequencing experiments (Duthie and Kalve, 2002). It was demonstrated that deep red cyanine dye-labeled terminators, when combined with a suitable DNA polymerase, could produce uniform band patterns and yield high-quality sequencing data. Structures of a Cy5.5 labeled ddATP is shown in Figure 7.2. The Cy-dye series (Cy 3, 5, 5.5) can also be used to produce energy-transfer (ET) cassettes for sequencing (Kumar, 2002). With a donor being Cy3 that is excited by a Nd:YAG laser (532 nm), Cy3, 3.5, 5, and 5.5 can be used as acceptor dyes for preparing ET primers. Other near-IR dyes, such as IRDye 700 and IRDye 800, are available from LiCOR Inc., and they have also been successfully employed in multi-lane sequencing in slab gel formats (http://www.licor.com/bio/IRDyes/index.jsp). Molecular Probes has also developed the AlexaFluor series of dyes that span across the visible and near-IR. AlexaFluor 680 and 750 have spectral properties almost identical to Cy5.5 and Cy7 dyes, respectively, but with greater brightness and better photostability (http://probes.invitrogen.com/handbook/sections/0103.html). Another class of cyanine-based near-IR dyes is called the WellRED dye set, which has been introduced by Beckman Coulter Inc. for use with their CEQ™ Series Genetic Analysis system (http://www.beckman.com/products/specifications/geneticanalysis/ceq/wellreddyes_con_stat.asp). Excitation of these dyes is achieved with 650-nm and 750-nm diode lasers. The dye set is available in the form of dideoxy terminator conjugates and facilitates four-color sequencing on Beckman's CEQ capillary array machine.

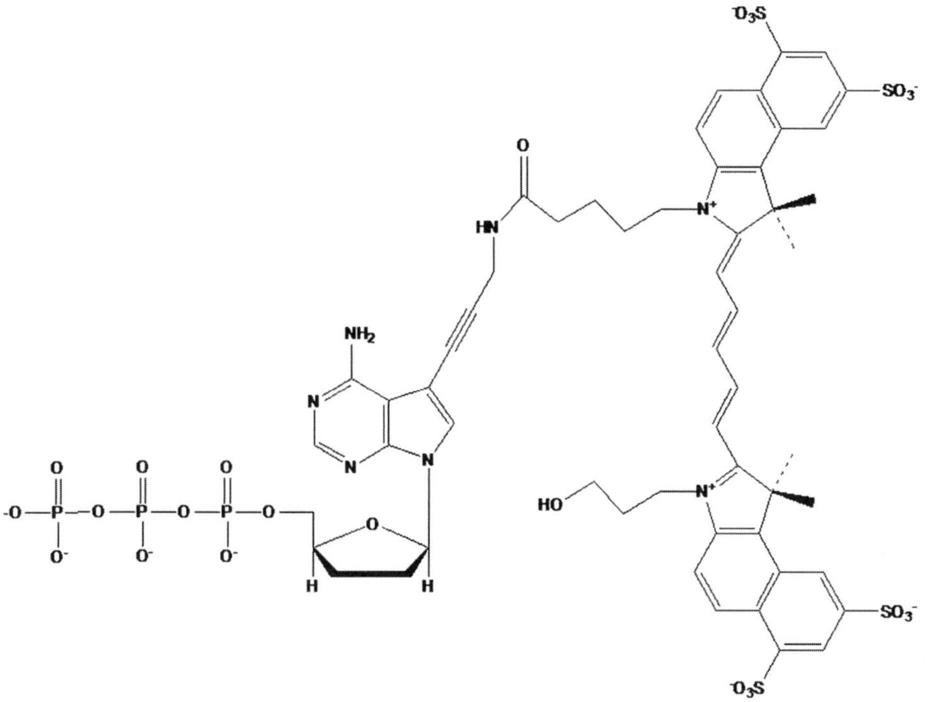

Figure 7.2. Near-IR labeled dye terminator. In this case, the terminator labeled is a ddATP. The linker is a propagyl amine, which contains a triple bond to provide rigidity to the linker to minimize dye interaction with the DNA polymerase during incorporation events. The dye is attached to the nucleobase on a non-hydrogen-bonding site of the ddNTP. The dye used here is a tricarbocyanine, which contains four water-solubilizing groups (sulfonates).

7.4. FLUORESCENCE-BASED DNA SEQUENCING STRATEGIES

One of the primary advantages of using fluorescence detection for DNA sequencing is its multiplexing capability—that is, its ability to identify multiple analytes in a single assay. The fluorescent probes attached to each nucleotide base can be identified simultaneously based on their characteristic fluorescence properties. Two properties associated with fluorescent probes have been used for identification purposes: spectral characteristics (color discrimination) and fluorescence lifetimes (lifetime discrimination). In each method, several different strategies can be implemented to reconstruct the sequence. These strategies may vary in terms of the number of dyes used, the number of detection channels required, or the need for running 1–4 parallel electrophoresis lanes. For example, if only one dye is used to label all four bases, the electrophoresis must be run in four different lanes, one for each base comprising the DNA molecule. However, if four different dyes are used, the electrophoresis can be reduced to one lane; as a consequence, the throughput of the instrument goes up by a factor of 4.

While color or spectral discrimination is the most commonly used method in DNA sequencing in which spectrally distinct reporters are attached either to a sequencing primer or to a dideoxynucleotide and are identified by their unique emission maxima, we will not discuss color or spectral techniques for multiplexed DNA sequencing in this context. We will, instead, focus on the use of time-resolved fluorescence detection strategies for DNA sequencing and the use of hybrid techniques, in which color and time-resolved methods are used to increase multiplexing capabilities.

7.4.1. Lifetime Discrimination Methods

While most sequencing applications using fluorescence utilize spectral discrimination to identify the terminal bases during electrophoretic sizing, an alternative or complementary approach is to use the fluorescence lifetimes of the labeling dyes to identify each of the bases. The fluorescence lifetime (τ_f) is an intrinsic photophysical property of fluorophores that measures the average time difference between electronic excitation and fluorescence emission. The monitoring and identification of multiple dyes by lifetime discrimination during a gel separation can allow for an additional identification protocol when combined with color discrimination to provide high multiplexing capabilities. Several principal advantages associated with fluorescence lifetime identification protocols include the following:

1. The calculated lifetime is immune to concentration differences. As such, dye-labeled terminators can potentially be used as well as dye-labeled primers with a wide choice in polymerase enzymes to suit the particular sequencing application; the base identification can be accomplished with high accuracy irrespective of the intensity of an electrophoretic band.
2. Lifetimes can be determined with higher precision than fluorescence intensities under appropriate conditions, improving base-calling accuracy.
3. Lifetime determinations do suffer from spectral leakage due to broad fluorophore emission profiles.
4. Multi-dye fluorescence can potentially be processed on a single detection channel without the need for spectral sorting to multiple detection channels (Soper et al., 1995).

Several problems do arise when considering lifetime discrimination for DNA sequencing. One potential difficulty is the poor photon statistics (low number of photocounts in a decay profile from which the lifetime is extracted), especially when utilizing micro-separation techniques such as capillary and microchip gel electrophoresis, which have low sample loading masses and short residence time of fluorophores within the excitation beam (1–5 s). Basically, the low number of photocounts acquired to construct the decay profile during electrophoresis can produce low precision in lifetime measurements since the fluorescence lifetime is extracted and calculated directly from the decay. In addition, the high scattering medium in which the fluorescence is measured (gel matrix) can produce large levels of background photons that would be included into the decay profile, lowering precision in the measurement. The poor precision would consequently affect the accuracy in the base call. An additional concern with lifetime measurements for calling bases in DNA sequencing is the complex instrumentation required for lifetime determinations as well as the complex algorithms that are required for extracting the lifetimes from the decay profiles. Nevertheless, many of these concerns have been addressed using near-IR fluorescence. The increased availability of pulsed diode lasers and single-photon avalanche photodiodes has had a tremendous impact on the ability to assemble simple time-resolved instruments appropriate for sequencing applications, with performance characteristics comparable to those using visible wavelengths. The use of near-IR fluorescence also reduces the number of background and scattering photons processed during detection, potentially increasing the sensitivity of the instrument and improving the photon statistics.

There are two different methods for measuring fluorescence lifetimes; frequency-resolved (He and McGown, 2000; He et al., 1998; Li et al., 1997; Li and McGown, 1996, 1999, 2000, 2001; McIntosh et al., 2000; Nunnally et al., 1997) and time-resolved (Chang and Force, 1993; Flanagan et al., 1998; Lassiter et al., 2000, 2002; Lieberwirth et al., 1998; Neumann et al., 2000; Sauer et al., 1998, 1999, 2001; Seeger et al., 1993; Soper et al., 1996; Waddell et al., 2000a,b; Zhang et al., 1999; Zhu et al., 2003, 2004). In frequency-resolved methods, also called phase-modulation techniques, the excitation source is intensity-modulated at a high frequency, typically employing sine wave modulation. When the fluorescent sample is excited by the modulated light, the emission responds at the same modulation frequency, but with a time delay with reference to the excitation with the phase delay related to the lifetime of the sample. This time delay is characterized by a phase shift, which is then used to calculate the fluorescence lifetime. In addition, the fluorescence lifetime can be measured from the intensity of the demodulated signal. The ability to measure short lifetimes depends on the frequency of the modulation and the efficiency of the phase-sensitive measuring electronics.

In the case of time-resolved fluorescence, the fluorophore is excited by a pulsed light source at a relatively high repetition rate. The time duration of the light pulse needs to be as short as possible, preferably much shorter than the fluorescence lifetime being measured. The emitted photons from the sample are time-correlated to the excitation pulse from which the lifetime can be determined. At present, most time-resolved measurements are performed using the time-correlated single-photon counting (TCSPC) technique. A typical TCSPC device consists of a pulsed excitation source, a fast detector, and timing electronics that include a constant fraction discriminator (CFD), an analog-to-digital converter (ADC), a time-to-amplitude converter (TAC), and a multi channel analyzer (MCA). The time difference between the excitation and the arrival of the resulting fluorescence photon to the detector is recorded electronically. The recorded time differences over many excitation–emission cycles are placed in the appropriate time channels of the MCA, and a statistical histogram is constructed representing the decay profile of the fluorophore. A calculation algorithm is then applied to the decay to extract the fluorescence lifetime. The shortest lifetime that can be measured reliably by this method is determined by the response time of the instrument (Instrument Response Function, IRF), which depends on the width of the excitation pulse, the spread of the travel times of photoelectrons in the photon detector, and the jitter in the measuring electronics. The IRF can be deconvolved from the collected decay profile using a deconvolution algorithm to provide a more accurate representation of the fluorescent lifetime. The time-resolved mode is a digital (photon counting) method; therefore it typically shows better signal-to-noise ratio than a frequency-resolved measurement, making it more attractive for separation platforms that deal with minute amounts of sample. In addition, the use of time-resolved methods allows for the use of time-gated detection in which background photons, which are coincident with the laser pulse (scattered photons), can be gated out electronically improving the signal-to-noise ratio in the measurement. However, TCSPC typically shows a limited dynamic range due to pulse pile up effects at high counting rates.

A device that has been used for making on-line lifetime measurements during CGE is shown in Figure 7.3 (Legendre et al., 1996). The light source consisted of an actively pulsed solid-state GaAlAs diode laser lasing at 780 nm with a repetition rate of 80 MHz and an average power of 5.0 mW. The pulse width of the laser was ~50 ps (FWHM). The detector

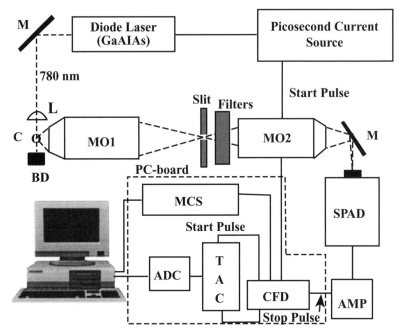

Figure 7.3. Block diagram of a near-IR time-correlated single-photon counting (TCSPC) detector for CGE. The laser is focused onto a capillary column with the emission collected using a 40× microscope objective (N.A. = 0.85). The fluorescence is imaged onto a slit and then spectrally filtered and focused onto the SPAD. L, laser singlet focusing lens; C, capillary; BD, beam dump; MO1, collecting microscope objective; MO2, focusing microscope objective; SPAD, single-photon avalanche diode; AMP, amplifier; CFD, constant fraction discriminator; TAC, time-to-amplitude converter; ADC, analog-to-digital converter; and MCS, multichannel scaler (Legendre et al., 1996).

selected for this instrument was a single-photon avalanche diode (SPAD) possessing an active area of 150 μm offering a high single-photon detection efficiency (>60% above 700 nm). The counting electronics were situated on a single TCSPC board, which was plugged directly into a PC-bus exhibiting a dead time of <260 ns, allowing efficient processing of single-photon events at counting rates exceeding 2×10^6 counts/s. This set of electronics allowed for the collection of 128 sequential decay profiles with a timing resolution of 9.77 ps per channel. The instrument possessed a response function of approximately 275 ps (FWHM), adequate for measuring fluorescence lifetimes in the subnanosecond regime.

One of the most important aspects associated with lifetime measurements in sequencing applications is the processing or calculation algorithm used to extract the lifetime value from the resulting decay. The accuracy of the base call depends directly on the lifetime differences between fluorophores in the dye set and the relative precision in the measurement. Algorithms used in this application require not only the calculation of the lifetimes precisely, even under the situation of poor photon statistics, but also the ability to perform on-line measurements during the electrophoresis. The typical calculation algorithm for lifetime determinations, nonlinear least squares (NLLS), can deconvolve the IRF from the overall decay and provide a more accurate lifetime value. Unfortunately, this algorithm is calculation intensive and it produces large errors in cases where photon statistics are poor. Moreover, it is more suitable for static measurements rather than dynamic (on-line). Two other simple algorithms for on-the-fly fluorescence lifetime determinations that have been evaluated are the maximum likelihood estimator (MLE) and the rapid lifetime determination method (RLD) (Soper and Legendre, 1994).

MLE calculates the lifetime via the following relation (Hall and Selinger, 1981):

$$1 + \left(e^{T/\tau_f} - 1\right) - m\left(e^{mT/\tau_f} - 1\right)^{-1} = N_t^{-1} \sum_{i-1}^{m} i N_i \qquad (7.1)$$

where m is the number of time bins within the decay profile, N_t is the number of photocounts in the decay spectrum, N_i is the number of photocounts in time bin i, and T is the width of each time bin. A table of values using the left-hand side (LHS) of the equation is calculated by setting m and T to the experimental values and using lifetime values (τ_f) ranging over the anticipated values. The right-hand side (RHS) of the equation is constructed from actual decay data over the appropriate time range. The fluorescent lifetime is determined by matching the value of the RHS obtained from the data with the table entry from the LHS. The relative standard deviations in the MLE can be determined using $N^{-1/2}$.

Fluorescence lifetimes are calculated using the RLD method by integrating the number of counts within the decay profile over a specified time interval and using the following relationship (Ballew and Demas, 1989):

$$\tau_f = -\Delta t / \ln(D_1/D_0) \qquad (7.2)$$

where Δt is the time range over which the counts are integrated, D_0 is the integrated counts in the early time interval of the decay spectrum, and D_1 represents the integrated number of counts in the later time interval. Both the MLE and RLD methods can extract only a single lifetime value from the decay, which in the case of multiexponential profiles would represent a weighted average of the various components comprising the decay.

Wolfrum and co-workers have developed a special pattern recognition technique for calling bases using lifetime discrimination methods (Koellner et al., 1996). Basically, the method involves comparing a simulated decay pattern to the measured decay and searches for the pattern that best fits the measurement. This algorithm is equivalent to the minimization of a log-likelihood ratio, where fluorescence decay profiles serve as the pattern. Since the pattern recognition algorithm uses the full amount of information present in the data, it potentially has the lowest error or misclassification probability.

Soper et al. (1995) demonstrated the feasibility of performing on-line lifetime determinations during capillary gel electrophoresis (CGE) separation of DNA sequencing ladders. C-terminated fragments produced from Sanger chain-termination protocols and labeled with a near-IR fluorophore at the 5' end of the sequencing primer were electrophoresed and the lifetimes of various components within the electropherogram were determined. The average lifetimes determined using the MLE method was found to be 581 ps with a standard deviation of ±9 ps (RSD = 1.9%). This result indicated that MLE could produce high precision, even for ultra-dilute conditions. The favorable accuracy and precision was aided by the use of near-IR fluorescence detection, which minimized scatter contributions into the decay as well as background fluorescence.

7.4.1.1. Four-Lifetime/One-Lane DNA Sequencing

In the four-lifetime/one-lane method, four DNA ladders of differently labeled fragments generated by Sanger chain termination reactions are separated in one lane, typically in a single capillary gel column, and the base calling is done with lifetime discrimination as opposed to spectral discrimination. Obviously, the dye sets suitable for color discrimination are not necessarily appropriate for the use in lifetime discriminations. New dye sets must be developed that suit this identification method. In lifetime discrimination, it is not necessary to use dyes with discrete emission maxima; therefore, structural variations in the dye set can be relaxed. For instance, Flanagan et al. (1998) developed a dye set that consisted of a series of structurally similar near-IR tricarbocyanines that possessed identical absorption (765 nm) and emission (794 nm) maxima. The lifetimes of the dyes were varied by incorporating a single halide (I, Br, Cl, or F) into the molecular structure. This dye set, with lifetimes ranging from 947 ps to 843 ps when measured in a polyacrylamide gel, have been suggested for use in a four-lifetime/one-lane sequencing experiment. Sequencing primers labeled with this dye set demonstrated uniform mobilities in gel electrophoresis applications, irrespective of the linker structures (see Table 7.2). However, since these were tricarbocyanine dyes, the lifetimes were found to be <1.0 ns and the lifetime differences among the dye set was somewhat small ($\Delta \tau_f = 70$ ps, $\sim$8% relative difference). Lassiter et al. (2002) optimized the experimental conditions for using these dyes for DNA sequencing by lifetime discrimination.

Wolfrum and co-workers demonstrated the use of a four-lifetime/one-lane approach for DNA sequencing using CGE (Lieberwirth et al., 1998). In their work, four dye labels selected from the rhodamine, cyanine, and oxazine families were covalently tethered to the 5' end of oligonucleotide primers. The dyes exhibited similar absorption and emission maxima and were excited efficiently with a 630-nm pulsed diode laser operated at a repetition rate of 22 MHz. The labels exhibited fluorescence lifetime values of 1.6, 2.4, 2.9, and 3.7 ns, with the difference between the dyes adequate for efficient identification in sequencing applications (see Figure 7.4a). This dye set allowed for the use of a simple detection system that was equipped with a single laser and a single avalanche photodiode. The instrumental response function of the entire system was measured to be $\sim$600 ps (FWHM). The time-resolved data were managed using the TCSPC technique. Using appropriate linker structures, dye-dependent mobility shifts were minimized, eliminating the need for post-electrophoresis corrections. The dye-labeled sequencing fragments were identified by both MLE and pattern recognition algorithms, with the latter method providing higher overall base-calling accuracy. Using an M13 template, Wolfrum and co-workers were able to demonstrate a read length of 660 bases with a probability of correct classification >90% (see Figure 7.4a–d).

7.4.1.2. Two-Lifetime/Two-Lane

To show that fluorescence lifetimes could also be obtained from multiple electrophoretic lanes, a scanning system for measuring fluorescence lifetimes from multi-lane slab gels has been reported by Lassiter et al. (2000). In that report, a modified microscope head was inserted into an automated slab gel sequencer, which consisted of a near-IR time-resolved scanning imager and implemented a two-lifetime/two-lane sequencing approach. Two dyes in each lane were identified by their lifetimes on-line during gel electrophoresis. Two commercially available cyanine-based near-IR dyes, IRD700 and Cy5.5 were chosen as fluorescence reporters and were labeled at the 5' end of a sequencing primer. A-terminated bases were labeled with IRD700, and T-terminated bases labeled with Cy5.5 were electrophoresed in one lane, while C (IRD700) and G (Cy5.5) tracts occupied an adjacent lane. The similar absorption and emission properties of these two dyes allowed efficient processing of the emission on a single detection channel and excitation with a single source. The lifetimes for these two dyes were calculated by the MLE algorithm and determined to be 718 ± 5 ps and 983 ± 13 ps for IRD700 and Cy5.5, respectively. Figure 7.5 shows an

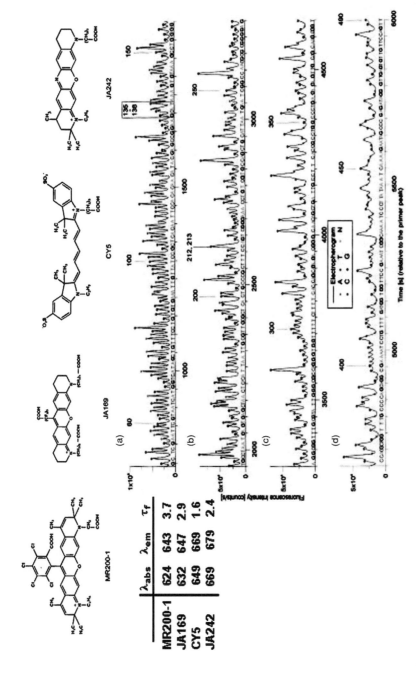

Figure 7.4. Chemical structures of the four fluorescent dyes used for DNA sequencing with time-resolved, fluorescence lifetime discrimination. Shown in the accompanying table are the fluorescence properties of these dyes, including their absorption (λ_{abs}) and emission (λ_{em}) wavelengths (maximum) and their fluorescence lifetime (τ_f). The figure also shows a sequencing trace that was run using capillary gel electrophoresis with fluorescence lifetimes used for identifying the constituent bases. The sequencing was performed using dye-labeled primers of an M13mp18 template. The sieving matrix was a 5% polyacrylamide gel run at a field strength of 160 V/cm. The excitation was supplied by a 630 nm diode laser operated at 22 MHz that generated 500 ps (FWHM) pulses at a peak power of 200 mW. The fluorescence was processed using time-correlated single-photon counting electronics and an actively quenched single-photon avalanche diode. Lifetimes were calculated using MLE, and the data used for constructing the decay histogram are presented as a point (apex of the electrophoretic peak) along the electrophoretic trace (Lieberwirth et al., 1998). See insert for color representation of this figure.

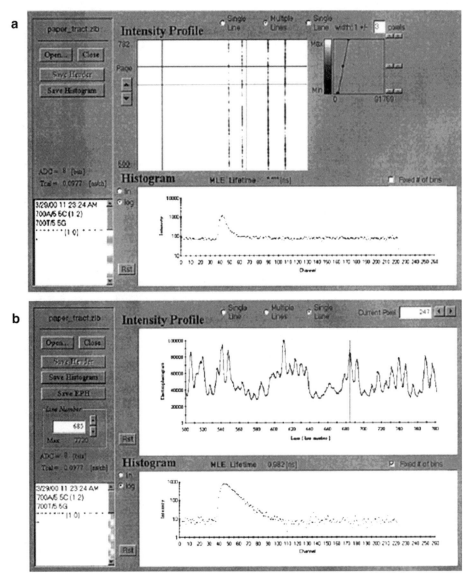

Figure 7.5. Front-panel displays of the data analysis in analyzing time-resolved data from gel images. (a) The upper portion of the screen shows the gel intensity image obtained from a time-resolved fluorescence scanner system. The lower half of the screen shows the instrument response function profile constructed from a region on the gel that was defined by the vertical and horizontal cursors (location free from any dye). (b) The top portion of the screen shows the intensity electropherogram of a single base DNA sequencing tract. The bottom portion displays a decay profile representing a single image pixel selected by the cursor (single vertical line) from the electropherogram (Lassiter et al., 2000).

intensity electropherogram and the data analysis screen for this time-resolved fluorescence experiment. The base calling accuracy for an M13 template using this approach was 99.7% for 670 bases, better than the 95.7% calling accuracy obtained using a single-color/four-lane strategy carried out on the same instrument. The improved base calling accuracy using lifetime identification was a consequence of the increased information content in the electrophoretic bands, particularly in those which were experiencing poor electrophoretic resolution.

7.4.1.3. Combination of Color-Discrimination and Time-Resolved Methods

Using fluorescence for readout has proven to be a viable multiplexed method for DNA sequencing. In this method, multiple analytes can be run in one track, increasing the information content (multiplexing) that can be processed simultaneously. Up to four reporters, one for each nucleotide base, have been identified in a single electrophoresis lane using either color (spectral) or lifetime multiplexing. Indeed, many commercially available automated DNA sequencers are equipped with four-color capabilities. However, the use of only color multiplexing limits the number of probes that can be identified simultaneously due to the broad emission profiles associated with most molecular labeling dyes. The coupling of lifetime discrimination with traditional color discrimination allows increased fluorescence multiplexing, enabling DNA sequencing to be performed with greater throughput. The basic reason is that during time-resolved measurements, all intensity-based data are preserved, while lifetime data adds an additional layer of information. For example, in a color and lifetime hybrid instrument, different dyes can be identified by color; in each color channel, different labels can be distinguished through their characteristic fluorescent lifetimes. Based on this concept, Soper and co-workers developed a two-color, time-resolved fluorescence microscope using near-IR fluorescence and first demonstrated the combination of color and lifetime discrimination schemes (Zhu et al., 2003).

The hybrid microscope built for color/lifetime discrimination could acquire both intensity and time-resolved fluorescence data on-line during gel electrophoresis in either a capillary or a microchip format for DNA sequencing applications (Zhu et al., 2003, 2004). A diagram of the optical fiber-based, dual-color, time-resolved microscope is shown in Figure 7.6. It consisted of two pulsed diode lasers (680 and 780 nm), both operated at a repetition frequency of 40 MHz, which were coupled to the microscope head using single-mode fibers. The laser light was directed onto a focusing objective using a dichroic mirror; the resulting emission was collected by the same objective and focused onto a multimode fiber, which transported the luminescence to two SPADs. The emission was sorted spectrally using a second dichroic beam splitter and isolated by appropriate interference filters before reaching one of the two SPADs (710-nm channel or 810-nm channel). A PC-based card in a computer contained all the photon counting electronics to process the time-resolved data. A router was used to properly register photons generated from different detectors, simultaneously allowing measurements in two channels. The dual-color microscope demonstrated a time response of 450 ps (FWHM) and 510 ps for the 710- and 810-nm channels, respectively. The use of near-IR fluorescence detection greatly simplified the hardware and allowed superior detection limits. The mass limits of detection were determined to be 7.1×10^{-21} and 3.2×10^{-20} mol (SNR = 3) for the two detection channels by electrophoresing two near-IR dye-labeled sequencing primers through a capillary gel column.

One of the benefits of applying time-resolved measurements is the ability to eliminate the cross-talk between different color channels. For example, using the two-color, time-resolved microscope for DNA sequencing, the problem of cross-talk accompanying almost all color discrimination methods was successfully solved by implementing optical delay filtering. Briefly, an extra length of optical fiber was inserted into the 680-nm laser pulse train to introduce a phase shift of this laser relative to the 780-nm laser. The time delay

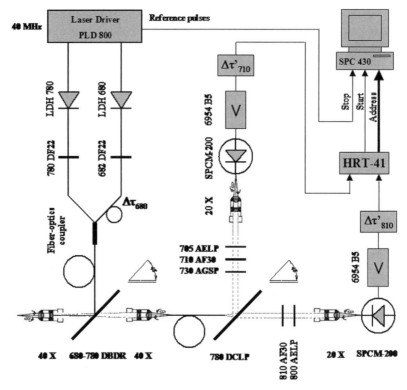

Figure 7.6. Schematic diagram of a dual-color, time-resolved, near-IR hybrid microscope. The microscope used two diode lasers providing excitation at 680 and 780 nm and a pair of SPADs for photon transduction (Zhu et al., 2003).

of the 680-nm pulse train with respect to the 780-nm pulse train resulted in a phase shift of the corresponding fluorescence emission. As such, spectral leakage of a dye in the wrong channel was separated from the fluorescence emission resulting from the other dyes (see Figure 7.7). Then, by setting the counting electronics for each color channel to process photons only over a selected time frame while excluding the delayed spectral leakage signals, the elimination of cross-talk was achieved. The fluorescence intensity data, appearing as normal electropherograms, were constructed from the time-resolved data by integrating all photoevents for a preselected time interval (e.g., 1 s) and plotting versus electrophoresis time.

The common method for increasing system throughput is to add additional electrophoresis lanes, which requires additional sample processing pipelines in the system with the limit of lanes that can be added determined by the amount of real estate that can be interrogated by the fluorescence detector without sacrificing signal-to-noise. For example, work has demonstrated the ability of performing 384 capillary gel separations simultaneously on a single device with the fluorescence readout accomplished using a rotary scanner (Emrich et al., 2002). The throughput of such a device could be doubled without adding more separation lanes if eight fluorescent dyes could be efficiently distinguished, allowing the ability to run two different samples in the same gel lane. The use of color discrimination exclusively is difficult when processing multiple dye sets due to the broad emission profiles associated with most molecular systems. Therefore, the ability to build hybrid systems that provide

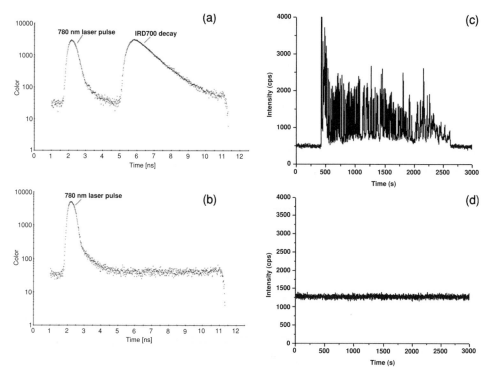

Figure 7.7. Time-filtering using an optical delay line (i.e., single-mode optical fiber) to remove spectral leakage of fluorescence into the inappropriate color channel. The time-resolved response is shown in parts a and b, while the fluorescence intensity response, for the 810-nm color channel, is shown in parts c and d, both without (a, c) and after (b, d) the time filtering. The labeling dye used in these measurements was IRD700 that was conjugated to a sequencing primer. The dual-color, time-resolved microscope contained two lasers, operating at 680 and 780 nm that were operated in a pulsed mode with a repetition frequency of 40 MHz. The data was acquired during a capillary gel electrophoresis run using a sieving matrix consisting of POP6. The electrophoretic track is from a single base (G) produced via Sanger sequencing with an IRD700-labeled primer.

fluorescence information via both color and lifetime is attractive for high multiplexing opportunities in DNA sequencing.

A two-color/two-lifetime sequencing run was carried out as an initial test of the color and time-resolve hybrid system approach (Zhu et al., 2003). Two dye pairs, IRD700 and AlexaFluor680, were processed in a 710-nm color channel, and IRD800 and near-IR-Br dye (see Table 7.2) [93] were used in an 810-nm color channel. Each pair of dyes showed minimal differences in their excitation and emission profiles, but possessed distinguishable lifetimes in the sequencing matrix used for electrophoretic sorting of the DNA. The decay profiles for all four dyes as well as the instrument response function in each color channel are shown in Figure 7.8. Four single-base tracts were run prior to the sequencing experiment to determine the lifetime value of each dye label under typical electrophoresis conditions. The average lifetime values, calculated using the MLE algorithm, for each dye label were 913 ± 9 ps, 1493 ± 11 ps, 454 ± 7 ps, and 744 ± 16 ps, for IRD700, Alexa680, IRD800, and NIR-Br, respectively. Based on the predetermined lifetime values, an automatic peak recognition algorithm was applied to assist the identification of the terminal bases. The raw data were subjected to mobility shift corrections but without going through any other data

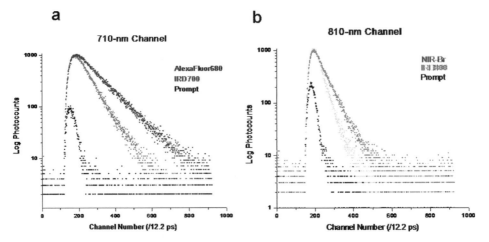

Figure 7.8. Instrument response function and fluorescence decay profiles for a two-color/two lifetime hybrid instrument. (a) Fluorescence decay profiles for IRD700-labeled G- and AlexaFluor680-labeled T fragments as well as the instrument response function for the 710-nm color channel. (b) Fluorescence decay profiles for IRD800-labeled C- and NIR-Br-labeled A-sequencing fragments and the instrument response function generated in the 810-nm color channel. Individual dye-labeled DNA ladders were analyzed using CGE with POP6 as the sieving matrix. Decays were constructed by integrating photocounts over 5 pixels (integration time of 5 s) centered on individual electrophoretic peaks from each sequencing trace. The instrument response functions were accumulated over 5 pixels from the gel track prior to migration of the DNA fragments into the detection volume (Zhu et al., 2003). See insert for color representation of this figure.

manipulations that are normally involved in many automated sequencing machines, such as removal of cross-talk, baseline adjustment, or signal normalization. The calling accuracy was 95.1% over a read length of 650 base pairs, with the majority of errors occurring at late times within the electrophoresis trace due to poor separation efficiency at the end of the run.

The same strategy (two-color/two-lifetime) was also tested on a microchip electrophoresis format (Zhu et al., 2004). The sequencing samples were required to be purified and pre-concentrated prior to electrophoretic sorting to improve the precision of the lifetime determinations due to the low loading levels associated with the microchip format. The fluorescence lifetimes of all dyes were determined with favorable precisions on the microchips.

7.4.2. Potential Applications

The primary motivation for coupling time-resolved measurements with color discrimination is to increase the fluorescence multiplexing capability to increase the information content obtainable in a single gel lane, thereby increasing the data throughput. With appropriate dye sets, a two-color, four-lifetime sequencing strategy could be envisioned to allow the identification of eight unique reporters. This scheme could, for example, allow simultaneous forward and reverse reads from both ends of a double-stranded DNA. The final steps of most shotgun sequencing strategies involve sequencing from both ends of selected M13 clones to build a scaffold (map) and fill in map gaps using directed reads Chen et al., 1993). The two-color/four-lifetime strategy could be effectively used here for front- and back-end reads in a single lane, where eight probes need to be analyzed simultaneously. Sequencing from both ends also yields important positional information since the distance between the read pair is known (Edwards and Caskey, 1991).

As multiplexed dye systems using color discrimination are further developed, lifetime identification methods could be incorporated into any color method to further increase the multiplexing. The potential extension of the hybrid strategy may also involve the use of FRET systems. For example, a measurement of four different fluorescence lifetimes in four spectrally distinct channels that use FRET would allow 16 different parameters to be measured in a single lane and still, using a single excitation source, efficiently excite the donor dye. Such a dye set, however, remains to be developed.

7.5. INSTRUMENTAL FORMATS FOR FLUORESCENCE-BASED DNA SEQUENCING USING LIFETIMES

Proper instrumentation design of a detector system is the cornerstone of any successful implementation of DNA sequencing, since the performance of the detector will dramatically affect read length and calling accuracy. Reading fluorescence during the electrophoretic separation of DNA sequencing ladders and accurately and efficiently identifying each terminal base offers a challenging task. A good sequencing instrument should possess the following characteristics: high sensitivity, high base-calling accuracy, high-throughput ability, and system robustness. High sensitivity is of particular importance because separation platforms used to fractionate DNA are being developed with miniaturization as a major goal. Loading amounts as low as the attomole range (10^{-18}) are often demanded. The detector must be able to read fluorescence signatures with reasonably high signal-to-noise ratio in order to accurately call the base. As systems are further scaled-down, increases in sensitivity of the fluorescence reader will have to occur.

A successful DNA sequencer also requires excellent base identification capabilities. This has been carried out by accurately processing the fluorescence by either their emission wavelength (color) or their fluorescence lifetimes. For color identification, the hardware must provide efficient excitation of the dye set and effectively sort the color into appropriate channels without producing significant amounts of cross-talk. For lifetime discrimination, the hardware must possess a favorable instrument response function so as to eliminate the need for deconvolution to simplify data processing. Common to both color and lifetime discrimination are the ability to provide excellent discrimination of different fluorescence reporters. The dye sets used should possess broad spacing between their emission maxima for color sorting and broad spacing between their fluorescence lifetimes for lifetime discrimination.

In addition, system throughput is another important criterion by which to evaluate a sequencer. High throughput requires many separation channels or lanes incorporated into a separation platform to process samples in parallel. As noted previously, system throughput can also be expanded by incorporating a higher level of multiplexing capability into the system as well. For multi-lane systems, all lanes need to be illuminated simultaneously with acceptable laser power and detected with high sensitivity at high duty cycles, since signal aliasing can be introduced that can effect electrophoretic performance. Numerous designs and improvements have been made toward developing comprehensive instruments that are capable of fast, automated, and sensitive operation even when operated in a multichannel format.

DNA sequencing was initially performed using slab gels, in which up to 96 lanes could be run simultaneously with loading volumes in the tens of microliters. With the success of capillary-based separation technologies and the desire for increased automation, capillary-based instruments have become the second-generation sequencing instruments. To achieve high-throughput production of DNA sequencing information for CGE, parallel sample processing in multiple capillaries, termed capillary array electrophoresis (CAE), has been introduced. With the recent advances in miniaturizing capillary dimensions on planar microfabricated devices, chip-based DNA sequencing has offered increases in speed and throughput compared to capillary array electrophoresis. The increase in throughput using microchips has been realized not only by speeding up individual runs but also by implementing array formats on a single chip. As such, fluorescence detection systems for chip-based DNA sequencing need to be developed. For example, a four-color rotary confocal scanner has been demonstrated for a 96-lane microchip array for DNA sequencing (Paegel, 2002).

The evolution of instruments for fluorescence-based DNA sequencing has led to the realization of high-throughput, high-performance automated DNA sequencers. For instance, the successful development of a 96-capillary CAE instrument, such as the MegaBACE and ABI PRISM® 3700 DNA analyzer, greatly contributed to the early completion of the Human Genome Project (HGP). The throughput of a slab gel machine was estimated to be ~600 bases/hour (Prober et al., 1987). Today, automated CAE sequencing instruments that use fluorescent dyes and laser scanners or imagers for production-scale sequencing allow processing 50,000 to 60,000 bases of sequence information in only a few hours.

Irrespective of the separation platform used, the fluorescence detection systems required for high-throughput arrays can be categorized into two types—scanning or imaging. The following sections provide a few typical examples of each type when implementing time-resolved identification of the constituent nucleotide bases.

7.5.1. Time-Resolved Fluorescence Scanning Detectors

Neumann and co-workers have described a time-resolved scanner for reading 16 capillary arrays (Neumann et al., 2000). This scanner consisted of a pulsed diode laser, operating at 640 nm with a repetition rate of 50 MHz and an avalanche diode detector. A confocal imager configured in a time-correlated single-photon-counting arrangement was kept stationary. The capillary array was linearly translated through the detection zone. Up to 16 capillaries were interrogated at scan rates approaching 0.52 Hz.

Lassiter and co-workers reported on the integration of a near-IR time-solved fluorescence scanner with an automated slab gel sequencer for lifetime-based DNA sequencing. Due to the size of the slab gel and the gel plates, the gel medium could not be translated underneath the relay optics of the detector. Therefore, a scanner containing the entire detection optics that could move in a linear fashion over the gel plate was constructed (see Figure 7.9). A pulsed diode laser operating at 680 nm was mounted on a microscope head at a 56° angle with respect to the boro-float gel plates to minimize reflected radiation from being coupled into the optical system. The laser diode was driven by an electrical short-pulse generator, which supplied a repetition rate of up to 80 MHz. The laser radiation was focused onto the surface of the gel plates using an $f/1.4$ lens that produced a spot on the gel of approximately 20 μm × 30 μm (elliptical beam shape of diode laser). The emission generated from the gel was subsequently collected by $f/1.2$ optics mounted in the microscope head, filtered with a single band-pass filter and then focused onto the face of an actively quenched avalanche photodiode with a large photoactive area. All photon-counting electronics were integrated into a PC board and situated in a computer to process the time-resolved data. The simple

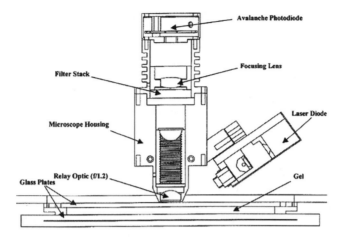

Figure 7.9. Block diagram of a time-resolved near-IR laser-induced scanner. The laser was mounted at 56° with respect to the scanning surface to minimize light reflection from the glass surface (Lassiter et al., 2000).

instrumental reconfiguration implemented in this work demonstrated that many existing machines, which use steady-state fluorescence, could be easily modified to do time-resolved measurements as well, acquiring both steady-state and the time-resolved data.

7.5.2. Time-Resolved Fluorescence Imaging Detectors

The ability to construct imaging time-resolved systems has been made feasible for DNA sequencing applications with the advent of multichannel detectors, such as CCDs, that demonstrate ultrafast time resolution (Becker et al., 2004; Kress et al., 2003; Mitchell et al., 2002; Stortelder et al., 2004; Urayama et al., 2003). These multichannel detectors are similar to conventional CCD cameras used for steady-state measurements, except that they also consist of gated-image intensifiers that provide the ultrafast time resolution. The intensifier is a phosphor plate generating nearly 200 photons per incident photo-event, thus providing gain (~200-fold) for high-sensitivity measurements. The phosphor plate acts as an optical shutter as well, providing the prerequisite time resolution. For a time-resolved measurement to construct the decay profile from which the lifetime could be extracted, the image intensifier is triggered with the excitation pulse from the laser to initiate image acquisition. Therefore, the system also requires a laser operating in a pulsed-mode. This acquisition can be delayed and shifted in time with respect to the excitation laser pulse by 25-ps steps (for 25-ps steps over a range of 12 ns, each decay profile would consist of 480 data points) to construct the entire decay curve. The timing resolution of such a system would depend on the speed of the gate, the pulse width of the excitation laser, and the bin width of the MCA used to accumulate the individual photon events of the decay profile. The optical arrangement would be similar to that used for most steady-state measurements required for spectral discrimination, in which the pulsed laser would need to simultaneously irradiate either an array of capillaries or a series of microchannels configured on a chip. The resulting fluorescence could then be collected by the appropriate relay optic and imaged onto the multichannel, time-resolved detector.

REFERENCES

Ambroz, M, Beeby, A, MacRobert, AJ, Simpson, MSC, Svensen, RK, and Phillips, D (1991). Preparative, analytical and fluorescence spectroscopic studies of sulfonated aluminium phthalocyanine photosensitizers. *Photochem Photobiol B: Biology* **9**:87–95.

Ansorge, W, Sproat, BS, Stegemann, J, and Schwager, C (1986). A non-radioactive automated method for DNA sequence determination. *J Biochem Biophys Methods* **13**:315–323.

Ansorge, W, Sproat, B, Stegemann, J, Schwager, C, and Zenke, M (1987). Automated DNA sequencing: Ultrasensitive detection of fluorescent bands during electrophoresis. *Nucleic Acids Res* **15**:4593–4602.

Ballew, RM, and Demas, JN (1989). An error analysis of the rapid lifetime determination method for the evaluation of single exponential decays. *Anal Chem* **61**:30–33.

Becker, W, Bergmann, A, Hink, MA, Konig, K, Benndorf, K, and Biskup, C (2004). Fluorescence lifetime imaging by time-correlated single-photon counting. *Microsc Res Tech* **63**:58–66.

Bradbrook, EF, and Linstead, RP (1936). Preparation of the ten dicyanonaphthalenes and the related naphthalenedicarboxylic acids. *J Chem Soc, Abstr,* 1739–1744.

Carson, S, Cohen, AS, Belenkii, A, Ruiz-Martinez, MC, Berka, J, and Karger, BL (1993). DNA sequencing by capillary electrophoresis: Use of a two-laser–two-window intensified diode array detection system. *Anal Chem* **65**:3219–26.

Casay, GA, Lipowski, J, Czuppon, T, Narayanan, N, and Patonay, G (1994). Spectroscopic investigations of a tetrasubstituted aluminum naphthalocyanine near-infrared compounds. *Spectrosc Lett* **27**:417–437.

Chang, K, and Force, RK (1993). Time-resolved laser-induced fluorescence study on dyes used in DNA sequencing. *Appl Spectrosc* **47**:24–29.

Chen, EY, Schlessinger, D, and Kere, J (1993). Ordered shotgun sequencing, a strategy for integrated mapping and sequencing of YAC clones. *Genomics* **17**:651–656.

Collins, FS, Green, ED, Guttmacher, AE, and Guyer, MS (2003). A vision for the future of genomics research. *Nature (London, United Kingdom)* **422**:835–847.

Duthie, RS, and Kalve, IM (2002). Novel cyanine dye-labeled dideoxynucleoside triphosphates for DNA sequencing. *Bioconjug Chem* **13**:699–706.

Edwards, A, and Caskey, CT (1991). Closure strategies for random DNA sequencing. *Methods (San Diego, CA, United States)* **3**:41–47.

Emrich, CA, Tian, HJ, Medintz, IL, and Mathies, RA (2002). Microfabricated 384-lane capillary array electrophoresis bioanalyzer for ultrahigh-throughput genetic analysis. *Anal Chem* **74**:5076–5083.

Flanagan, JH, Jr, Owens, CV, Romero, SE, Waddell, E, Kahn, SH, Hammer, RP, and Soper, SA. (1998). Near-infrared heavy-atom-modified fluorescent dyes for base-calling in DNA-sequencing applications using temporal discrimination. *Anal Chem* **70**:2676–2684.

Giesecke, H, Obermaier, B, Domdey, H, and Neubert, WJ (1992). Rapid sequencing of the sendai virus 6.8 Kb large (L) gene through primer walking with an automated DNA sequencer. *J Vir Methods* **38**:47–60.

Gilson, TR, and Hendra PJ (1970). *Laser Roman Spectroscopy: A Survey of Interest Primarily to Chemists.* Wiley, Chichester, UK.

Hall, P, and Selinger, B (1981). Better estimates of exponential decay parameters. *J Phys Chem* **85**:2941–2946.

Hammer, RP, Owens, CV, Hwang S-H, Sayes CM, and Soper, SA (2002). Asymmetrical, water-soluble phthalocyanine dyes for covalent labeling of oligonucleotides. *Bioconjug Chem* **13**:1244–1252.

Hardison, RC (2003). Comparative genomics. *PLoS Biology* **1**:156–160.

He, H, and McGown, LB (2000). DNA sequencing by capillary electrophoresis with four-decay fluorescence detection. *Anal Chem* **72**:5865–5873.

He, H, Nunnally, BK, Li, L-C, and McGown, LB (1998). On-the-fly fluorescence lifetime detection of dye-labeled DNA primers for multiplex analysis. *Anal Chem* **70:**3413–3418.

Karger, AE, Harris, JM, and Gesteland, RF (1991). Multiwavelength fluorescence detection for DNA sequencing using capillary electrophoresis. *Nucleic Acids Res* **19:**4955–4962.

Kieleczawa, J, Dunn, JJ, and Studier FW (1992). DNA sequencing by primer walking with strings of contiguous hexamers. *Science* **258:**1787–1791.

Kobayashi, N, Ogata, H, Nonaka, N, and Luk'yanets, EA (2003). Effect of peripheral substitution on the electronic absorption and fluorescence spectra of metal-free and zinc phthalocyanines. *Chem Eur J* **9:**5123–5134.

Koellner, M, Fischer, A, Arden-Jacob, J, Drexhage, KH, Mueller, R, Seeger, S, and Wolfrum, J (1996). Fluorescence pattern recognition for ultrasensitive molecule identification: comparison of experimental data and theoretical approximations. *Chem Phys Lett* **250:**355–360.

Kovshev, EI, and Luk'yanets, EA (1972). Phthalocyanines and related compounds. XI. Substituted 2,3-naphthalocyanines. *Zh Obshch Khim* **42:**1593–1597.

Kress, M, Meier, T, Steiner, R, Dolp, F, Erdmann, R, Ortmann, U, and Ruck, A (2003). Time-resolved microspectrofluorometry and fluorescence lifetime imaging of photosensitizers using picosecond pulsed diode lasers in laser scanning microscopes. *J Biomed Opt* **8:**26–32.

Kumar, S (2002). Fluorescent dye nucleotide conjugates for DNA sequencing. *Modified Nucleosides: Synth Appl*; 87–110.

Lander, ES, Linton, LM, Birren, B, Nusbaum, C, Zody, MC, Baldwin, J, Devon, K, Dewar, K, Doyle, M, FitzHugh, W, Funke, R, Gage, D, Harris, K, Heaford, A, Howland, J, Kann, L, Lehoczky, J, LeVine, R, McEwan, P, McKernan, K, Meldrim, J, Mesirov, JP, Miranda, C, Morris, W, Naylor, J, Raymond C, Rosetti, M, Santos, R, Sheridan, A, Sougnez, C, Stange-Thomann, N, Stojanovic, N, Subramanian, A, Wyman, D, Sulston, J, Ainscough, R, Beck, S, Bentley, D, Burton, J, Clee, C, Carter, N, Coulson, A, Deadman, R, Deloukas, P, Dunham, A, Dunham, I, Durbin, R, French, L, Grafham, D, Gregory, S, Hubbard, T, Humphray, S, Hunt, A, Jones, M, Lloyd, C, McMurray, A, Matthews, L, Mercer, S, Milne, S, Mullikin, JC, Mungall, A, Plumb, R, Ross, M, Shownkeen, R, Sims, S, Waterston, RH, Wilson, RK, Hillier, LW, McPherson, JD, Marra, MA, Mardis, ER, Fulton, LA, Chinwalla, AT, Pepin, KH, Gish, WR, Chissoe, SL, Wendl, MC, Delehaunty, KD, Miner, TL, Delehaunty, A, Kramer, JB, Cook, LL, Fulton, RS, Johnson, DL, Minx, PJ, Clifton, SW, Harkins, T, Branscomb, E, Predki, P, Richardson, P, Wenning, S, Slezak, T, Doggett, N, Cheng, J-F, Olsen, A, Lucas, S, Elkin, C, Uberbacher, E, Frazier, M, Gibbs, RA, et al. (2001). Initial sequencing and analysis of the human genome. *Nature (London, United Kingdom)* **409:**860–921.

Lassiter, SJ, Stryjewski, W, Legendre, BL, Jr, Erdmann, R, Wahl, M, Wurm, J, Peterson, R, Middendorf, L, and Soper, SA (2000). Time-resolved fluorescence imaging of slab gels for lifetime base-calling in DNA sequencing applications. *Anal Chem* **72:**5373–5382.

Lassiter, SJ, Stryjewski, W, Owens, CV, Flanagan, JH, Jr, Hammer, RP, Khan, S, and Soper, SA (2002). Optimization of sequencing conditions using near-infrared lifetime identification methods in capillary gel electrophoresis. *Electrophoresis* **23:**1480–1489.

Lee, LG, Connell, CR, Woo, SL, Cheng, RD, McArdle, BF, Fuller, CW, Halloran, ND, and Wilson, RK (1992). DNA sequencing with dye-labeled terminators and T7 DNA polymerase: effect of dyes and dNTPs on incorporation of dye-terminators and probability of termination fragments. *Nucleic Acids Res* **20:**2471–2483.

Legendre, BL, Jr, Williams, CC, Soper, SA, Erdmann, R, Ortmann, U, and Enderlein, J. (1996). An all solid-state near-infrared time-correlated single photon counting instrument for dynamic lifetime measurements in DNA sequencing applications. *Rev Sci Instrum* **67:**3984–3989.

Li, L-C, He, H, Nunnally, BK, and McGown, LB (1997). On-the-fly fluorescence lifetime detection of labeled DNA primers. *J Chromatogr B: Biomed Sci Appli* **695:**85–92.

Li, L-C, and McGown, L,B. (1996). On-the-fly frequency-domain fluorescence lifetime detection in capillary electrophoresis. *Anal Chem* **68:**2737–2743.

Li, L, and McGown, LB (1999). Effects of gel material on fluorescence lifetime detection of dyes and dye-labeled DNA primers in capillary electrophoresis. *J Chromatogr A* **841**:95–103.

Li, L-C, and McGown, LB (2000). Improving signal to background ratio for on-the-fly fluorescence lifetime detection in capillary electrophoresis. *Electrophoresis* **21**:1300–1304.

Li, L, and McGown, LB (2001). Comparison of sieving matrices for on-the-fly fluorescence lifetime detection of dye-labeled DNA fragments. *Fresenius' Jour Anal Chem* **369**:267–272.

Lieberwirth, U, Arden-Jacob, J, Drexhage, KH, Herten, DP, Mueller, R, Neumann, M, Schulz, A, Siebert, S, Sagner, G, Klingel, S, Sauer, M, and Wolfrum, J (1998). Multiplex dye DNA sequencing in capillary gel electrophoresis by diode laser-based time-resolved fluorescence detection. *Anal Chem* **70**:4771–4779.

Lipowska, M, Patonay, G, and Strekowski, L (1993). New near-infrared cyanine dyes for labeling of proteins. *Synth Commun* **23**:3087–94.

McCubbin, I, and Phillips, D. (1986). The photophysics and photostability of zinc(II) and aluminum(III) sulfonated naphthalocyanines. *J Photochem* **34**:187–195.

McIntosh, SL, Nunnally, BK, Nesbit, AR, Deligeorgiev, TG, Gadjev, NI, and McGown, LB (2000). Fluorescence lifetime for on-the-fly multiplex detection of DNA restriction fragments in capillary electrophoresis. *Anal Chem* **72**:5444–5449.

Middendorf, LR, Bruce, JC, Bruce, RC, Eckles, RD, Grone, DL, Roemer, SC, Sloniker, GD, Steffens, DL, Sutter, SL, et al. (1992). Continuous, on-line DNA sequencing using a versatile infrared laser scanner/electrophoresis apparatus. *Electrophoresis* **13**:487–494.

Middendorf, LR, Bruce, JC, Bruce, RC, Eckles, RD, Roemer, SC, and Sloniker, GD (1993). A versatile infrared laser scanner/electrophoresis apparatus. *Proc SPIE: Int Soc Opt Eng* **1885**:423–434.

Mitchell, AC, Wall, JE, Murray, JG, and Morgan, CG (2002). Measurement of nanosecond time-resolved fluorescence with a directly gated interline CCD camera. *J Micros—Oxford* **206**:233–238.

Mujumdar, RB, Ernst, LA, Mujumdar, SR, and Waggoner, AS (1989). Cyanine dye labeling reagents containing isothiocyanate groups. *Cytometry* **10**:11–19.

Neumann, M, Herten, DP, Dietrich, A, Wolfrum, J and Sauer, M (2000). Capillary array scanner for time-resolved detection and identification of fluorescently labelled DNA fragments. *J Chromatogr A* **871**:299–310.

Nunnally, BK, He, H, Li, L,-C, Tucker, SA, and McGown, LB (1997). Characterization of visible dyes for four-decay fluorescence detection in DNA sequencing. *Anal Chem* **69**:2392–2397.

Onyango, P (2004). The role of emerging genomics and proteomics technologies in cancer drug target discovery. *Curr Cancer Drug Targets* **4**:111–124.

Paegel, BME, Wedemayer, GJ, Scherer, JR, and Mathies, RA (2002). High throughput DNA sequencing with a microfabricated 96-lane capillary array electrophoresis bioprocessor. *Proc Natl Acad Sci USA* **99**:574–579.

Patonay, G, and Antoine, MD (1991). Near-infrared fluorogenic labels: New approach to an old problem. *Anal Chem* **63**:321A–322A, 324A–327A.

Prober, JM, Trainor, GL, Dam, RJ, Hobbs, FW, Robertson, CW, Zagursky, RJ, Cocuzza, AJ, Jensen, MA, and Baumeister, K (1987). A system for rapid DNA sequencing with fluorescent chain-terminating dideoxynucleotides. *Science (Washington, DC, United States)* **238**:336–341.

Sauer, M, Arden-Jacob, J, Drexhage, KH, Gobel, F, Lieberwirth, U, Muhlegger, K, Muller, R, Wolfrum, J, and Zander, C (1998). Time-resolved identification of individual mononucleotide molecules in aqueous solution with pulsed semiconductor lasers. *Bioimaging* **6**:14–24.

Sauer, M, Angerer, B, Han, KT, and Zander, C (1999). Detection and identification of single dye labeled mononucleotide molecules released from an optical fiber in a microcapillary: First steps towards a new single molecule DNA sequencing technique. *Phys Chem Chem Phys* **1**:2471–2477.

Sauer, M, Angerer, B, Ankenbauer, W, Foldes-Papp, Z, Gobel, F, Han, KT, Rigler, R, Schulz, A, Wolfrum, J, and Zander, C (2001). Single molecule DNA sequencing in submicrometer channels: state of the art and future prospects. *J Biotechnol* **86**:181–201.

Seeger, S, Bachteler, G, Drexhage, KH, Arden-Jacob, J, Deltau, G, Galla, K, Han, KT, Mueller, R, Koellner, M, et al. (1993). Biodiagnostics and polymer identification with multiplex dyes. *Ber Bunsen-Ges* **97**:1542–1548.

Sharman, WM, Kudrevich, SV, and van Lier, JE (1996). Novel water-soluble phthalocyanines substituted with phosphonate moieties on the benzo rings. *Tetrahedron Lett* **37**:5831–5834.

Shealy, DB, Lipowska, M, Lipowski, J, Narayanan, N, Sutter, S, Strekowski, L, and Patonay, G (1995). Synthesis, chromatographic separation, and characterization of near-infrared labeled DNA oligomers for use in DNA sequencing. *Anal Chem* **67**:247–251.

Shen, T, Yuan, Z, and Xu, H. (1989). Fluorescent properties of phthalocyanines. *Dyes and Pigments* **11**:77–80.

Smith, LM, Sanders, JZ, Kaiser, RJ, Hughes, P, Dodd, C, Connell, CR, Heiner, C, Kent, SBH, and Hood, LE (1986). Fluorescence detection in automated DNA sequence analysis. *Nature (London, United Kingdom)* **321**: 674–679.

Soper, SA, and Legendre, BL, Jr (1994). Error analysis of simple algorithms for determining fluorescence lifetimes in ultradilute dye solutions. *Appl Spectrosc* **48**: 400–405.

Soper, SA, and Mattingly, QL (1994). Steady-state and picosecond laser fluorescence studies of nonradiative pathways in tricarbocyanine dyes: Implications to the design of near-IR fluorochromes with high fluorescence efficiencies. *J Am Chem Soc* **116**:3744–3752.

Soper, SA, Legendre, BL, Jr, and Williams, DC (1995). Online fluorescence lifetime determinations in capillary electrophoresis. *Anal Chem* **67**:4358–4365.

Soper, SA, Flanagan, JH, Jr, Legendre, BL, Jr, Williams DC, and Hammer RP (1996). Near-infrared, laser-induced fluorescence detection for DNA sequencing applications. *IEEE J Selected Top Quantum Electroni* **2**:1129–1139.

Stortelder, A, Buijs, JB, Bulthuis, J, Gooijer, C, and van der Zwan, G (2004). Fast-gated intensified charge-coupled device camera to record time-resolved fluorescence spectra of tryptophan. *Appl Spectros* **58**:705–710.

Strekowski, L, Lipowska, M, and Patonay, G (1992a). Facile derivatizations of heptamethine cyanine dyes. *Synth Commun* **22**:2593–2598.

Strekowski, L, Lipowska, M, and Patonay, G (1992b). Substitution reactions of a nucleofugal group in heptamethine cyanine dyes. Synthesis of an isothiocyanato derivative for labeling of proteins with a near-infrared chromophore. *J Org Chem* **57**: 4578–4580.

Strekowski, L, Lipowska, M, Gorecki, T, Mason, JC, and Patonay, G (1996). Functionalization of near-infrared cyanine dyes. *J Heterocyclic Chem* **33**:1685–1688.

Swerdlow, H, and Gesteland, R (1990). Capillary gel electrophoresis for rapid, high resolution DNA sequencing. *Nucleic Acids Res* **18**:1415–1419.

Urayama, P, Zhong, W, Beamish, JA, Minn, FK, Sloboda, RD, Dragnev, KH, Dmitrovsky, E, and Mycek, MA (2003). A UV-visible-NIR fluorescence lifetime imaging microscope for laser-based biological sensing with picosecond resolution. *Appl Phys B: Lasers Opt* **76**: 483–496.

Voss, H, Wiemann, S, Grothues, D, Sensen, C, Zimmermann, J, Schwager, C, Stegemann, J, Erfle, H, Rupp, T, and Ansorge, W (1993). Automated low-redundancy large-scale DNA-sequencing by primer walking. *Biotechniques* **15**:714–721.

Waddell, E, Lassiter, S, Owens, CV, Jr, and Soper, SA (2000a). Time-resolved near-IR fluorescence detection in capillary electrophoresis. *J Liq Chromatogr Related Technol* **23**:1139–1158.

Waddell, E, Wang, Y, Stryjewski, W, McWhorter, S, Henry, AC, Evans, D, McCarley, RL, and Soper, SA (2000b). High-resolution near-infrared imaging of DNA microarrays with time-resolved acquisition of fluorescence lifetimes. *Anal Chem* **72**:5907–5917.

Wheeler, BL, Nagasubramanian, G, Bard, AJ, Schechtman, LA, and Kenney, ME (1984). A silicon phthalocyanine and a silicon naphthalocyanine: Synthesis, electrochemistry, and electrogenerated chemiluminescence. *J Am Chem Soc* **106**:7404–7410.

Williams, DC, and Soper, SA (1995). Ultrasensitive near-IR fluorescence detection for capillary gel electrophoresis and DNA sequencing applications. *Anal Chem* **67**:3427–3432.

Wrobel, D, and Boguta, A (2002). Study of the influence of substituents on spectroscopic and photoelectric properties of zinc phthalocyanines. *J Photochem Photobiol, A: Chemistry* **150**:67–76.

Zhang, Y, Soper, SA, Middendorf, LR, Wurm, JA, Erdmann, R, and Wahl, M (1999). Simple near-infrared time-correlated single photon counting instrument with a pulsed diode laser and avalanche photodiode for time-resolved measurements in scanning applications. *Appl Spectrosc* **53**:497–504.

Zhu, L, Stryjewski, W, Lassiter, S, and Soper, SA (2003). Fluorescence multiplexing with time-resolved and spectral discrimination using a near-IR detector. *Anal Chem* **75**:2280–2291.

Zhu, L, Stryjewski, WJ, and Soper, SA (2004). Multiplexed fluorescence detection in microfabricated devices with both time-resolved and spectral-discrimination capabilities using near-infrared fluorescence. *Anal Biochem* **330**:206–218.

CHAPTER

8

ULTRASENSITIVE MICROARRAY DETECTION OF DNA USING ENZYMATICALLY AMPLIFIED SPR IMAGING

HYE JIN LEE, ALASTAIR W. WARK, AND ROBERT M. CORN

8.1. INTRODUCTION

The use of surface bioaffinity interactions in a microarray format has become an indispensable tool for modern biological research. Measurements of DNA–DNA hybridization, protein–DNA binding, and antibody–antigen interactions can all be performed now with biopolymer microarrays that permit the simultaneous characterization of multiple bioaffinity binding processes. In particular, DNA microarrays currently provide researchers with a simple and systematic method for the simultaneous identification and quantification of multiple nucleic acid sequences in a sample for a variety of applications, including gene expression analysis (Lockhart and Winzeler, 2000; Stears et al., 2003; Nelson et al., 2004; Thomson et al., 2004), single nucleotide polymorphism (SNP) profiling (Gerry et al., 1999; Kwok, 2001; Gerion et al., 2003; Giusto and King, 2003; Zhong et al., 2003), and the detection of specific viral and bacterial species (Wang et al., 2002; Sengupta et al., 2003; Volokhov et al., 2003; Gonzalez et al., 2004; Lin et al., 2004; Chen et al., 2005). The selectivity and reliability of DNA microarrays are usually determined by the sequence specificity of the surface hybridization of target DNA or RNA from solution onto the various DNA or RNA microarray elements.

Fluorescence microscopy is the most commonly used approach for biosensing with DNA microarrays (van Hal et al., 2000; Zammatteo et al., 2000; Epstein et al., 2002; Lehr et al., 2003; Livache et al., 2003; Stears et al., 2003; Stoughton, 2005). If a fluorescent tag is attached to one end of a single-stranded DNA (ssDNA) molecule, the hybridization of this target ssDNA onto a microarray element that contains specific complementary probe oligonucleotides can be detected with a fluorescent imaging experiment. For example, this type of measurement is the basis for the analysis of the large-scale photolithographically generated microarrays employed by Affymetrix (Pease et al., 1994; Schena et al., 1995; Winzeler et al., 1998; Brenner et al., 2000; Berchuck et al., 2005).

For cases where labeling is not possible or not desired, the surface-sensitive optical technique of surface plasmon resonance imaging (SPRI) is an excellent alternative for the detection of surface bioaffinity measurements with biopolymer microarrays (Lyon et al., 1999; Brockman et al., 2000; Smith and Corn, 2003; Wegner et al., 2004a; Lee et al., 2005d). SPRI measures biopolymer adsorption onto a surface by detecting any changes in the local index of refraction that occur upon binding events. SPR imaging has been applied to the

---

New Frontiers in Ultrasensitive Bioanalysis. Edited by Xiao-Hong Nancy Xu
Copyright © 2007 John Wiley & Sons, Inc.

detection of DNA–DNA (Lee et al., 2001; Nelson et al., 2001; Smith et al., 2002). RNA–DNA (Nelson et al., 2002; Goodrich et al., 2004a,b), DNA–protein (Smith et al., 2003a; Kyo et al., 2004; Shumaker-Parry and Campbell, 2004; Shumaker-Parry et al., 2004; Okumura et al., 2005), peptide–protein (Wegner et al., 2002; Wegner et al., 2004b), carbohydrate–protein (Smith et al., 2003b), and antibody–antigen (Wegner et al., 2003; Kanda et al., 2004; Wilkop et al., 2004) binding in a microarray format. SPRI is also potentially a superb candidate for the direct analysis of genomic DNA and RNA in an array format without labeling or PCR amplification. However, to date the application of SPRI to genomic DNA samples has not been possible due to a concentration limit of 1 nM for the detection of short (16–20mer) oligonucleotides; genomic DNA samples typically have DNA concentrations of approximately 20 fM (Goodrich et al., 2004a).

In fact, a target DNA concentration of 20 fM is also too small to be detected directly in a typical fluorescent imaging experiment. An amplification method must be used to increase the sensitivity of the surface bioaffinity measurement; typically a small amount of a particular DNA fragment in a genomic sample is enzymatically amplified using the polymerase chain reaction (PCR) with a fluorescent tag attached to one of the PCR primers (Sengupta et al., 2003; Gonzalez et al., 2004). For large sets of DNA molecules, simultaneous PCR amplification is not possible. For these systems, sandwich assays involving three oligonucleotides and either enzymes or nanoparticles have been proposed as a means of amplifying the signal (Kawai et al., 1993; Morata et al., 2003; Storhoff et al., 2004; Bao et al., 2005). Some of these amplification methods have also been used in conjunction with SPRI (He et al., 2000).

Recently, we have developed a novel surface-based enzymatic process that can be used to lower the limit of detection for SPRI measurements of ssDNA to a concentration of 1 fM. This enzymatically amplified SPR imaging method can be used for the direct detection of specific sequences from human genomic DNA. In this method, RNase H is used in conjunction with RNA microarrays for the enzymatically amplified detection of DNA. In this chapter we describe in detail this method for the ultrasensitive detection of DNA with enzymatically amplified SPRI.

This chapter is divided into four sections. In Section 8.2, we describe the SPRI measurement in detail and discuss how to make DNA and RNA microarrays on chemically modified gold surfaces. In Section 8.3, we examine the use of SPRI for the direct detection of DNA, and what limits the sensitivity of this method to 1 nM. In Section 8.4, we introduce the surface RNase H hydrolysis reaction that serves as the core of the enzymatic amplification process, and we show what experiments and analyses are required to completely characterize surface enzyme reactions. Finally, in Section 8.5, we show how enzymatically amplified SPRI can be used to detect femtomolar concentrations of oligonucleotides, PCR products, and unamplified genomic DNA directly.

8.2. OVERVIEW OF SPRI METHODOLOGY

Surface plasmon resonance (SPR) is a surface-sensitive optical technique that can be applied to the real-time monitoring of biomolecular adsorption and/or desorption events at biopolymer layers formed on a thin gold film or other noble metal surfaces. Surface plasmons are electromagnetic waves that propagate along a metal/dielectric interface. The optical field intensity of the surface plasmon waves decays exponentially from the surface of the metal into the dielectric layer. For a gold film, this decay length is about 200 nm, thus defining a region where the SPR response is sensitive to localized changes in refractive index due to

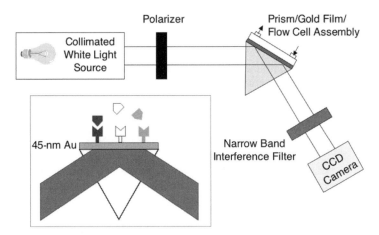

Figure 8.1. Schematic overview of the SPRI apparatus. Inset is a representation of the analyte/thin gold film/prism assembly. See insert for color representation of this figure.

the adsorption or desorption of molecules. Surface plasmons cannot be excited directly by light at planar air–metal or water–metal interfaces because momentum-matching conditions are not satisfied. Instead, a diffraction grating or prism arrangement, such as that shown in Figure 8.1, is required to convert incident p-polarized light photons into surface plasmons (Hanken et al., 1998; Knoll, 1998; Homola et al., 1999). Changes in the SPR response can be measured using three different instrumental formats: (1) scanning angle SPR, (2) scanning wavelength SPR, and (3) SPR imaging. Angle shift measurements (Szabo et al., 1995; Chinowsky et al., 2003) are the most commonly undertaken technique and form the basis of the commercial instruments available from Biacore and Texas Instruments. In scanning angle SPR, the reflectivity of monochromatic incident light is monitored as a function of incident angle, while in scanning wavelength SPR (Frutos et al., 1999; Corn and Weibel, 2001), it is the incident angle that remains fixed. The latter approach is the principle behind the Fourier Transform SPR instrument available from GWC Technologies. Both of these techniques are used to study a single region on a gold surface. In contrast, SPR imaging measurements simultaneously monitor spatially resolved changes in reflectivity at a fixed angle and wavelength due to biomolecular adsorption onto an array surface. Commercial SPR imaging instruments are available from GWC Technologies and HTS Biosystems.

The high-throughput capabilities of SPRI make it an attractive tool for studies involving the screening of multiple biomolecular interactions on a single chip surface. Figure 8.1 shows a schematic representation of the SPR imaging instrument used to detect the adsorption of biopolymers in solution to surface-immobilized biomolecules such as DNA (Nelson et al., 2001; Kyo et al., 2004), RNA (Goodrich et al., 2004a,b), peptides (Wegner et al., 2002, 2004b), carbohydrates (Smith et al., 2003b), and proteins (Wegner et al., 2003; Kanda et al., 2004).

The output from a collimated white light source is first passed through a polarizer before being directed through a high-index prism/sample assembly at an optimal incident angle. The p-polarized light impinges on the back of a gold thin film, whose surface is chemically modified with an array of biomolecules. The reflected light is then collected via a narrow band-pass filter centered at 830 nm onto a CCD camera. Molecules are delivered to the gold surface using a flow cell with adsorption at a particular array element, resulting in an increase in reflectivity. This is detected by subtracting images acquired before and after the surface binding event. For measured changes in percentage reflectivity ($\Delta\%R$) under 10%, a linear relationship exists between $\Delta\%R$ and the corresponding change in refractive index (Nelson et al., 2001). This relationship can also be applied to determine the surface coverage of biomolecules adsorbed onto the surface, thus allowing the use of SPRI to quantitatively evaluate the binding affinity between the biomolecules in solution and the multiple probes immobilized on the array.

8.2.1. Surface Attachment Chemistry

The use of well-characterized and robust surface chemistries to tether biological molecules onto gold surfaces in an array format is an essential component of a successful SPRI experiment. Because noble metal films are required for the propagation of surface plasmons, commercially available DNA arrays created on glass substrates, such as those provided by Affymetrix or inkjet printing processes (Winzeler et al., 1998; Hughes et al., 2001; Berchuck et al., 2005), cannot be utilized. Consequently, we have developed a strategy using self-assembled alkanethiol monolayers (SAMs) containing an ω-terminated amine functional group as the foundation of the array. Thiol-modified biomolecules such as thiol-modified DNA (Brockman et al., 1999), RNA (Goodrich et al., 2004a,b), carbohydrates (Smith et al., 2003b), and cysteine-containing peptides (Wegner et al., 2002) can then be covalently attached to the surface through the use of heterobifunctional linker molecules such as SSMCC (sulfosuccinimidyl 4-(N-maleimidomethyl) cyclohexane-1-carboxylate) and SPDP (N-succinimidyl 3-(2-pyridyldithio)-propionamido). This approach has been successfully applied to create equally robust DNA, RNA, and peptide microarrays.

Figure 8.2 shows two different reaction schemes for the surface immobilization of thiol-modified probe molecules. In the first approach, the N-hydroxysulfosuccinimide (NHS)

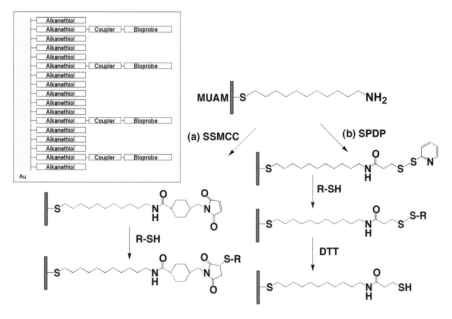

Figure 8.2. Reaction schemes outlining two different surface attachment chemistries [(a) SSMCC and (b) SPDP] for the surface immobilization of thiol-modified probe biomolecules.

ester moiety of SSMCC is reacted with a densely packed SAM of 11-mercaptoundecylamine (MUAM) to form an amide bond, leaving the free maleimide group to react with a thiol-modified biomolecule (Figure 8.2a). Alternatively, a MUAM monolayer is reacted with the molecule SPDP, resulting in the creation of a surface terminated with pyridyl disulfide groups (Figure 8.2b). Thiol-modified oligonucleotides (both DNA and RNA) are then attached via a thiol-disulfide exchange reaction with pyridine-2-thione as the leaving group. The SPDP chemistry has the advantage of being reversible as compared to SSMCC; the disulfide bond can be cleaved in the presence of dithiothreitol to regenerate the sulfhydryl-terminated surface. The thiol-modified probe oligonucleotides used are typically single-stranded and consist of 16–20 bases with a spacer between the C6-thiol modifier and the probe sequence. This spacer is designed to improve the accessibility of the surface attached molecule to the target molecule in solution. Examples of spacers employed include 15 thymine bases, $(T)_{15}$, or 12 ethylene glycol molecules, $(EG)_{12}$, for DNA and eight uracil bases, $(U)_8$, for RNA probes. In addition, surfaces fabricated using these approaches are very stable and can be used for several assay cycles as well as allowing good control over the surface density of probe molecules. For example, the hybridization of target DNA onto ssDNA microarrays (16mer short oligonucleotides) followed by regeneration of the array by washing with 8 M urea can be repeated up to 20 times (Lee et al., 2001; Nelson et al., 2001; Wark et al., 2005). Both attachment chemistries (SSMCC and SPDP) provide a typical ssDNA monolayer surface coverage of approximately 5×10^{12} molecules/cm^2.

8.2.2. Array Fabrication

The fabrication of arrays containing multiple, independently addressable elements on a single gold surface for use in SPRI measurements can be achieved using a combination of self-assembly, surface attachment chemistry, and array patterning. Two different array fabrication methods have been developed in our laboratory: (i) UV photopatterning in combination with a series of chemical modification steps allowing the production of relatively large numbers (> 100) of array elements (Brockman et al., 1999) and (ii) the use of polydimethylsiloxane (PDMS) microfluidic networks which are physically sealed onto the chemically modified gold surface (Lee et al., 2001).

Spotted Microarrays. (Brockman et al., 1999). The first step of the fabrication method is the reaction of the MUAM modified gold surface with the *N*-hydroxysuccinimide ester of 9-fluorenylmethoxycarbonyl (Fmoc-NHS), which serves as a hydrophobic protecting group. By exposing the Fmoc surface to UV radiation through a quartz mask containing 500-μm square features, patterns of bare gold spots surrounded by the hydrophobic background were created. The slide is then immersed again in an ethanolic MUAM solution for 2 hours, with the resulting MUAM patches then reacted with a heterobifunctional linker, SSMCC or SPDP, followed by thiol-modified oligonucleotides. Each of the array elements is separately addressed using a manually operated picopump to spot SSMCC or SPDP and thiol-modified probes (DNA or RNA) onto the surface. This approach is successful because the hydrophobic Fmoc background ensures that each droplet is contained on a particular array element, thus avoiding cross contamination of other sequences on neighboring elements. The Fmoc background can be completely removed with a mildly basic solution and the regenerated MUAM surface reacted with an NHS derivative of polyethylene glycol (PEG-NHS) to create a background resistant to nonspecific adsorption of biomolecules. Figure 8.3a shows an example of a raw SPR image of an array fabricated using this procedure with over

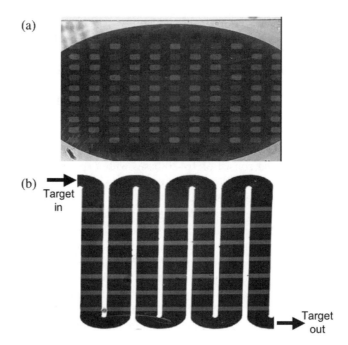

Figure 8.3. SPR raw images showing (a) an array fabricated using a multistep chemical protection/deprotection process in conjunction with UV photopatterning using a mask featuring a 500-μm square pattern. (b) The line array was created using a set of parallel PDMS microfluidic channels to deliver linker and probe molecules before being replaced by a second serpentine PDMS channel to create a continuous-flow cell.

110 individual elements. Furthermore, if a 250-μm square photomask is used instead, it is possible to attain over 300 elements on a single microarray surface.

Microfluidic Line Arrays. (Lee et al., 2001). A second approach involves the coupling of PDMS microfluidic channels to the gold surface for use in SPRI measurements with the aim of reducing chemical consumption and sample volume as well as speeding up analysis times. First, a set of parallel microchannels (300-μm width, 14.2-mm length, 35-μm depth) with 700-μm spacing between channels is created by replication from a 3-D silicon wafer master using soft lithography methods (Lee et al., 2001). The PDMS microchannels are then physically attached onto a MUAM modified gold surface before the heterobifunctional cross-linker (SSMCC or SPDP) and thiol-modified DNA or RNA are sequentially passed through using a simple differential pumping system to create a series of individual line elements on the gold array surface. After the surface immobilization of probe molecules is complete, the channels are removed and the background MUAM monolayer that surrounds the channels is reacted with PEG-NHS, with no other chemical protection and deprotection steps required. By changing the spacing and widths of the microchannels, a single chip can contain a maximum of 100 different probe elements.

In order to achieve a continuous flow of sample to the surface array for kinetics measurements, the large flow cell (100 μl) used in Figure 8.3a is replaced with a second PDMS microchannel (see Figure 8.3b). The serpentine design (670-μm width, 9.5-cm length, 200-μm depth, ~10-μL total volume) facilitates well-controlled and reproducible sample delivery to each array element as well as significantly reducing the sample volume. Discrete SPR imaging probing regions are formed by orienting the microchannel perpendicular to the probe line array. The microchannel is created by replication from a 3-D aluminum master (Wegner et al., 2004b). SPR imaging kinetics experiments are performed using a continuous flow of solution through the serpentine microchannel to prevent mass transport limitations, while equilibrium measurements are obtained under stopped-flow conditions using the large flow cell. Additionally, a constant temperature sample holder encased in a specially designed water jacketed cell, allowing the system temperature to be controlled to within 0.1°C, is used to reduce fluctuations in SPR signal due to temperature variations (Goodrich et al., 2004b; Lee et al., 2005a). Finally, if the simultaneous injection of multiple samples is desired, a set of parallel PDMS microchannels placed perpendicular to the line array can be used to deliver 1–2 µL of target sample per channel (Lee et al., 2001).

8.3. DIRECT DETECTION OF DNA BY HYBRIDIZATION ADSORPTION

The direct analysis of genomic DNA and RNA in an array format without labeling or PCR amplification would be extremely advantageous for a variety of biological applications. The detection and identification of ssDNA oligonucleotides by sequence specific adsorption onto an ssDNA microarray element to form a double-stranded duplex is called hybridization adsorption. If the target oligonucleotide is labeled with a fluorophore, fluorescence imaging can be used to detect DNA with the microarray. If the DNA has no fluorescent tag, then SPRI can be used to directly detect DNA hybridization adsorption. SPR imaging measurements of DNA microarrays has been demonstrated previously by a number of research groups (Guedon et al., 2000; Nelson et al., 2001; Livache et al., 2003; Rella et al., 2004; Shumaker-Parry et al., 2004; Okumura et al., 2005; Wark et al., 2005). However, the detection limit for DNA or RNA analysis via hybridization adsorption of untagged target molecules typically lies in the nanomolar range (Guedon et al., 2000; Nelson et al., 2001; Livache et al., 2003; Okumura et al., 2005; Wark et al., 2005). A typical sample of genomic DNA (35 µg/mL) has a ssDNA concentration of around 20 fM. What can be done to improve the SPRI methodology to increase its sensitivity into the femtomolar range?

To answer this question, we must carefully examine the thermodynamic and kinetic limitations of surface bioaffinity measurements. An example of the detection of DNA by hybridization adsorption onto DNA microarrays with SPRI is shown in Figure 8.4. A four-component ssDNA microarray is exposed to two of the complementary DNA target sequences, with only the perfectly matched array elements forming duplexes via hybridization adsorption (Figure 8.4b). A positive increase in percent reflectivity ($\Delta\%R$) due to selective hybridization adsorption is observed; Fresnel calculations and experimental evidence show that if the SPRI response remains below 10%, then it is directly proportional to the relative surface coverage (θ) of complementary DNA (Nelson et al., 2001), where $\theta = \Gamma/\Gamma_{tot}$ and Γ represents the molecular surface density. θ is related to the bulk concentration (C) of ssDNA by the Langmuir adsorption isotherm:

$$\theta = K_{Ads}C/(1 + K_{Ads}C) \tag{8.1}$$

where K_{Ads} is the Langmuir adsorption coefficient. A typical Langmuir isotherm for 16mer adsorption is shown in Figure 8.4c. At a bulk concentration of $C_{0.5} = 1/K_{Ads}$, the relative surface coverage is 0.5. For a 16mer DNA, $K_{Ads} = 2 \times 10^7$ M^{-1}, so that $C_{0.5}$ is at 50 nM.

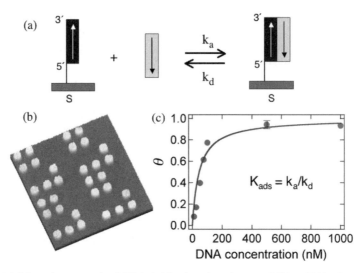

Figure 8.4. (a) Schematic representing DNA hybridization adsorption onto DNA or RNA microarrays. (b) An example of an SPR difference image showing the sequence-specific hybridization/adsorption of target DNA onto a four-component ssDNA microarray. This was obtained by exposing the ssDNA array to a 500 nM solution containing two different target DNA sequences with duplex formation occurring only at the complementary probe array elements. A maximum of 130 individually addressable array elements on a single chip with a total surface area of 0.8 cm² can be created using a 500-μm² photopatterning mask. (c) An example showing a plot of the relative surface coverage (θ) as a function of target complementary DNA concentration. The solid line represents a Langmuir isotherm fit to the data. A value of $K_{\text{ads}} = 2.0\ (\pm 0.2) \times 10^7$ M^{-1} was determined from the fit.

Comparable K_{Ads} values have been observed with both fluorescence imaging and SPR measurements (Liebermann et al., 2000; Nelson et al., 2001; Peterson et al., 2002; Lee et al., 2005a; Levicky and Horgan, 2005; Wark et al., 2005).

At low surface coverages, the Langmuir isotherm depends linearly on the target DNA concentration:

$$\theta = K_{\text{Ads}} C \qquad (8.2)$$

Equation (8.2) shows one reason why it is difficult to measure 1 fM target concentrations with a surface bioaffinity measurement: As C goes to zero, θ goes to zero as well! If one wants to detect ssDNA at a bulk concentration of 1 fM, then one must be able to detect a relative surface coverage of $\theta = 2 \times 10^{-8}$! As mentioned in the previous section, the surface density of DNA probes on a microarray element is typically 5×10^{12} molecules/cm², so that the surface density of dsDNA is $\Gamma = 10^5$ molecules/cm² at 1 fM. For a 500-μm array element, this corresponds to 250 molecules; for a 50-μm array element, this surface density corresponds to 2.5 molecules!! SPRI typically has a detection limit of 1 nM for 16mer oligonucleotide adsorption, which corresponds to a lowest relative surface coverage of $\theta \sim 2 \times 10^{-2}$ or $\Gamma \sim 10^{11}$ molecules/cm². If a sandwich assay is used with a DNA-coated nanoparticle, a relative surface coverage of $\theta \sim 2 \times 10^{-4}$ corresponding to a detection limit of approximately 10 pM is observed (He et al., 2000). For comparison, a typical lowest surface coverage that can be observed above background for the detection of DNA in a sandwich assay with a fluorescently tagged DNA molecule is $\Gamma \sim 10^8$ molecules/cm², corresponding to a detection limit of approximately 1 pM (Zammatteo et al., 2000; Lehr et al., 2003; Livache et al., 2003).

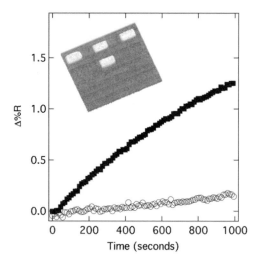

Figure 8.5. Representative real-time SPRI curves showing sequence specific hybridization adsorption of complementary DNA onto a surface probe element at concentrations of 1 nM (O) and 10 nM (■). The figure inset is an SPR difference image obtained by subtracting images acquired before and after the hybridization adsorption of 100 nM target DNA onto DNA array elements corresponding to a final Δ%R change of 2%. [Reprinted with permission from *Anal. Chem.* **77**:5096–5100, Copyright 2005, American Chemical Society.]

A second reason that it is difficult to detect a 1 femtomolar solution of ssDNA with SPRI hybridization adsorption measurements is the kinetics of hybridization adsorption. The hybridization adsorption rate (k_a) for a 16mer was measured to be approximately $10^5 M^{-1} s^{-1}$ from the data in Figure 8.5. Since the velocity of the adsorption reaction is $k_a C$, this rate decreases significantly as the concentration is lowered. Moreover, there is also a diffusional contribution that becomes significant at lower concentrations due to the time required for the molecules to reach the surface. These two effects are observed in Figure 8.5, which shows the time-dependent SPRI responses upon exposure of an ssDNA array to 10 and 1 nM solutions of complementary 16mer DNA. The initial adsorption rate, as determined from the initial slopes, decreases by a factor of 10 as the target DNA concentration is decreased from 10 to 1 nM, as expected from simple Langmuir kinetics. Thus, due to both Langmuir adsorption equilibrium arguments and adsorption kinetics, the detection limit for the direct detection of ssDNA oligonucleotides with SPRI is estimated to be 1 nM. To analyze genomic DNA at a concentration of 1 fM with SPRI, an amplification methodology is required. While an ELISA-based amplification scheme may provide sufficient sensitivity (Kawai et al., 1993; Morata et al., 2003), we will choose a different strategy for DNA detection, employing RNase H in conjunction with RNA microarrays to create a new type of surface enzyme amplification scheme.

8.4. RNase H SURFACE ENZYME KINETICS

RNase H is an endoribonuclease that selectively hydrolyzes the RNA strand of an RNA–DNA heteroduplex (Nakamura et al., 1991; Katayanagi et al., 1993; Zamaratski et al., 2001). Figure 8.6 shows a schematic drawing of the two possible ways that RNase H can react with surface-bound RNA–DNA heteroduplexes. The heteroduplex is first formed on

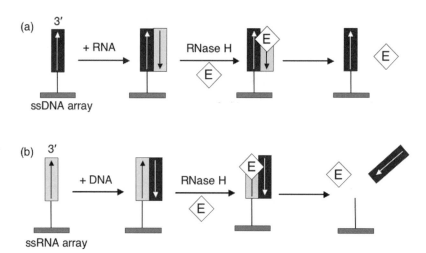

Figure 8.6. Schematic representation of RNase H activity at surface-bound RNA–DNA heteroduplexes formed by hybridization adsorption of (a) target RNA onto ssDNA microarrays and (b) target DNA onto ssRNA microarrays.

the surface of either a DNA or RNA microarray (see Figures 8.6a and 8.6b, respectively) by hybridization adsorption of the complementary oligonucleotide from solution. After hybridization, the RNA strand of the heteroduplex will be hydrolyzed and destroyed by RNase H. In the case of the DNA microarray (Figure 8.6a), hydrolysis leaves the original ssDNA still attached to the surface and capable of hybridization adsorption. In contrast, the RNase H hydrolysis of the RNA microarray results in both the destruction of the surface array element and the release of the target DNA back into solution (Figure 8.6b).

To demonstrate this surface RNase H activity, a series of real-time SPR imaging measurements were obtained for the reaction of RNase H with RNA–DNA heteroduplexes formed on DNA and RNA microarrays. Figure 8.7 plots the real-time SPRI signal for both cases (labeled a and b, respectively). A rapid rise in $\Delta\%R$ to a steady-state value of $+1.6\%$ was observed upon the hybridization adsorption of DNA or RNA onto the surfaces from 500 nM complement solutions. Subsequent rinsing of the DNA and RNA microarrays with buffer followed by exposure to RNase H resulted in a rapid decrease in $\Delta\%R$ from the hydrolysis of the RNA on the surface. For the case of the DNA microarray, a $\Delta\%R$ of -1.6% was observed; this decrease brought the SPRI signal back to the original reflectivity level prior to hybridization adsorption and suggests that the RNase H completely removed all surface-bound RNA. In contrast, for the case of the heteroduplexes formed using RNA microarrays, a decrease of -3.2% was observed corresponding to the complete loss of both the target DNA and probe RNA from the surface. No hybridization adsorption was observed upon subsequent exposure of the RNA microarray to complementary ssDNA. Moreover, if the experiments were repeated after the creation of either surface-bound dsDNA or dsRNA, no surface enzymatic hydrolysis was observed. These additional measurements confirmed the selectivity and specificity of the surface RNase H reaction to the RNA–DNA heteroduplex.

How does one quantify the reaction rate of the RNase H surface hydrolysis process? The characterization of the surface enzyme reactions utilized in biosensing processes is extremely important for optimization of the array-based biosensor, because an enzyme reaction can be orders of magnitude slower on a surface as compared to solution due to

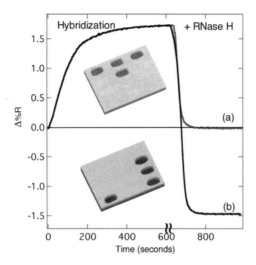

Figure 8.7. Real-time SPRI curves representing the sequence specific hybridization adsorption of (a) a 500 nM solution of complementary RNA onto an ssDNA microarray and (b) a 500 nM solution of complementary DNA onto an ssRNA microarray. The surface hydrolysis of both DNA–RNA heteroduplexes (a) and (b) was monitored at an RNase H concentration of 8 nM. The figure insets are representative SPR difference images obtained by subtracting images acquired before and after RNase H hydrolysis of heteroduplexes formed in parts a and b.

a variety of transport and energetic factors. We have recently shown (Fang et al., 2005; Lee et al., 2005b) that a simple model can be used to describe surface enzyme reaction dynamics, such as the surface RNase H hydrolysis process that forms the foundation of our enzymatically amplified SPRI measurements.

A general reaction scheme for the enzymatic processing of a biopolymer microarray that couples diffusion, adsorption, desorption and surface catalysis is shown in Figure 8.8.

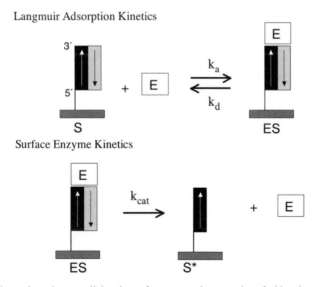

Figure 8.8. A reaction scheme outlining the surface enzymatic processing of a biopolymer microarray.

The enzyme (E) first adsorbs from solution onto the surface-bound substrate (S) to create a 1:1 surface complex (ES). The surface complex then reacts to form the surface-bound product (S^*) and releases the enzyme back into solution. This reaction scheme differs from the typical ELISA-based enzymatic method for detecting species on surfaces, which uses a sandwich assay in which an enzyme–protein conjugate binds to an adsorbed molecule and then reacts with a substrate in solution to create an amplified detection signal (Crowther, 1995). For the case of surface enzyme kinetics, the substrate is attached to the surface, and upon reaction of the surface complex, the enzyme is released back into solution.

The model shown in Figure 8.8 can be quantified with three rate constants (k_a, k_d, and k_{cat}) and an additional parameter (β) which describes the steady-state diffusion of enzyme to the surface. We rewrite the model for the surface enzyme reaction here in Equations (8.3)–(8.5):

$$E_{(x=\infty)} \xrightarrow{k_m} E_{(x=0)} \tag{8.3}$$

$$S + E_{(x=0)} \underset{k_d}{\overset{k_a}{\rightleftarrows}} ES \tag{8.4}$$

$$ES \xrightarrow{k_{cat}} S^* + E_{(x=0)} \tag{8.5}$$

where $E_{(x=\infty)}$ and $E_{(x=0)}$ are the bulk and surface enzyme species respectively, k_m is the steady-state mass transport coefficient, S is the RNA–DNA surface-bound substrate (the RNA–DNA heteroduplex), ES is the surface enzyme–substrate complex (the RNase H–heteroduplex complex), k_a and k_d are the Langmuir adsorption and desorption rate constants, S^* is the surface product (ssDNA), and k_{cat} is the surface reaction rate for the enzyme complex (Fang et al., 2005). The ratio k_a/k_d is the Langmuir adsorption coefficient K_{Ads}. The steady-state mass transport coefficient (k_m) can also be written as D/δ, where D is the diffusion coefficient for the enzyme and δ is the steady-state diffusion layer thickness (Bourdillon et al., 1999).

If the surface coverages for the three surface species S, ES, and S^* are denoted as Γ_S, Γ_{ES}, and Γ_{S*}, respectively, and the total number of surface sites is Γ_{tot}, then the surface kinetics equations can be expressed in terms of the relative surface coverages $\theta_x = \Gamma_x/\Gamma_{tot}$, where $x = S$, ES, or S^*:

$$\theta_S + \theta_{ES} + \theta_{S*} = 1 \tag{8.6}$$

$$\frac{d\theta_{ES}}{dt} = \frac{k_a[E]^b(1 - \theta_{ES} - \theta_{S*}) - (k_d + k_{cat})\theta_{ES}}{1 + \beta(1 - \theta_{ES} - \theta_{S*})} \tag{8.7}$$

$$\frac{d\theta_{S*}}{dt} = k_{cat}\theta_{ES} \tag{8.8}$$

In Eq. (8.7), $[E]^b$ is the bulk enzyme concentration and β is the dimensionless diffusion parameter (Schuck and Minton, 1996; Bourdillon et al., 1999) mentioned previously and defined by Eq. (8.9):

$$\beta = \frac{k_a \Gamma_{tot}}{k_m} = \frac{k_a \Gamma_{tot} \delta}{D} \tag{8.9}$$

These equations were derived in a series of recent papers (Fang et al., 2005; Lee et al., 2005b), and can be solved by simple Euler integration methods with the initial conditions $\theta_S = 1$ and $\theta_{ES} = \theta_{S^*} = 0$ to yield three time-dependent surface coverages $\theta_{ES}(t)$, $\theta_{S^*}(t)$, and $\theta_S(t)$. We have used these computer-generated solutions to Eqs. (8.6)–(8.8) to fit our SPRI measurements and obtain the four constants k_a, k_d, k_{cat}, and β (Fang et al., 2005; Lee et al., 2005b).

SPRI measurements are just one of few possible experimental methods that have been employed to monitor surface enzyme processes in real-time. While most research efforts have focused on the use of fluorescence based detection methods (Gaspers et al., 1994, 1995; Jervis et al., 1997; Tachi-iri et al., 2000; Bosma et al., 2003; Tawa and Knoll, 2004), SPR-based techniques are an alternative tool for measuring time-dependent surface coverages of untagged adsorbing species that provides excellent discrimination against possible bulk signal contributions (Peterson et al., 2000, 2002; Goodrich et al., 2004b; Kanda et al., 2004; Shumaker-Parry and Campbell, 2004; Shumaker-Parry et al., 2004; Wegner et al., 2004b; Fang et al., 2005; Lee et al., 2005b).

Perhaps the most effective experimental method for the quantification of surface enzyme processes is to use a combination of SPR and fluorescence experiments. For example, Knoll and co-workers have recently used the combination of SPR and surface plasmon fluorescence spectroscopy (SPFS), which is a very sensitive fluorescence method for detecting labeled surface biochemical species (Yu et al., 2003; Tawa and Knoll, 2004; Yao et al., 2004; Stengel and Knoll, 2005). Kim et al. (2002) have also employed a combination of SPR and SPFS to create separate profiles of the enzyme adsorption and substrate cleavage steps. Moreover, we have recently demonstrated that the combination of time-resolved SPRI and SPFS measurements can be used for the study of RNase H hydrolysis of heteroduplexes on DNA microarrays (Fang et al., 2005).

Figure 8.9 plots the theoretical curves for $\theta_{ES}(t)$, $\theta_S(t)$, and $\theta_{S^*}(t)$ for the reaction of a 2 nM RNase H solution with a monolayer of RNA–DNA heteroduplexes formed previously by the hybridization adsorption of ssRNA onto a ssDNA microarray element. These curves were obtained by fitting SPRI data to the solutions of Eqs. (8.6)–(8.8) (Fang et al., 2005).

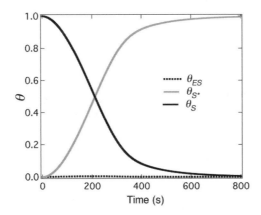

Figure 8.9. Theoretical analysis for the RNase H hydrolysis reaction of a surface immobilized RNA–DNA heteroduplex. The solid lines represent the simulated curves for $\theta_{ES}(t)$, $\theta_S(t)$, and $\theta_{S^*}(t)$ obtained using Eqs. (8.6)–(8.8) with the parameters $k_a = 3.4 \times 10^6$ M^{-1} s^{-1}, $[E]^b = 2$ nM, $k_d = 0.1$ s^{-1}, $k_{cat} = 1.0$ s^{-1}, and $\beta = 180$.

The values of the four reaction parameters in the model obtained from this fit are $k_a = 3.4 \times 10^6$ M^{-1} s^{-1}, $k_d = 0.1$ s^{-1}, $k_{cat} = 1.0$ s^{-1}, and a diffusion parameter $\beta = 180$ for the SPRI measurements (Fang et al., 2005). Note that the enzyme–substrate complex coverage (θ_{ES}) remains very small throughout the course of the surface reaction. This is because the surface catalysis rate constant (k_{cat}) is much larger than the surface adsorption rate ($k_a[E]^b$), so that the enzyme reacts very quickly and is released from the surface. In fact, if we look at the Langmuir adsorption coefficient obtained from k_a and k_d ($K_{Ads} = 3.4 \times 10^7$ M^{-1}), we find that $C_{0.5}$ is approximately 30 nM, so that even the equilibrium enzyme surface coverage would be low at a bulk target DNA concentration of 1 nM. However, the k_{cat} value of 1 s^{-1} for RNase H is approximately 10 times slower than that observed for the RNase H reaction in solution (Hogrefe et al., 1990; Keck et al., 1998), so improvements in the surface attachment chemistry are still possible.

8.5. ULTRASENSITIVE DNA DETECTION WITH ENZYMATICALLY AMPLIFIED SPRI

General Methodology. Now that we have characterized the surface RNase H reaction, we are in the position to use this surface enzyme process to enhance the sensitivity of the SPRI measurements. A generalized surface destruction/amplification process that can be used on microarrays is depicted schematically in Figure 8.10. A surface containing attached probe molecules is exposed to a solution containing target molecules and an enzyme (e.g., RNase H, Exo III). After a target molecule binds to a probe molecule, the enzyme recognizes and binds to the surface-bound target–probe complex. The enzyme then destroys the probe, releasing both itself and the target molecule back into solution. The released target is now free to bind to another surface probe molecule, which can subsequently be destroyed by an enzyme molecule. The repeated target binding/enzymatic destruction/target release will result in a large amount of surface probe removal with a very small number of target molecules. This probe loss can be observed in the SPRI signal, although, in principle, this amplification methodology can work with a number of different surface-based detection techniques on various substrates.

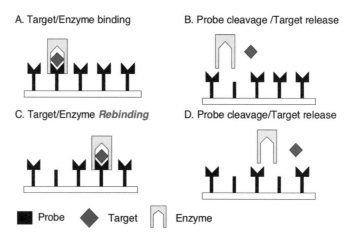

Figure 8.10. Schematic presentation of a general surface enzymatic amplification procedure.

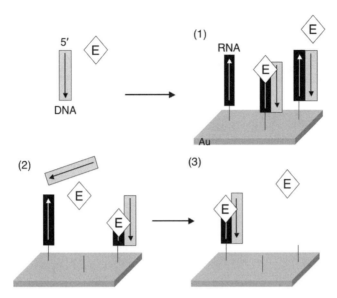

Figure 8.11. Schematic outlining the RNase H amplified SPRI detection of DNA using RNA microarrays. See insert for color representation of this figure.

This amplification method has been specifically applied to the RNase H hydrolysis of RNA microarrays (Goodrich et al., 2004a,b; Lee et al., 2005c), as well as to the Exo III hydrolysis of dsDNA microarrays (Lee et al., 2005a). Figure 8.11 shows the specific amplification scheme for the RNase H system. When the ssRNA microarray is exposed to a solution containing both target DNA and RNase H, the DNA will first bind to a complementary RNA molecule in a specific microarray element on the surface and form an RNA–DNA heteroduplex (step 1). RNase H will then bind to the heteroduplex and selectively hydrolyze the RNA probe. This hydrolysis process releases both the enzyme and the target DNA back into solution (step 2). The released DNA molecule is then free to bind to another RNA probe molecule in the same array element. Either the same RNase H molecule or a different one can bind to the new surface bound heteroduplex and again hydrolyze the probe RNA molecule (step 3). After a sufficient amount of time, a very small amount of target DNA can remove all of the RNA probe molecules from the surface with this cyclic binding and hydrolysis process.

Femtomolar Oligonucleotide Detection. To determine the sensitivity of enzymatically amplified SPRI measurements, we performed a number of experiments to detect ssDNA 16mers at various concentrations. Figure 8.12 shows an SPRI image (a) and line profile (c) obtained upon the hybridization adsorption of a target 16mer ssDNA (100 nM) onto ssRNA microarray elements in a two-component array (the second type of array element was a 16mer ssDNA negative control). A 1.7% increase in $\Delta\%R$ was observed upon hybridization adsorption as expected from the Langmuir adsorption isotherm. Also shown in Figure 8.12 are an SPRI image (b) and a line profile (c) taken after the microarray was exposed for 30 minutes to a solution containing 100 pM ssDNA and 8 nM of RNase H. A significant (−3.4%) loss in the SPRI signal was observed that almost equaled the loss observed from the solution of 500 nM ssDNA and RNase H as previously shown in Figure 8.7. Note that the sample signal loss obtained from the 100 pM DNA solution is substantially larger

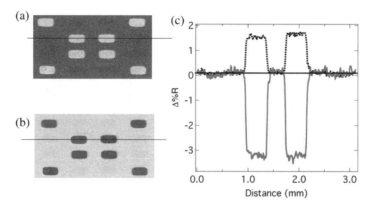

Figure 8.12. (a) An SPR difference image showing sequence specific hybridization of a 100 nM solution of target DNA onto ssRNA microarray elements. (b) An SPR image obtained by subtracting images taken before and after the ssRNA array was exposed to a solution containing 100 pM complementary target DNA and 8 nM RNase H for 30 minutes. (c) Comparison of line profiles in both images demonstrates the advantage of enzymatic amplification. The upper line profile shows an increase in percent reflectivity due to hybridization adsorption of target DNA, while a larger decrease in percent reflectivity is observed in the lower line profile using a much lower concentration (100 pM) of target DNA. The decrease in percent reflectivity corresponds to a complete removal of the ssRNA probes due to repeat cycles of hybridization adsorption of target DNA followed by enzyme hydrolysis.

than the meager $\Delta\%R$ gain observed from the 1 nM ssDNA solution without RNase H shown in Figure 8.5. Given enough time, both the 100 nM DNA solution and the 100 pM solution can completely remove all of the ssRNA from the surface in the presence of RNase H; however, the loss observed in the 500 nM DNA solution occurred at a substantially faster rate.

The initial rate of SPRI signal loss during the enzymatically amplified SPRI experiment can be used as a measure of the ssDNA target concentration. Figure 8.13 plots the time required to reach a 0.2% drop in percent reflectivity for an ssRNA microarray element as a

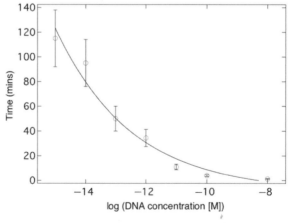

Figure 8.13. Plot of time required to reach an initial 0.2% change in reflectivity during RNase H hydrolysis versus a target DNA concentration ranging from 1 fM to 10 nM. [Reprinted with permission from *J Am Chem Soc* supporting materials **126**:4086–4087. Copyright 2004, American Chemical Society.]

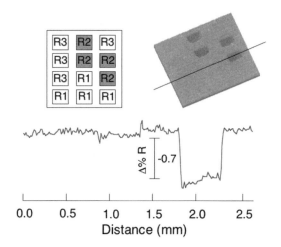

Figure 8.14. An SPR difference image showing the RNase H amplified detection of 1 fM 20mer target DNA using an RNA microarray. The top left is a schematic of the three-component RNA microarray. The line profile taken across the difference image shows a decrease in percent reflectivity ($-0.7 \, \Delta\%R$) at the R2 probe elements only with no changes in the SPR signal observed at the R1 and R3 array elements. [Reprinted with permission from *J Am Chem Soc* **126**:4086–4087. Copyright 2004, American Chemical Society.]

function of target DNA concentration from 10 nM down to 1 fM in the presence of RNase H. As the concentration of target DNA decreases, the time required to reach this SPRI signal level becomes longer due to the drop in the surface concentration of heteroduplexes. A linear relationship between target DNA concentration and time to reach a given loss in SPR signal was observed for concentrations from 1 fM to 10 pM. As expected, lower concentrations of DNA required more time to reach a desired SPR signal level due to the fact that more binding–hydrolysis cycles per DNA molecule were required. For a 10 pM ssDNA solution, 12 ± 2 minutes is required to observe a $-0.2 \, \Delta\%R$; for 1 fM the time required was 120 ± 25 minutes. The lowest concentration detected with enzymatically amplified SPRI was an amazing 1 fM. To verify this detection limit, an SPRI reflectivity difference image was obtained from a three-component RNA microarray after approximately a 4-hour exposure to a solution containing a 1 fM 20mer target DNA complementary to only one of the three RNA array elements in the presence of RNase H (see Figure 8.14). A loss of $-0.7 \, \Delta\%R$ was observed for this array element from this 1 fM solution of target DNA; no loss in SPRI was observed from the other two ssRNA microarray elements. This experiment attests to the very high selectivity and specificity of the surface RNase H reaction. Also note that the reaction time for the hydrolysis reaction is not determined by DNA diffusion to the surface, since the bulk solution is never depleted of target ssDNA. Instead, the time required for the signal loss is limited by the rate of ssDNA diffusion on the surface and the rates of hybridization adsorption and surface catalysis. The improvement of the detection limit from 1 nM for hybridization adsorption to 1 fM for enzymatically amplified SPR is a remarkable factor of 10^6! An estimate of the amount of enzymatic amplification achieved in this experiment was obtained from the stoichiometry of the experiment; 0.5 attomoles of complementary DNA were required to remove approximately 3 femtomoles of RNA probes, so on average every target molecule in solution removed approximately 6000 RNA probes from the surface.

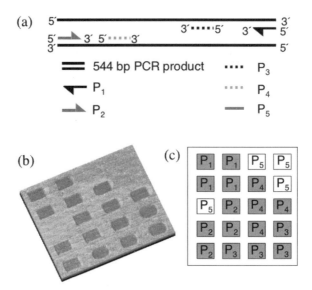

Figure 8.15. The RNase H amplified detection of a 544 bp PCR product of the TSPY gene. (a) Schematic showing the relative binding sites for probe molecule hybridization to the PCR product. (b) SPR difference image obtained by subtracting images taken before and 30 minutes after a 100 pM solution of the PCR product was exposed to the RNA array surface. A decrease in percent reflectivity was observed for the array elements (P_1, P_2, P_3, and P_4) designed to bind to the PCR product. P_5 serves as a negative control. (c) Schematic showing the five-component RNA microarray. [Reprinted with permission from *Anal Chem* **76**:6173–6178. Copyright 2004, American Chemical Society.] See insert for color representation of this figure.

PCR Product Detection. In addition to single-stranded oligonucleotides, enzymatically amplified SPRI can also be used to detect longer strands of dsDNA such as those created by the PCR amplification of a genomic DNA sample. A 544-bp segment of the TSPY gene was obtained from human genomic DNA by PCR amplification (Goodrich et al., 2004b). A five-component RNA array was constructed with probe sequences P_1–P_4 designed to specifically bind to the TSPY PCR product, and P_5 was used as a negative control (see Figure 8.15c). The location of the binding sites for the four probes on the PCR product is shown in Figure 8.15a. A sized and purified sample of the PCR product was then diluted to a concentration of 100 pM and heated to 95°C for 5 minutes to denature the duplexed DNA strands. RNase H was then added to the DNA solution, which was then injected into the 100 μL cell in contact with the RNA microarray surface. Figure 8.15b shows the SPRI differential reflectance image obtained from this experiment; a decrease in percent reflectivity can be seen for all of the probes that were designed to specifically bind to this PCR product, while the P_5 negative control array element showed no change. This experiment illustrates that the RNase H surface enzymatic process can be used to detect longer dsDNA PCR products; the SPRI image obtained from a 100 pM solution of unpurified products gave similar results (Goodrich et al., 2004b).

Direct Detection of Genomic DNA. Finally, the sensitivity of enzymatically amplified SPRI is actually sufficient to directly detect genomic DNA *without* PCR amplification. A three-component RNA microarray was constructed with two RNA probe sequences designed to bind to the TSPY gene (R1 and R2 in Figure 8.16), and a third negative control RNA sequence (R3). Male and female genomic DNA samples were obtained commercially,

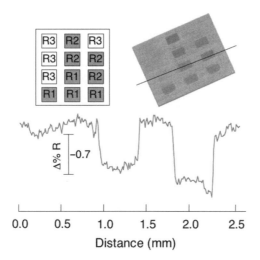

Figure 8.16. An SPR difference image obtained for the detection of 7 fM male genomic DNA using the RNase H amplification method. The upper left inset shows the pattern of the three-component array with the RNA probes R1 and R2 designed to selectively bind to the TSPY gene on the Y chromosome with R3 used as a negative control. The line profile displays a decrease in percent reflectivity for the R1 and R2 array elements where hybridization adsorption followed by RNA probe hydrolysis occurred. [Reprinted with permission from *J Am Chem Soc* **126**:4086–4087. Copyright 2004, American Chemical Society.] See insert for color representation of this figure.

sheared by vortexing and sonication, diluted in buffer, heated to 95°C for 5 min, and then mixed with RNase H. The concentration of the TSPY gene sequences in the male genomic DNA used was estimated to be 7 fM (Goodrich et al., 2004a). The array was then exposed to the genomic DNA for approximately 4 hours and the SPR reflectance difference image in Figure 8.16 was obtained. A significant decrease in SPRI signal (–0.7%) was observed for the two probes specific for the TSPY gene, while no SPRI signal loss was observed in either the negative control or background. As a control, enzymatically amplified SPRI experiments were performed on female genomic DNA under the same conditions; no decrease in SPRI signal from any of the array elements was observed. These experiments unequivocally demonstrate that enzymatically amplified SPRI can be used to directly detect specific DNA sequences from genomic samples.

8.6. SUMMARY AND OUTLOOK

The ability to sensitively detect and identify multiple DNA or RNA sequences in a microarray format without the use of labeling can greatly assist in many ongoing biosensing applications including gene expression analysis, biowarfare detection, and viral identification. The enzymatically amplified SPRI detection described in this chapter highlights the potential of enzymes as invaluable surface bioengineering tools for use in the development of novel microarray-based biosensing methodologies. Other examples of surface enzyme reactions implemented with multielement surface biosensors include ligation (Gerry et al., 1999; Broude et al., 2001; Zhong et al., 2003; Deng et al., 2004), primer extension (Hultin et al., 2005), and single base extension reactions (Giusto and King, 2003) for single nucleotide polymorphism detection.

The most widely used method for the detection and identification of DNA or RNA using DNA microarrays is fluorescence imaging, which has a detection limit typically in the region of 1–10 pM (Zammatteo et al., 2000; Lehr et al., 2003; Livache et al., 2003). Therefore, in order to use this method for the detection of femtomolar DNA concentrations—for example, in genomic DNA studies—enzymatic amplification methods such as PCR are commonly used prior to array exposure (Sengupta et al., 2003; Gonzalez et al., 2004). Alternatively, coupled enzymatic labeling methods such as an ELISA sandwich assay can be used after target adsorption (Kawai et al., 1993; Morata et al., 2003). However, these enzymatic amplification methods are not readily adapted to a multiplexed array format; PCR requires multiple primers for multiple target DNA and is often unable to reproduce the relative concentrations of nucleic acid sequences from complex mixtures, while the enzymatic product detected in ELISA measurements is a solution species that can diffuse into neighboring array elements. An alternative strategy for the ultrasensitive detection of DNA using DNA microarrays is the use of nanoparticle tagging in a sandwich assay format (Storhoff et al., 2004; Bao et al., 2005). In these cases, a 50–200 fM detection limit was achieved only after a second silver staining amplification step.

In this chapter, we have demonstrated that the sensitivity of SPRI for the direct detection and identification of DNA can be lowered down to 1 fM by monitoring the repeated enzymatic hydrolysis of surface RNA–DNA heteroduplexes by RNase H. This method of "enzymatically amplified SPRI" is an extraordinary improvement compared to the nanomolar DNA detection limit of SPRI based solely on hybridization adsorption. This method is successful because one target DNA molecule can destroy multiple surface bound RNA molecules. The technique can be used to directly detect oligonucleotides, PCR fragments, and sequences in unamplified genomic DNA, and will work with all DNA sequences that can form DNA–RNA heteroduplexes (Goodrich et al., 2004a,b). Furthermore, this enzymatic amplification method can be used in conjunction with other detection methods such as fluorescence and electrochemical measurements for even more sensitive DNA detection, perhaps with the ability to detect single molecular binding events.

Although the RNase H amplification technique offers excellent sensitivity and selectivity, it requires the use of an RNA microarray. At present, there are only a few reports on the creation of RNA microarrays currently available in the literature (McCauley et al., 2003; Goodrich et al., 2004a,b; Collett et al., 2005; Lee et al., 2005c). Special care is required during RNA microarray fabrication in order to avoid degradation of the RNA either by nucleases from environmental sources or during the chemical processing of the surface. The design of new array fabrication strategies that require very little *ex situ* handling or chemical processing of the RNA remains a significant challenge that, if met, will greatly promote the application of RNA microarrays in a number of research areas.

Finally, the surface enzyme reaction can be orders of magnitude slower on a surface as compared to that in solution. The surface density, orientation, and spacing from the chip surface of the immobilized substrate as well as the enzyme–substrate binding affinity are some of the many energetic and transport factors that can greatly affect enzymatic activity on a surface. The application of powerful *in situ* surface analytical surface techniques that can directly monitor enzyme activity in combination with a good theoretical understanding is essential for the optimization of these new surface enzymatic based techniques. To date, there are very few quantitative studies considering the kinetics and thermodynamics of enzyme-catalyzed reactions on surfaces; further investigation into these fundamental research areas will undoubtedly accelerate the harnessing of the unique properties of enzymes for ultrasensitive biosensing applications.

ACKNOWLEDGMENT

This research was funded by the National Institute of Health (2RO1 GM059622-04) and the National Science Foundation (CHE-0551935).

REFERENCES

Bao, YP, Martin Huber, M, Wei, T, Marla, SS, Storhoff, JJ, and Müller, UR (2005). SNP identification in unamplified human genomic DNA with gold nanoparticle probes. *Nucleic Acids Res* **33**:e15.

Berchuck, A, Iversen, ES, Lancaster, JM, Pittman, J, Luo, J, Lee, P, Murphy, S, Dressman, HK, Febbo, PG, West, M, Nevins, JR, and Marks, JR (2005). Patterns of gene expression that characterize long-term survival in advanced stage serous ovarian cancers. *Clin Cancer Res* **11**:3686–3696.

Bosma, AY, Ulijn, RV, McConnell, G, Girkin, J, Hallingc, PJ, and Flitsch, SL (2003). Using two photon microscopy to quantify enzymatic reaction rates on polymer beads. *Chem Commun* **22**:2790–2791.

Bourdillon, C, Demaille, C, Moiroux, J, and Saveant, J (1999). Activation and diffusion in the kinetics of adsorption and molecular recognition on surfaces. Enzyme-amplified electrochemical approach to biorecognition dynamics illustrated by the binding of antibodies to immobilized antigens. *J Am Chem Soc* **121**:2401–2408.

Brenner, S, Johnson, M, Bridgham, J, Golda, G, Lloyd, DH, Johnson, D, Luo, S, McCurdy, S, Foy, M, Ewan, M, Roth, R, George, D, Eletr, S, Albrecht, G, Vermaas, E, Williams, SR, Moon, K, Burcham, T, Pallas, M, DuBridge, RB, Kirchner, J, Fearon, K, Mao, J, and Corcoran, K (2000). Gene expression analysis by massively parallel signature sequencing (MPSS) on microbead arrays. *Nat Biotechnol* **18**:630–634.

Brockman, JM, Frutos, AG, and Corn, RM (1999). A multistep chemical modification procedure to create DNA arrays on gold surfaces for the study of protein–DNA interactions with surface plasmon resonance imaging. *J Am Chem Soc* **121**:8044–8051.

Brockman, JM, Nelson, BP, and Corn, RM (2000). Surface plasmon resonance imaging measurements of ultrathin organic films. *Annu Rev Phys Chem* **51**:41–63.

Broude, NE, Woodward, K, Cavallo, R, Cantor, CR, and Englert, D (2001). DNA microarrays with stem–loop DNA probes: Preparation and applications. *Nucleic Acids Res* **29**:e92.

Chen, CC, Teng, LJ, Kaiung, S, and Chang, TC (2005). Identification of clinically relevant viridans streptococci by an oligonucleotide array. *J Clin Microbiol* **43**:1515–1521.

Chinowsky, TM, Quinn, JG, Bartholomew, DU, Kaiser, R, and Elkind, JL (2003). Performance of the Spreeta 2000 integrated surface plasmon resonance affintiy sensor. *Sens Actuators B* **91**:266–274.

Collett, JR, Cho, EJ, Lee, JF, Levy, M, Hood, AJ, Wan, C, and Ellington, AD (2005). Functional RNA microarrays for high-throughput screening of antiprotein aptamers. *Anal Biochem* **338**:113–123.

Corn, RM, and Weibel, SC (2001). Fourier transform surface plasmon resonance. In: Chalmers, J, and Griffiths, PR (eds.), *Handbook of Vibrational Spectroscopy*, Vol. 2. John Wiley & Sons, New York, pp. 1057–1064.

Crowther, JR (1995). ELISA: Theory and practice. In: Crowther, JR (ed.), *Methods in Molecular Biology*, Vol. 42. Humana Press, Totowa, NJ.

Deng, J, Zhang, X, Manga, Y, Zhang, Z, Zhou, Y, Liu, Q, Lu, H, and Fu, Z (2004). Oligonucleotide ligation assay-based DNA chip for multiplex detection of single nucleotide polymorphism. *Biosens Bioelectron* **19**:1277–1283.

Epstein, JR, Biran, I, and Walt, DR (2002). Fluorescence-based nucleic acid detection and microarrays. *Anal Chim Acta* **469**:3–36.

Fang, S, Lee, HJ, Wark, AW, Kim, HM, and Corn, RM (2005). Determination of ribonuclease H surface enzyme kinetics by surface plasmon resonance imaging and surface plasmon fluorescence spectroscopy. *Anal Chem* **77**:6528–6534.

Frutos, AG, Weibel, SC, and Corn, RM (1999). Near infrared surface plasmon resonance measurements of ultrathin films. 2. Fourier transform SPR spectroscopy. *Anal Chem* **71**:3935–3940.

Gaspers, PB, Robertson, CR, and Gast, AP (1994). Enzymes on immobilized substrate surfaces: Diffusion. *Langmuir* **10**:2699–2704.

Gaspers, PB, Gast, AP, and Robertson, CR (1995). Enzymes on immobilized substrate surfaces: reaction. *J Colloid Interface Sci* **172**:518–529.

Gerion, D, Chen, F, Kannan, B, Fu, A, Parak, WJ, Chen, DJ, Majumdar, A, and Alivisatos, AP (2003). Room-temperature single-nucleotide polymorphism and multiallele DNA detection using fluorescent nanocrystals and microarrays. *Anal Chem* **75**:4766–4772.

Gerry, NP, Witowski, NE, Day, J, Hammer, RP, Barany, G, and Barany, F (1999). Universal DNA microarray method for multiplex detection of low abundance point mutations. *J Mol Biol* **292**:251–262.

Giusto, DD, and King, GC (2003). Single base extension (SBE) with proofreading polymerases and phosphorothioate primers: Improved fidelity in single-substrate assays. *Nucleic Acids Res* **31**:e7.

Gonzalez, SF, Krug, MJ, Nielsen, ME, Santos, Y, and Call, DR (2004). Simultaneous detection of marine fish pathogens by using multiplex PCR and DNA microarrays. *J Clin Microbiol* **42**:1414–1419.

Goodrich, TT, Lee, HJ, and Corn, RM (2004a). Direct detection of genomic DNA by enzymatically amplified SPR imaging measurements of RNA microarrays. *J Am Chem Soc* **126**:4086–4087.

Goodrich, TT, Lee, HJ, and Corn, RM (2004b). Enzymatically amplified surface plasmon resonance imaging method using RNase H and RNA microarrays for the ultrasensitive detection of nucleic acids. *Anal Chem* **76**:6173–6178.

Guedon, P, Livache, T, Martin, F, Lesbre, F, Roget, A, Bidan, G, and Levy, Y (2000). Characterization and optimization of a real-time, parallel, label-free, polypyrrole-based DNA sensor by surface plasmon resonance imaging. *Anal Chem* **72**:6003–6009.

Hanken, DG, Jordan, CE, Frey, BL, and Corn, RM (1998). Surface plasmon resonance measurements of ultrathin organic films at electrode surfaces. In: Bard, A, and Rubinstein, I (eds.), *Electroanalytical Chemistry: A Series of Advances*, Vol. 20, Marcel Dekker, New York, pp. 141–225.

He, L, Musick, MD, Nicewarner, SR, Salinas, FG, Benkovic, SJ, Natan, MJ, and Keating, CD (2000). Colloidal Au-enhanced surface plasmon resonance for ultrasensitive detection of DNA hybridization. *J Am Chem Soc* **122**:9071–9077.

Hogrefe, HH, Hogrefe, RI, Walder, RY, and Walder, JA (1990). Kinetic analysis of *Escherichia coli* RNase H using DNA–RNA–DNA/DNA substrates. *J Biol Chem* **265**:5561–5566.

Homola, J, Yee, SS, and Gauglitz, G (1999). Surface plasmon resonance sensors: Review. *Sens Actuators B* **54**:3–15.

Hughes, TR, Mao, M, Jones, A, Burchard, J, Marton, MJ, Shannon, KW, Lefkowitz, SM, Ziman, M, Schelter, JM, Meyer, MR, Kobayashi, S, Davis, C, Dai, H, He, YD, Stephaniants, SB, Cavet, G, Walker, WL, West, A, Coffey, E, Shoemaker, DD, Stoughton, R, Blanchard, AP, Friend, SH, and Linsley, PS (2001). Expression profiling using microarrays fabricated by an ink-jet oligonucleotide synthesizer. *Nat Biotechnol* **19**:342–347.

Hultin, E, Kaller, M, Ahmadian, A, and Lundeberg, J (2005). Competitive enzymatic reaction to control allele-specific extensions. *Nucleic Acids Res* **33**:e48.

Jervis, EJ, Haynes, CA, and Kilburn, DG (1997). Surface diffusion of cellulases and their isolated binding domains on cellulose. *J Biol Chem* **272**:24016–24023.

Kanda, V, Kariuki, JK, Harrison, DJ, and McDermott, MT (2004). Label-free reading of microarray-based immunoassays with surface plasmon resonance imaging. *Anal Chem* **76**:7257–7262.

Katayanagi, K, Okumura, M, and Morikawa, K (1993). Crystal structure of escherichia coli RNase HI in complex with Mg^{2+} at 2.8 Å resolution: Proof for a single Mg^{2+} binding site. *Proteins: Struct Funct Genet* **17**:337–346.

Kawai, S, Maekawajiri, S, and Yamane, A (1993). A simple method of detecting amplified DNA with immobilized probes on microtiter wells. *Anal Biochem* **209**:63–69.

Keck, JL, Goedken, ER, and Marqusee, S (1998). Activation/attenuation model for RNase H. *J Biol Chem* **273**:34128–34133.

Kim, J-H, Roy, S, Kellis, JT, Jr., Poulose, AJ, Gast, AP, and Robertson, CR (2002). Protease adsorption and reaction on an immobilized substrate surface. *Langmuir* **18**:6312–6318.

Knoll, W (1998). Interfaces and thin films as seen by bound electromagnetic waves. *Annu Res Phys Chem* **49**:569–638.

Kwok, P-Y (2001). Methods for genotyping single nucleotide polymorphisms. *Annu Rev Genomics Hum Genet* **2**:235–258.

Kyo, M, Yamamoto, T, Motohashi, H, Kamiya, T, Kuroita, T, Tanaka, T, Engel, JD, Kawakami, B, and Yamamoto, M (2004). Evaluation of MafG interaction with Maf recognition element arrays by surface plasmon resonance imaging technique. *Genes to Cells* **9**:153–164.

Lee, HJ, Goodrich, TT, and Corn, RM (2001). SPR imaging measurements of 1-D and 2-D DNA microarrays created from microfluidic channels on gold thin films. *Anal Chem* **73**:5525–5531.

Lee, HJ, Li, Y, Wark, AW, and Corn, RM (2005a). Enzymatically amplified SPR imaging detection of DNA by exonuclease III digestion of DNA microarrays. *Anal Chem* **77**:5096–5100.

Lee, HJ, Wark, AW, Goodrich, TT, Fang, S, and Corn, RM (2005b). Surface enzyme kinetics for biopolymer microarrays: A combination of Langmuir and Michaelis–Menten concepts. *Langmuir* **21**:4050–4057.

Lee, HJ, Wark, AW, Li, Y, and Corn, RM (2005c). Fabricating RNA microarrays with RNA-DNA surface ligation chemistry. *Anal Chem* **77**:7832–7837.

Lee, HJ, Yan, Y, Marriott, G, and Corn, RM (2005d). Quantitative functional analysis of protein complexes on surfaces. *J Physiol* **563**:61–71.

Lehr, H-P, Reimann, M, Brandenburg, A, Sulz, G, and Klapproth, H (2003). Real-time detection of nucleic acid interactions by total internal reflection fluorescence. *Anal Chem* **75**:2414–2420.

Levicky, R, and Horgan, A (2005). Physicochemical perspectives on DNA microarray and biosensor technologies. *Trends Biotechnol* **23**:143–149.

Liebermann, T, Knoll, W, Sluka, P, and Herrmann, R (2000). Complement hybridization from solution to surface-attached probe-oligonucleotides observed by surface-plasmon-field-enhanced fluorescence spectroscopy. *Colloids Surf A* **169**:337–350.

Lin, B, Vora, GJ, Thach, D, Walter, E, Metzgar, D, Tibbetts, C, and Stenger, DA (2004). Use of oligonucleotide microarrays for rapid detection and serotyping of acute respiratory disease-associated adenoviruses. *J Clin Microbiol* **42**:3232–3239.

Livache, T, Maillart, E, Lassalle, N, Mailley, P, Corso, B, Guedon, P, Roget, A, and Levy, Y (2003). Polypyrrole based DNA hybridization assays: Study of label free detection processes versus fluorescence on microchips. *J Pharm Biomed Anal* **32**:687–696.

Lockhart, DJ, and Winzeler, EA (2000). Genomics, gene expression and DNA arrays. *Nature* **405**:827–836.

Lyon, LA, Holliway, WD, and Natan, MJ (1999). An improved surface plasmon resonance imaging apparatus. *Rev Sci Instrum* **70**:2076–2081.

McCauley, TG, Hamaguchi, N, and Stanton, M (2003). Aptamer-based biosensor arrays for detection and quantification of biological macromolecules. *Anal Biochem* **319**:244–250.

Morata, P, Queipo-Ortuno, MI, Reguera, JM, Garcia-Ordonez, MA, Cardenas, A, and Colmenero, JD (2003). Development and evaluation of a PCR-enzyme-linked immunosorbent assay for diagnosis of human brucellosis. *J Clin Microbiol* **41**:144–148.

Nakamura, H, Oda, Y, Iwai, S, Inoue, H, Ohtsuka, E, Kanaya, S, Kimura, S, Katsuda, C, Katayangi, K, Morikawa, K, Miyashiro, H, and Ikehara, M (1991). How does RNase H recognize a DNA–RNA hybrid. *Proc Natl Acad Sci USA* **88**:11535–11539.

Nelson, BP, Grimsrud, TE, Liles, MR, Goodman, RM, and Corn, RM (2001). Surface plasmon resonance imaging measrements of DNA and RNA hybridization adsorption onto DNA microarrays. *Anal Chem* **73**:1–7.

Nelson, BP, Liles, MR, Frederick, K, Goodman, RM, and Corn, RM (2002). Label-free detection of 16s rRNA hybridization onto DNA arrays using surface plasmon resonance imaging. *Environ Microbiol* **4**:735–743.

Nelson, PT, Baldwin, DA, Scearce, LM, Oberholtzer, JC, Tobias, JW, and Mourelatos, Z (2004). Microarray-based, high-throughput gene expression profiling of microRNAs. *Nature Methods* **1**:155–161.

Okumura, A, Sato, Y, Kyo, M, and Kawaguchi, H (2005). Point mutation detection with the sandwich method employing hydrogel nanospheres by the surface plasmon resonance imaging technique. *Anal Biochem* **339**:328–337.

Pease, AC, Solas, D, Sullivan, EJ, Cronin, MT, Holmes, CP, and Fodor, SPA (1994). Light-generated oligonucleotide arrays for rapid DNA sequence analysis. *Proc Natl Acad Sci USA* **91**:5022–5026.

Peterson, AW, Heaton, RJ, and Georgiadis, RM (2000). Kinetic control of hybridization in surface immobilized DNA monolayer films. *J Am Chem Soc* **122**:7837–7838.

Peterson, AW, Wolf, LK, and Georgiadis, RM (2002). Hybridization of mismatched or partially matched DNA at surfaces. *J Am Chem Soc* **124**:14601–14607.

Rella, R, Spadavecchia, J, Manera, MG, Siciliano, P, Santino, A, and Mita, G (2004). Liquid phase SPR imaging experiments for biosensors applications. *Biosens Bioelectron* **20**:1140–1148.

Schena, M, Shalon, D, Davis, RW, and Brown, PO (1995). Quantitative monitoring of gene expression patterns with a complementary DNA microarray. *Science* **270**:467–470.

Schuck, P, and Minton, AP (1996). Analysis of mass transport-limited binding kinetics in evanescent wave biosensors. *Anal Biochem* **240**:262–272.

Sengupta, S, Onodera, K, Lai, A, and Melcher, U (2003). Molecular detection and identification of influenza viruses by oligonucleotide microarray hybridization. *J Clin Microbiol* **41**:4542–4550.

Shumaker-Parry, JS, and Campbell, CT (2004). Quantitative methods for spatially resolved adsorption/desorption measurements in real time by surface plasmon resonance microscopy. *Anal Chem* **76**:907–917.

Shumaker-Parry, JS, Zareie, MH, Aebersold, R, and Campbell, CT (2004). Microspotting streptavidin and double stranded DNA arrays on gold for high-throughput studies of protein–DNA interactions by surface plasmon resonance microscopy. *Anal Chem* **76**:918–929.

Smith, EA, Kyo, M, Kumasawa, H, Nakatani, K, Saito, I, and Corn, RM (2002). Chemically induced hairpin formation in DNA monolayers. *J Am Chem Soc* **124**:6810–6811.

Smith, EA, and Corn, MR (2003). Surface plasmon resonance imaging as a tool to monitor biomolecular interactions in an array based format. *Appl Spectrosc* **57**:320A–332A.

Smith, EA, Erickson, MG, Ulijasz, AT, Weisblum, B, and Corn, RM (2003a). Surface plasmon resonance imaging of transcription factor proteins: Interactions of bacterial response regulators with DNA arrays on gold films. *Langmuir* **19**:1486–1492.

Smith, EA, Thomas, WD, Kiessling, LL, and Corn, RM (2003b). Surface plasmon resonance imaging studies of protein–carbohydrate interactions. *J Am Chem Soc* **125**:6140–6148.

Stears, RL, Martinsky, T, and Schena, M (2003). Trends in microarray analysis. *Nat Med* **9**:140–145.

Stengel, G, and Knoll, W (2005). Surface plasmon field-enhanced fluorescence spectroscopy studies of primer extension reactions. *Nucleic Acids Res* **33**:e69.

Storhoff, JJ, Marla, SS, Bao, P, Hagenow, S, Mehta, H, Lucas, A, Garimella, V, Patno, T, Buckingham, W, Cork, W, and Müller, UR (2004). Gold nanoparticle-based detection of genomic DNA targets on microarrays using a novel optical detection system. *Biosens Bioelectron* **19**:875–883.

Stoughton, RB (2005). Applications of DNA microarrays in biology. *Annu Rev Biochem* **74**:53–82.

Szabo, A, Stolz, L, and Granzow, R (1995). Surface plasmon resonance and its use in biomolecular interaction analysis (BIA). *Curr Opin Struct Biol* **5**:699–705.

Tachi-iri, Y, Ishikawa, M, and Hirano, K.-i. (2000). Investigation of the hydrolysis of single DNA molecules using fluorescence video microscopy. *Anal Chem* **72**:1649–1656.

Tawa, K, and Knoll, W (2004). Mismatching base-pair dependence of the kinetics of DNA–DNA hybridization studied by surface plasmon fluorescence spectroscopy. *Nucleic Acids Res* **32**:2372–2377.

Thomson, JM, Parker, J, Perou, CM, and Hammond, SM (2004). A custom microarray platform for analysis of microRNA gene expression. *Nature Methods* **1**:47–53.

van Hal, NLW, Vorst, O, van Houwelingen, AMML, Kok, EJ, Peijnenburg, A, Aharoni, A, van Tunen, AJ, and Keijer, J (2000). The application of DNA microarrays in gene expression analysis. *J Biotechnol.* **78**:271–280.

Volokhov, D, Chizhikov, V, Chumakov, K, and Rasooly, A (2003). Microarray-based identificaiton of thermophilic *Campylovacter jejuni*, *C. coli*, *C. lari*, and *C. upsaliensis*. *J Clin Microbiol* **41**:4071–4080.

Wang, D, Coscoy, L, Zylberberg, M, Avila, PC, Boushey, HA, Ganem, D, and DeRisi, JL (2002). Microarray-based detection and genotyping of viral pathogens. *Proc Natl Acad Sci USA* **99**:15687–15692.

Wark, AW, Lee, HJ, and Corn, RM (2005). Long-range surface plasmon resonance imaging for bioaffinity sensors. *Anal Chem* **77**:3904–3907.

Wegner, GJ, Lee, HJ, and Corn, RM (2002). Characterization and optimization of peptide arrays for the study of epitope–antibody interactions using surface plasmon resonance imaging. *Anal Chem* **74**:5161–5168.

Wegner, GJ, Lee, HJ, Marriott, G, and Corn, RM (2003). Fabrication of histidine-tagged fusion protein arrays for surface plasmon resonance imaging studies of protein–protein and protein–DNA interactions. *Anal Chem* **75**:4740–4746.

Wegner, GJ, Lee, HJ, and Corn, RM (2004a). Surface plasmon resonance imaging measurements of DNA, RNA, and protein interactions to biomolecular arrays. In: Kambhampati, D. (ed.), *Protein Microarray Technology*. Wiley-VCH, Weinheim, pp. 107–129.

Wegner, GJ, Wark, AW, Lee, HJ, Codner, E, Saeki, T, Fang, S, and Corn, RM (2004b). Real-time SPR imaging measurements for the multiplexed determination of protein adsorption/desorption kinetics and surface enzymatic reactions on peptide microarrays. *Anal Chem* **76**:5677–5684.

Wilkop, T, Wang, Z, and Cheng, Q (2004). Analysis of μ-contact printed protein patterns by SPR imaging with a LED light source. *Langmuir* **20**:11141–11148.

Winzeler, EA, Richards, DR, Conway, AR, Goldstein, AL, Kalman, S, McCullough, MJ, McCusker, JH, Stevens, DA, Wodicka, L, Lockhart, DJ, and Davis, RW (1998). Direct allelic variation scanning of the yeast genome. *Science* **281**:1194–1197.

Yao, D, Yu, F, Kim, J, Scholz, J, Nielsen, PE, Sinner, E, and Knoll, W (2004). Surface plasmon field-enhanced fluorescence spectroscopy in PCR product analysis by peptide nucleic acid probes. *Nucleic Acids Res* **32**:e177.

Yu, F, Yao, D, and Knoll, W (2003). Surface plasmon field-enhanced fluorescence spectroscopy studies of the interaction between an antibody and its surface-coupled antigen. *Anal Chem* **75**:2610–2617.

Zamaratski, E, Pradeepkumar, PI, and Chattopadhyaya, J (2001). A critical survey of the structure–function of the antisense oligo/RNA heteroduplex as substrate for RNase H. *J Biochem Biophys Methods* **48**:189–208.

Zammatteo, N, Jeanmart, L, Hamels, S, Courtois, S, Louette, P, Hevesi, L, and Remacle, J (2000). Comparison between different strategies of covalent attachment of DNA to glass surfaces to build DNA microarrays. *Anal Biochem* **280**:143–150.

Zhong, X, Reynolds, R, Kidd, JR, Kidd, KK, Jenison, R, Marlar, RA, and Ward, DC (2003). Single-nucleotide polymorphism genotyping on optical thin-film biosensor chips. *Proc Natl Acad Sci USA* **100**:11559–11564.

CHAPTER 9

ULTRASENSITIVE ANALYSIS OF METAL IONS AND SMALL MOLECULES IN LIVING CELLS

RICHARD B. THOMPSON

9.1. INTRODUCTION

Metal ions are essential nutrients and play a host of roles in living organisms. While the participation of metal ions such as iron in respiration and other metals in enzymatic reactions has been known for a century, many other functions are just beginning to be elucidated. This is particularly true for so-called "trace" metal ions such as zinc, copper, selenium, vanadium, molybdenum, and others. The explosion in understanding of the signaling functions of calcium has been achieved due to the invention of fluorescent probes both sensitive and selective enough to permit free calcium ion concentrations to be quantitatively imaged in the microscope (Tsien, 1989). The quantitative imaging of metal ions (and other analytes) at even lower levels requires a new generation of probes with different selectivity and higher sensitivity. Indeed, much remains to be done in the study of other classes of analyte. The goal of this chapter is to introduce some of the approaches emerging in this field.

9.2. BOUNDARY CONDITIONS

To understand the trafficking in the various metal ions in cells (e.g., their distribution, what carriers they may be bound to, and their oxidation state), ideally one would like to be able to image the positions and concentrations of the ions specifically in their various states throughout the cell, with infinite time resolution and single atom sensitivity. Of course, doing all this simultaneously is infeasible at the current state of the art, but one can approach some of these desiderata individually already. Ultimately, this is a problem in chemical analysis. For our purposes we will focus on optical methods of analysis, because at present NMR does not yet offer image resolution comparable to optical microscopy (~ 100 nm), nor do most of the higher-energy spectroscopies such as x-ray fluorescence, Mössbauer, and the like. While spectroscopies coupled to electron microscopy offer potentially sub-Ångstrom resolution, the need to operate *in vacuo* makes them less desirable for biologically active specimens. Fascinating work has appeared recently on the use of mass spectroscopy for chemical analysis in single cells (Hummon et al., 2003), but its application to high-resolution imaging in cells remains to be achieved.

Among optical methods, Raman scattering microscopes are already available, but Raman (with the important exceptions of resonance and surface-enhanced Raman spectroscopies)

New Frontiers in Ultrasensitive Bioanalysis. Edited by Xiao-Hong Nancy Xu
Copyright © 2007 John Wiley & Sons, Inc.

is intrinsically insensitive and thus less useful for most trace analytes. Substantial recent progress has been made with second harmonic generation in imaging, but probes have not yet been developed which couple the recognition of a trace analyte to the nonlinear optical properties of a probe (Milliard et al., 2003); however, membrane potentials have been measured with this technique. Certainly, absorption spectroscopy has been coupled with microscopy for decades but, except in special circumstances, does not offer high sensitivity. The development of means to manufacture conducting particles with dimensions of light wavelength and smaller has opened up many new approaches. In particular, 100-nm particles of gold are excellent scatterers, and they have been developed for array applications (Trulson et al., 1998); coupling molecular recognition of a trace analyte to a perturbation of the scattering within the cellular milieu would appear to be a challenge. Other nanoparticles made of semiconductors exhibit size-dependent luminescence and are much less prone to photobleaching, fueling interest in them as fluorescent labels and sensors (Brauns and Murphy, 1997). Infrared and longer wavelengths [e.g., the so-called THz regime of hundreds of micron wavelength (Appleby et al., 2004)] also appear difficult to selectively couple to trace analytes and necessarily offer poorer resolution. Optical tomographies, while currently focused on imaging mesoscopic targets such as breast tumors, seem to have been little explored for analytical applications, probably because they rely for contrast on bulk properties of the tissue (Chance et al., 2005; Tuchin et al., 2005). All of the foregoing is not intended to minimize the utility of these techniques nor preclude their use; we only wish to indicate potential problems in employing them for ultrasensitive analysis.

By comparison, the dominant technique in the field is fluorescence microscopy using molecular probes that couple recognition of the analyte (metal ion, or whatever) to some fluorescence observable. The archetypical analytes are pH and calcium, but literally hundreds of probes have been developed over decades for scores of different analytes (Wolfbeis, 1991; Haugland, 2005; Thompson, 2005). The development of confocal and multiphoton microscopies (Denk et al., 1990; Egner and Hell, 2005) has enhanced the resolution of the technique to 80 nm, significantly better than visible light wavelength. The coupling of fluorescence labeling to biomolecularly based recognition by molecules such as antibodies (Coons et al., 1942) and nucleic acids has also had an enormous impact in detection of particular molecules on and within cells but, with notable exceptions, has been less useful for metal ion analysis (Darwish and Blake, 2002; Yang and Ellington, 2005).

9.3. THE CELLULAR MILIEU

From an analytical standpoint, the interior of the cell remains an enormous challenge: Even an organism as simple as *E. coli* contains a few hundred different small (e.g., nonpolymeric) molecules and ions, as well as a few thousand enzymes and a significant complement of nucleic acids. Even a simple eukaryotic cell is 10-fold more complex; furthermore, it is subdivided into compartments such as mitochondria, chloroplasts, lysosomes, and the nucleus. Not only do these intracellular compartments add complexity to the problem of introducing probes, but their interiors may be at a different pH (lysosomes) or propensity for oxidation than the rest of the cytoplasm. It is well to remember, moreover, that a typical cell cytoplasm is not a dilute, homogenous solution of low viscosity where translational diffusion happens rapidly. Rather, it has high concentrations (in g/mL) of proteins, including those that make up the cytoskeleton, in addition to nucleic acids and the aforesaid organelles. In extended cells like neurons, diffusive transport is otherwise so slow that molecular transport along axons is an important field of study. While cells evidently expend substantial energy maintaining intracellular conditions of pH, oxygen tension, and the levels of a host of other small molecules at a stable level, conditions in the cell or within a tissue can, nevertheless, change dramatically. For instance, under conditions of oxygen deprivation, brain and muscle tissue become substantially acidotic, the latter due to lactate buildup. Working in plant cells offers an additional degree of complexity due to the presence of the cell wall. In multicellular organisms, cells are differentiated and play a variety of roles, and thus they are chemically and morphologically distinct.

Finally, there is a dynamic element that must be taken into account—namely, that many of the most interesting questions in cell biology are those involving changes in the cell and its response to stimuli. The recent fascination with studies of gene expression in response to various stimuli largely overlooks other changes in the cell, because there are few tools to study them. For instance, activity levels of the enzyme HMG CoA reductase (which catalyzes the first committed step in the biosynthesis of cholesterol and is a main control point for cholesterol homeostasis) are modulated not only by expressing its gene to increase amounts of the protein, but also by changing the rate at which the enzyme is degraded, phosphorylating it, or competitively inhibiting it, none of which requires turning a gene on or off (Goldstein and Brown, 1990). Because of its topological as well as chemical complexity, the cell remains a very challenging target for study.

9.4. NEW PROBE TARGETS

As mentioned above, selective fluorescent probes exist for a relatively modest proportion of analytes, and in fact the metal ions have garnered more attention than many other interesting analytes. Table 9.1 lists several analytes for which fluorescent probes have been developed;

Table 9.1. Several Analytes for which Fluorescent Probes have been Developed

Target Analyte	Probe(s)	Reference(s)
pH	Scores	Haugland (2005)
Ca(II)	Scores	Grynkiewicz et al. (1985), Haugland (2005)
Zn(II)	Dozens	Hirano et al. (2004), Jiang and Guo (2004), Thompson et al. (2002b)
Oxygen	Long-lived fluorophores	Demas and DeGraff (2001), Lippitsch et al. (1988)
Mg(II)	Mag-Fura, others	Haugland (2005)
F-actin	Phallotoxin conjugates	Haugland (2005)
Na/K ATPase	Ouabain conjugates	Haugland (2005)
Liver alcohol dehydrogenase	Auramine O	Conrad et al. (1970)
Carbonic anhydrase	sulfonamides	Chen and Kernohan (1967), Fierke and Thompson (2001)
GABA receptors	Muscimol conjugates	Haugland (2005)
Glucose	Boronate derivatives	DiCesare and Lakowicz (2002)
Lactate	Protein biosensor	D'Auria et al. (2000)
Maltose	Protein biosensor	Li and Cass (1991)
Glutamine	Protein biosensor	Dattelbaum and Lakowicz (2001)

the list is not meant to be exhaustive. Certainly, there are many probes for analytes such as pH, O_2, and Cl; metals such as Ca, Mg, Zn, Cu; and lipids, nucleic acids (nonspecific and nonspecific), and protein (specific and nonspecific). Perhaps what is more interesting is the lack of probes for a host of very interesting small molecules, many of them involved in signaling, such as neurotransmitters, leukotrienes, and sterols; the cAMP and IP_3 probes stand out by comparison. Some of the key intermediary metabolites such as pyruvate, lactate, phosphoribosyl pyrophosphate, ribulose bis phosphate, and glucose would also be of interest in monitoring the energy and metabolic status of cells.

Among metal ions, most activity has been directed at Ca, Zn, Mg, Na, and K, because these are the most prevalent ions and exist predominantly in a single ionic state. Much less attention has been devoted to Fe, Cu, or Mn, which are (except for Fe) less prevalent and which undergo redox cycling between ionic states. Another issue that has been underappreciated [at least in biology (Prasad and Oberleas, 1968; Thompson et al., 2006) if not in marine chemistry (van den Berg, 1984)] has been the speciation of metal ions *in vivo*: What fraction of a particular element is in which valence state, and what fraction is bound to which ligands. For some metals such as Cu and Fe, there are believed to be essentially no "free" (or bound to rapidly exchangeable ligands) ions (Outten and O'Halloran, 2001), in part due to the likely toxicity of the metals due to Fenton reactions producing reactive oxygen species and, in the case of Fe(III), insolubility. In the case of Zn, which exists in a single ionic state and does not undergo the Fenton reaction, the essentially zero levels predicted for prokaryotic cells are not found in eukaryotic cells (Bozym et al., 2004).

A large fraction of potential targets are proteins that act as enzymes. While fluorogenic enzyme assays have been known for decades [e.g., the 4-methylumbelliferyl glycoconjugates, which produce highly fluorescent 4-methylumbelliferone upon hydrolysis with the cognate glycosidase (Loontiens and DeBruyne, 1965)], relatively few have been applied intracellularly in live cells. The new substrates that precipitate upon enzymatic activation offer the prospect of intracellular localization (Haugland, 2005) without having to fix or otherwise permeabilize the cell to permit entry of antibodies. While a number of fluorogenic substrates have been described for hydrolytic enzymes such as proteases, nucleases, and glycosidases, other classes of enzymes have received much less attention, such as those involved in intermediary metabolism and biosynthesis. Indeed, what one would like to know is not only the activity but the amount of a particular enzyme as well for example, what the specific activity is. Determining this inside a live cell with much precision seems challenging. Moreover, enzyme assay would appear to require consumption of the substrate, whose amount inside the cell may be limited and which may be difficult to load to a level approaching K_m.

For many years, the reagent of choice for identifying and localizing proteins and polysaccharides has been the antibody. Its high affinity and selectivity have made it the basis of (a) fluorescence-activated cell sorting (flow cytometry) immunofluorescent stains and (b) immunoassays of all kinds. However, antibodies classically have a number of drawbacks that imperil their use for quantitation of analytes in cells, particularly of dynamic processes therein. First is the difficulty of introducing an antibody into an intact cell; recently, means have been devised to solve this problem (Schwarze et al., 1999). Second, the antibody binding to antigens is often essentially irreversible, so the reaction is ill-suited to following rapidly changing events. Third is the difficulty of transducing antibody binding as a useful fluorescence signal. Most immunoassays (e.g., ELISA, RIA) incorporate a separation step before the immunocomplex is quantitated, which is evidently difficult to implement inside a cell. The well-known fluorescence polarization immunoassay avoids this separations step, as do energy-transfer-based immunoassays (Dandliker et al., 1973; Ullman and Schwarzberg, 1981), but these assays typically require known amounts of labeled antibodies to be present and are usually configured as competition assays with labeled antigen in addition. What are more desirable are small molecule fluorescent probes selective for particular proteins that exhibit useful changes in fluorescence upon binding, such as Auramine O for liver alcohol dehydrogenase (Conrad et al., 1970), phallotoxins for F-actin, and dansylamide for carbonic anhydrase (Chen and Kernohan, 1967). At present a fair number of these are known, but the majority are based on fluorescent-labeled inhibitors or agonists that only exhibit changes in polarization upon binding. While these latter have been very useful for localization, quantitation is difficult without a fluorescence polarization microscope. It might be fruitful to screen a library of potentially fluorescent compounds against key target proteins to see if one will serve as a probe. While pharmacophores will not have been developed to recognize many proteins of interest, other proteins (such as kinases and others involved in signaling cascades, as well as neurotransmitter receptors) have been the subjects of prolonged scrutiny to find tight, selective ligands and potential lead compounds that do not meet the stringent criteria for development as drugs may be perfectly adequate as the basis for fluorescent probes. Of course, such probes are overtly useful in further screening of potential pharmacophores, and they may be worth developing for that purpose alone. Certainly, getting small molecules into cells is easier than large molecules, and the diffusive and binding kinetics of small molecules often are faster than those of antibodies, for instance. Finally, if the fluorescence properties are suitable, quantitation by intensity ratio (if there is a shift upon binding), lifetime, or polarization becomes possible, and all of these have been demonstrated in the microscope.

One of the central issues here is quantitation, preferably coupled with an image of the cell to provide spatial information: e.g., a map of the analyte concentration. Recently, Dinely has highlighted the importance of at least knowing, and preferably controlling, the intracellular concentrations of fluorescent indicators of calcium, zinc, and other metals when using them to measure concentrations (Dinely et al., 2002). For instance, if a fluorescent indicator with an affinity constant of one nanomolar is present in a cell at a concentration of one micromolar where the analyte concentration is 10 nM, the fractional saturation of the indicator will be 1% instead of 90% and the analyte concentration will be underreported. This is clearly an issue irrespective of whether the signal change is in intensity, wavelength ratio, polarization, or lifetime. Accurately measuring the amount of indicator present, particularly at the single cell level, is not at all a trivial process. Certainly, the measures used in gene expression studies in comparison to levels of "housekeeping" genes are possible, but the accuracy, precision, and dynamic range have only recently been improved by the use of fluorescence ratio methods (Dmochowski et al., 2002).

Some of the "analytes" of interest are less chemical analytes than changes in chemical status. For instance, as mentioned above, activities of many proteins are modulated by phosphorylation. While antibodies to phosphorylated amino acids themselves exist and are commercially available, it is another turn of the wheel to (a) determine the proportion of a particular protein that is phosphorylated and (b) see how the proportion changes in real time. Similarly, while one can measure oxygen tension inside cells and tissues using long lifetime indicators, the oxidizing or reducing propensity within the cell is only just beginning to be measured (Hanson et al., 2004). Part of the issue here is mechanistic, in that oxidants such as oxygen may not be able to oxidize some compounds at a significant rate despite their E_0 values providing sufficient thermodynamic driving force. The redox probes of the Tsien group are an important first step in this direction. In other cases the activity of an enzyme (perhaps a participant in a cascade) is modulated by binding of a ligand, and thus it is of interest to know what proportion of the protein is ligand-bound. For signaling cascades, what one would like to be able to do is follow the activation of the cascade components at several steps simultaneously. While we may have individual probes for steps of a cascade, like for the appearance of phosphatidylserine on the exterior of the cell using Annexin V and activation of caspases in the apoptosis cascade, we should shortly be able follow the steps in concert. In cases where the cascade steps have been fully elucidated, it would be useful to know how many steps must be monitored to fully follow the process to be activated.

9.5. NEW APPROACHES TO INTRACELLULAR ANALYSIS

9.5.1. Expressed Probes

The Tsien group has developed several exciting new approaches to *in situ* analysis. Miyawaki et al. (1997) introduced the use of Green Fluorescent Protein variants fused to calmodulin, such that the dramatic conformational change induced in the dumbbell-shaped protein by binding calcium was transduced as a change in energy transfer efficiency (Figure 9.1). This approach combines the advantages of energy transfer assays with an *expressible* indicator: To measure the analyte, all one need do is express the gene within the relevant cell type (or even organism) to quantitate the analyte. Others have attempted to adapt this approach to measure other analytes such as zinc, with a range of success from modest (Barondeau et al., 2002; Pearce et al., 2000) to satisfactory (Bozym et al., 2004)(see below). Calmodulin is unusual, however, in that it undergoes a really substantial change in conformation upon calcium binding and thus donor–acceptor distance, which results in a significant change in energy

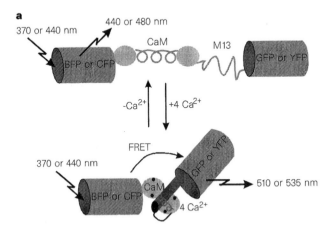

Figure 9.1. Expressible fluorescence energy transfer-based calcium sensor of Miyawaki et al. (1997). In the absence of Ca(II), the calmodulin (upper panel) is in an extended conformation such that the BFP on the left does not transfer energy, and one observes its emission. In the presence of calcium (lower panel), calmodulin adopts a more compact conformation, bringing the BFP in close proximity to the GFP, resulting in efficient energy transfer to the GFP, and reduced emission from BFP and enhanced emission from GFP. [Reproduced with permission from Miyawaki et al. (1997).]

transfer efficiency. The R^6 dependence of the efficiency on distance (Forster, 1948) means, however, that the system must be poised in terms of a carefully matched overlap integral and donor–acceptor distance [or distance distribution (Eis and Lakowicz, 1993)] to provide a significant change in transfer efficiency, and therefore signal. For instance, if the binding of the analyte changes the donor–acceptor distance from 20 to 30 Å but the donor and acceptor are chosen so that "Forster distance" where energy transfer is 50% efficient is 40 Å, analyte binding only results in a 14% change in transfer efficiency for a 50% increase in distance. With fluorescent donors and acceptors attached to proteins, the use of site-directed mutagenesis to position these moieties unambiguously and optimally is preferred (Thompson et al., 1996a). The change in energy transfer efficiency can be measured accurately as a change in intensity ratio, donor lifetime, or anisotropy (polarization) (Thompson et al., 1998). The flexibility of the energy transfer approach (i.e., one can induce a change in efficiency in principle by changing distance, spectra, quantum efficiency, refractive index, lifetime, and/or relative orientation) suggests that it will be even more widely used in the future.

9.5.2. Single-Molecule Detection in Cells

The development of single-molecule fluorescence detection techniques over the last several years (Moerner and Kador, 1989; Orrit and Bernard, 1990) (see Chapters 1, 2, and 4 in this book) has been very exciting in part because it potentially offers a way to look at molecules that may be very rare in the cell, and largely without having to average over an ensemble of molecules. The essence of the approach is to excite fluorescence within a relatively small (usually about one femtoliter) volume element of a medium containing fluorescent molecules at some low concentration (nanomolar), such that fluorescence above background levels will be readily collected and observed if a fluorophore happens to be within the volume element. Multiphoton excitation with a high numerical aperture objective is nearly ideal for this purpose. If the fluorescence collection is efficient, the fluorophore quantum yield is high, and the background is low, emission from individual molecules can be observed with statistically defined levels of confidence. It should be noted that this doesn't necessarily translate into determination of analytes at really low concentrations, since to ensure that a fluorophore will be observed in the tiny voxel within a reasonable time frame, one needs a nanomolar concentration of fluorophores. Nevertheless, in cells where the molecules of interest are likely to be localized to some structural feature such as the nucleus (e.g., you're looking a small fraction of the cell volume), this becomes a potentially fruitful endeavour.

Work has been ongoing at the single-molecule level in cells using the molecular beacon technology (Yao et al., 2003), but little has appeared using single indicator molecules to analyze metal ions. Several of the fluorescent probes for certain metal ions are good enough fluorophores that this should be possible in principle. At the single-molecule level, however, one is perforce no longer observing ensemble averages, but rather has entered the regime of stochastic sensing (Braha et al., 2000), where one measures concentration not by the fractional occupancy of an ensemble of sites, but by the time-averaged occupation of a single site. For low concentrations providing low fractional occupancy, this implies observing for a period of time, which at some point collides with the propensity of the fluorophore to photobleach. Use of fluorophores with much lower photobleaching propensity (like the semiconductor nanoparticles) would appear essential to such studies. The virtue of the patch-clamp-like techniques used in the existing stochastic sensors is that interrogation of the transducer is nondestructive and can be carried out for hours.

As mentioned above, an important issue in single-molecule work (indeed, in all fluorescence microscopy) is photobleaching, which is the often irreversible loss of fluorescence emission upon continuing excitation. The (photo)chemistry of the process differs for various fluorophores, but is often oxygen-dependent (Turro, 1978). It is important to remember that a fluorophore in the excited state has substantial additional energy and is thus more chemically reactive than in the ground state. Indeed, the fluorophore in the excited state often behaves entirely differently (from a chemical standpoint) from that in the ground state. For instance, it is well-known that some fluorophores that are weak bases in the ground state become strong acids in the excited state, and it is even more common that a molecule's redox properties change dramatically upon excitation. Reagents exist that counteract the effects of photobleaching, but these often are usable only in fixed and permeabilized cells. A rule of thumb in this regard is that a very good fluorophore (i.e., Rhodamine 6G) on the average will survive for a few thousand cycles of excitation and emission in solution before undergoing a destructive reaction. Since it is easy (by increasing the laser power and focusing in a small volume) to provide very intense excitation, the collection efficiency must be optimized since there is a limited opportunity to observe the fluorophore before it bleaches. In microscopes with high numerical aperture objectives, collection of fluorescence can be almost 50% efficient (Axelrod et al., 1984), whereas in ordinary fluorometers the efficiency is probably less than 10%. One answer to this is the use of metal-enhanced fluorescence, which under certain conditions is not emitted isotropically, but is emitted in a particular direction without the use of focusing optics (Lakowicz, 2004).

Finally, when studying trace analytes at the single-molecule level, one is concerned not only about the destructive effects of merely observing the fluorophore for prolonged periods, but also about the amount of time it will take the analyte to find the indicator molecule. For a small molecule present at nanomolar concentration with a diffusion-controlled association rate constant, this should take on the order of seconds in water, but with the slower (and anisotropic) diffusion within the cell and at lower analyte concentrations the binding process will be more prolonged. Within the cell, one imagines that small-molecule effectors seldom have to diffuse very far, and most likely only need to diffuse within a compartment to find their cognate binding site; effectors like steroids which must diffuse from the membrane into the nucleus perforce are not required to act quickly. Thus one imagines that it will be useful to localize the indicator within the organelle(s) of interest using chemical or molecular biological methods.

9.5.3. PEBBLES for Intracellular Measurements

An important development has been contributed by the Kopelman laboratory that promises to be extremely useful in understanding the biology of individual cells. Instead of dispersing a fluorescent indicator for a particular analyte throughout the cell or organelle of interest, the indicator is incorporated into very small (20–300 nm) beads and injected or otherwise introduced into cells. These beads are formed by emulsion polymerization and are named PEBBLES, an acronym short for "Probes Encapsulated by Biologically Localized Embedding" (Monson et al., 2003).

The incorporation of the fluorescent indicator into the PEBBLES offers a number of advantages. First, if the indicator itself is toxic or deleterious, the amount introduced into the cell is much smaller than if the indicator were present throughout the cytoplasm. A subtler, related advantage is that (for equilibrium binding indicators) the limited indicator amount means that the concentration of the analyte is not perturbed by excessive concentrations of indicator, as discussed by Dinely et al. (2002). The limited volume subtended by

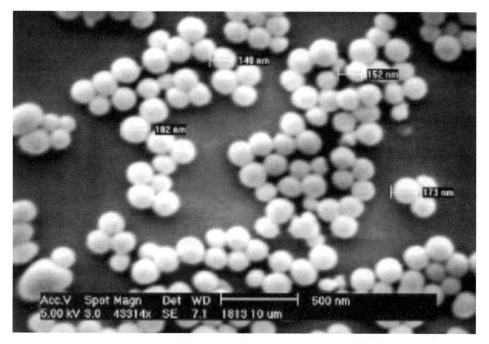

Figure 9.2. Scanning electron micrograph of sol–gel PEBBLES; the scale indicates that the average size is 160 nm. [Reproduced with permission from Monson et al. (2003).]

the PEBBLE (see Figure 9.2) means that background fluorescence due to autofluorescence from cell constituents can be accurately subtracted from the emission of the PEBBLE; for indicators dispersed throughout the cell, this is much harder to do accurately. Moreover, if different PEBBLES are present having indicators whose spectral properties overlap to some degree, the crosstalk interference between PEBBLES does not require correction since the emissions are spatially resolved as well; again, this is very difficult to do if the compounds are dispersed throughout the cell and not spatially resolved. Typically, the fluorescent indicator is sequestered within the PEBBLE matrix, such that the indicator is protected from adsorption to, for instance, protein(s) in the intracellular milieu which might change its response or degrade it (Burdette et al., 2001). Also, the sequestration limits any toxicity the indicator system might have. Finally, it is straightforward to include a second fluorescent dye for intensity ratiometric measurements, an approach whose advantages are widely appreciated.

PEBBLES have been made three different ways to date. The first and most widely used approach is to form ~40-nm polyacrylamide beads by polymerization in a microemulsion formulation. The polyacrylamide gel is similar to those used for analytical electrophoretic separations, being quite hydrophilic. The detergents used to form the microemulsion can be deleterious to protein-based indicators, however. PEBBLES can also be made from so-called sol–gel formulations, which comprise a rigid, porous, transparent, glass-like material containing the indicator and polyethylene glycol. The rugged nature of the sol–gel compared to the softer polyacrylamide suggests its use in more demanding environments. While the conditions needed to form sol–gels are comparatively mild compared to those required for regular glass, the conditions may be deleterious to protein-based sensors. Finally, for a more hydrophobic PEBBLE matrix suited to hydrophobic indicators, the use of decyl methacrylate has been developed.

PEBBLES may be introduced into cells by one of four routes: They may be introduced by liposome fusion or direct injection, propelled ballistically into the cell like a bullet using a device called a "gene gun," or taken up by endocytosis. The methods differ in their degree of efficacy with a given cell line and where within the cell the PEBBLE ends up. PEBBLES can be directly injected into the cell using a micropipette. While this procedure is used routinely for *in vitro* fertilization, there is reason to suspect that such injections may indeed disrupt the cell. PEBBLES may be phagocytosed like bacteria or other small particles; this approach is especially valuable to study the phagosomes or lysosomes, but by itself it may not be very useful for studying other organelles such as the mitochondria or nucleus. PEBBLES can be introduced into the cell by fusion of liposomes that contain them with the cell membrane. This technique is very benign but has little target specificity. Most ingenious of all is the use of the Gene Gun, a device that transfects cells with DNA by "spitting" (using a gas jet) gold nanoparticles coated with DNA into the cell. To a degree, the penetration of the gold nanoparticle into particular compartments in the cell can be influenced by the gun parameters, but the accuracy with which the nanoparticles can be placed is very modest. Of course, a virtue of the multiple means of introduction is that if one is unsatisfactory, one of the others might work.

Several analytes have been measured using PEBBLES, including oxygen, pH, Zn, Ca, and potassium. In principle, it would appear possible that anything measured by a dispersed indicator could also be measured with PEBBLEs. One of the advantages of PEBBLES is that if the fluorescent indicator exhibits just a simple intensity change upon binding (or colliding with) the analyte, it can be readily converted into a ratiometric indicator by inclusion of a spectrally distinct fluorophore that is unresponsive together with the indicator fluorophore: Co-location in the PEBBLE maintains the concentration relationship of the two fluorophores, as well as their spatial relationship. Figure 9.3 depicts PEBBLES in rat C6

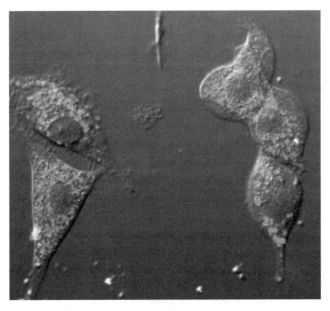

Figure 9.3. Fluorescence micrograph Rat C6 glioma cells with ballistically inserted PEBBLES. [Reproduced with permission from Monson et al. (2003).] See insert for color representation of this figure.

glioma cells containing a ratiometric oxygen indicator consisting of a mixture of ruthenium *tris* (dpp) together with Oregon Green; the oxygen quenches the metal complex efficiently, whereas the Oregon Green is quite insensitive. Thus the emission intensity of the complex declines with increasing oxygen concentration, whereas that of the organic fluorophore remains constant, and thus the observed ratio of the emissions is a simple function of the oxygen concentration (Monson et al., 2003).

One of the issues in the development of PEBBLE-based sensors is the means of their formation and its compatibility with biomolecules such as proteins. In particular, while the conditions are gentle, the emulsion polymerization can be deleterious to proteins since it contains substantial concentrations of surfactants, which are well known to denature proteins. Other components of this or the sol–gel or decyl methacrylate formulations may also be deleterious to certain indicator molecules. There seems to be no general solution, with individual sensor components requiring validation with a particular PEBBLE formulation on a case-by-case basis.

9.5.4. Metal Ion Biosensors

As mentioned above, while scores of low-molecular-weight fluorescent indicators have been described for a myriad of metal ions, in many cases this approach has appeared inadequate. For metal ions like Mg, Ca, K, and Na that are relatively abundant in the free form in cells, it has proven feasible to synthesize fluorescent indicator molecules (Haugland, 2005). For metal ions present at trace and ultratrace levels in the free form, such as Zn, Cu, Fe, and others, synthesizing indicators has proven more difficult (Thompson, 2005). While chelators are known that bind these metal ions very tightly, the need for selectivity in the complex matrices of the cytoplasm and extracellular medium is paramount: For a sensor, it is not enough to bind (and therefore respond to) free zinc at picomolar levels. Rather, it must bind zinc specifically in the presence of million-fold higher concentrations of calcium and magnesium. Classical chemical separations to remove interferents are of course problematic within an organism: The separation must be done *in situ* "on the fly" as it were. Some time ago we (Thompson and Jones, 1993) and others (Godwin and Berg, 1996; Regan and Clarke, 1990; Walkup and Imperiali, 1997) sought to adapt the extreme selectivity of biological molecules to this problem. In our case we exploited a human enzyme, carbonic anhydrase II, as a very selective ligand for zinc and other metal ions (Fierke and Thompson, 2001). We demonstrated that the protein would selectively bind these target metals at nanomolar to picomolar levels in the presence of 10 mM Ca and 50 mM Mg (the concentrations found in sea water). Looking at the structure of the active site (Christianson and Fierke, 1996) (Figure 9.4), it is easy to see that it would bind zinc with high affinity due to the position and orientation of the three histidine imidazole ligands, but would bind Ca and Mg with poor affinity, if at all. We believe that the high selectivity compared with classic chelators such as EDTA can be rationalized by considering that the ligands in the protein are held in position by hydrogen bonds to other amino acid residues, which must be displaced for other metals to bind and which therefore is energetically unfavorable (Kiefer et al., 1995b). By comparison, the chelator is much more flexible and can accommodate a greater variety of metal ions, which is undesirable in this case. We note that cyclophanes are considerably more rigid and have been adapted to some metallofluorescent indicators [notably for zinc (Kikuchi et al., 2004)], but these indicators exhibited very slow kinetics.

A unique advantage of the biosensor approach is the ease with which the protein [or other macromolecules such as RNA aptamers (Yang and Ellington, 2005)] can be modified

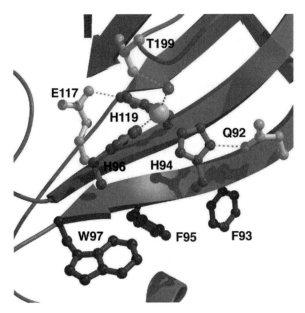

Figure 9.4. Active site of human carbonic anhydrase II indicating residues that are zinc ligands, as well as "second shell" residues which hold the ligand histidinyl imidazoles in position by hydrogen bonding. [Reproduced with permission of the copyright holder from Fierke and Thompson (2001).] See insert for color representation of this figure.

to improve its properties for a particular application. For instance, our collaborators, led by Professor Carol Fierke of the University of Michigan, have shown that the affinity, selectivity, and even kinetics of metal ion binding may be improved by mutagenesis of the protein, by either directed or combinatorial mutagenesis (Kiefer et al., 1995a; Huang et al., 1996; Hunt et al., 1999). Moreover, the use of site-directed mutagenesis has greatly simplified the labeling process, enabling us to optimize the response of fluorescent labels by positioning them with respect to the metal ion binding site (Thompson et al., 1996a). Finally, the use of macromolecules as transducers (as discussed above with the work of Miyawaki et al., 1997) enables the use of Forster transfer as a sensing approach, which has several advantages (Godwin and Berg, 1996; Thompson et al., 2002a; Thompson and Patchan, 1995).

We (and others) have found several means to transduce the metal recognition event as a change in fluorescence we can measure. Transduction as a simple intensity change is so prone to artifact and so hard to calibrate accurately (especially in the microscope) that it has been largely superseded by intensity ratios (Grynkiewicz et al., 1985), fluorescence lifetimes (Lippitsch et al., 1988; Szmacinski and Lakowicz, 1993) and fluorescence polarization (anisotropy) (Dandliker et al., 1973; Elbaum et al., 1996). Thus we use zinc-dependent binding of a fluorescent ligand to carbonic anhydrase and use the change in fluorescence intensities at differing wavelengths (Figure 9.5) (Thompson et al., 2000), fluorescence polarization (Elbaum et al., 1996), or fluorescence lifetime (Thompson and Patchan, 1995). For other metal ions that are colored when bound [such as Cu (II)], we can use quenching of nearby fluorescent labels due to Forster transfer if the label's emission overlaps the weak d–d absorbance band, resulting in reduced fluorescence lifetimes (Thompson et al., 1999a) (Figure 9.6). For uncolored ions, other proximity-dependent quenching mechanisms can be used; under certain conditions these mechanisms result in changes in fluorescence

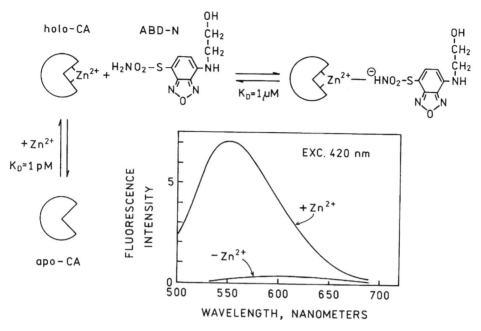

Figure 9.5. Principle for fluorescence detection of zinc by apocarbonic anhydrase and ABDN (left panel) and dependence of fluorescence emission intensity ratios on free zinc concentration (right panel). Binding of ABDN to carbonic anhydrase only occurs when zinc is bound to the active site, and results in a blue shift and increase in quantum yield; the proportion of blue-shifted emission reflects the fraction of protein with zinc bound, and thus the concentration. [Reproduced from Thompson et al. (2000), with permission.]

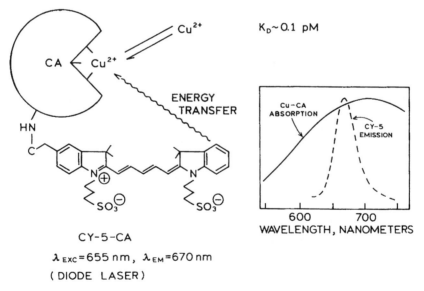

Figure 9.6. Schematic of fluorescence sensing of Cu(II) by carbonic anhydrase II. Energy transfer from the fluorescent label to protein-bound Cu(II) results in quenching and a decrease in lifetime; the fraction of the shorter lifetime corresponding to the Cu-bound form is proportional to the free Cu concentration. *Inset*: Spectral overlap of label fluorescence emission and bound Cu absorption. [Reproduced from Thompson et al. (1999), with permission of the copyright holder.]

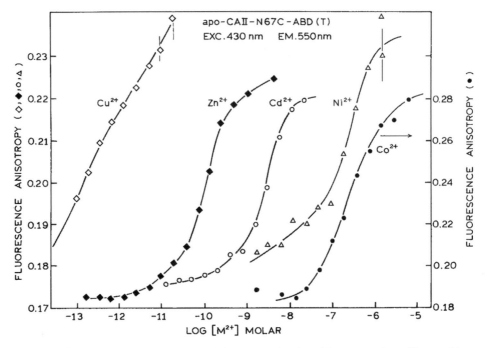

Figure 9.7. Fluorescence anisotropies of apoN67C-ABD-T as a function of the concentrations of five metal ions. [Reproduced with permission from Thompson, et al. (1999b).]

polarization which are more convenient to measure; an example of these data is shown in Figure 9.7.

Beginning with the work of Peterson, Hirschfeld, and Seitz, there has been interest in adapting fluorescence assays to fiber optics for purposes of determining analytes in remote or inaccessible places (Thompson, 1991; Thompson, 2005; Wolfbeis, 1991). Several workers have demonstrated sensors based on changes in intensity, intensity ratio (Goyet et al., 1992), and fluorescence lifetime (Bright, 1988). We also have adapted our metal ion biosensors to use with fiber optics (Thompson et al., 1996b; Thompson and Jones, 1993; Zeng et al., 2003, 2005), being particularly interested in measurements in the depths of the ocean (where sampling for analysis is slow, expensive, and prone to contamination) and *in vivo* in larger animals, where the skin is not transparent and the area of analytical interest may be deep within an organ or body cavity. Thus we have made measurements of Cu(II) at picomolar levels *in situ* in the ocean, in real time (Zeng et al., 2003), Similarly, we have measured free zinc in the brain of a large animal ischemia model *in situ* in real time. Recently, we replaced the off-axis parabolic mirrors in our fiber-optic apparatus (Thompson et al., 1990) with dichroic mirrors, which greatly simplified alignment of the system (Figure 9.8).

9.6. *IN VIVO* MEASUREMENTS

While isolated, often cultured cells have been essential tools in understanding the biology of multicellular organisms, ideally one would like to study the biology *in situ* in the intact, living organism. The question becomes, What fraction of the powerful tools usable on isolated cells

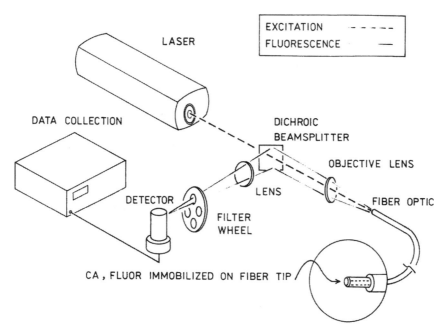

Figure 9.8. Schematic of fluorescence emission intensity ratio-based fiber-optic biosensor. [Reproduced with permission from Zeng, et al. (2005).]

can be applied to intact animals and plants? From the standpoint of fluorescence techniques, the issue becomes one of penetration: How far beneath the surface/skin/bark/shell can one see, and with how much resolution? In the cases of micro/meso organisms such as *C. elegans*, *Drosophila*, and zebrafish, the organisms are small and to a degree transparent, so this is less of an issue. For others, optical limitations such as scattering or absorption are an issue, which can be addressed to a degree by the use of infrared fluorescence or multiphoton excitation (Thompson, 1994). Thus transdermal measurements in nude mice have been made using infrared fluorophores (Fisher et al., 1996), and initial experiments looking through skin have been made using two-photon microscopy (Masters et al., 1997). For multiphoton excitation, however, the high NA microscope objectives that give the highest intensity also have the shortest focal lengths (<0.5 mm), making it difficult to penetrate far beneath the skin. For *in vivo* measurements deep in organisms (Peterson et al., 1980; Zeng et al., 2005), the insertion of fiber-optic sensors is another route, which may be more invasive.

9.7. CONCLUSIONS

We are in the midst of a very exciting time in the study of metal ions in biology, where the confluence of spectroscopic methods with chemical and molecular biology approaches offers unprecedented insight. Despite our improved techniques, important issues remain, including the difficulty of transdermal measurement in intact organisms, the need for new probes for many analytes, and the problems of quantitation and selectivity. Nevertheless, even with our current tools, much can be done to elucidate the biochemistry of metals in the myriad of organisms in the biosphere.

REFERENCES

Appleby, R, Chamberlain, JM, and Krapels, KA (2004). *Passive Millimeter Wave and Terahertz Imaging and Technology*. Society of Photooptical Instrumentation Engineers, 252 pp.

Axelrod, D, Burghardt, TP, and Thompson, N (1984). Total internal reflection fluorescence. *Annu Rev Biophys Bioeng* **13**:247–268.

Barondeau, DP, Kassman, CJ, Tainer, JA, and Getzoff, ED (2002). Structural chemistry of a green fluorescent protein Zn biosensor. *J Am Chem Soc* **124**(14):3522–3524.

Bozym, RA, Zeng, HH, Cramer, M, Stoddard, A, Fierke, CA, and Thompson, RB (2004). *In vivo* and intracellular sensing and imaging of free zinc ion. In: Cohn, GE, Grundfest, WS, Benaron, DA, and Vo-Dinh, T (eds.), *Proceedings of the SPIE Conference on Advanced Biomedical and Clinical Diagnostic Systems II*, SPIE, Bellingham, WA.

Braha, O, Gu, L, Zhou, L, Lu, X, Cheley, S, and Bayley, H (2000). Simultaneous stochastic sensing of divalent metal ions. *Nature Biotechnol* **18**:1005–1007.

Brauns, EB, and Murphy, CJ (1997). Quantum dots as chemical sensors. *Recent Res. Dev. Phys. Chem.* **1**:1–15.

Bright, FV (1988). Remote sensing with a multifrequency phase-modulation fluorometer. In Lakowicz SR (ed.) *Time-Resolved Laser Spectroscopy in Biochemistry* Los Angeles, CA, pp. 23–28.

Burdette, SC, Walkup, GK, Spingler, B, Tsien, RY, and Lippard, SJ (2001). Fluorescent sensors for Zn^{2+} based on a fluorescein platform: Synthesis, properties, and intracellular distribution. *J Am Chem Soc* **123**:7831–7841.

Chance, B, Alfano, RR, Tromberg, BJ, Tamura, M, and Sevick-Muraca, EM (2005). *Optical Tomography and Spectroscopy of Tissue VI*. Society of Photooptical Instrumentation Engineers, 550 pp.

Chen, RF, and Kernohan, J (1967). Combination of bovine carbonic anhydrase with a fluorescent sulfonamide. *J Biol Chem* **242**:5813–5823.

Christianson, DW, and Fierke, CA (1996). Carbonic anhydrase—Evolution of the zinc binding site by nature and by design. *Acc Chem Res* **29**:331–339.

Conrad, RH, Heitz, JR, and Brand, L (1970). Characterization of a fluorescent complex between Auramine O and horse liver alcohol dehydrogenase. *Biochemistry* **9**:1540–1546.

Coons, AM, Creeck, HJ, Jones, RN, and Berliner, E (1942). Fluorescent antibody technique. *J Immunol* **45**:159.

Dandliker, WB, Kelly, RJ, Dandliker, J, Farquhar, J, and Levin, J (1973). Fluorescence polarization immunoassay. Theory and experimental method. *Immunochemistry* **10**:219–227.

Darwish, IA, and Blake, DA (2002). Development and validation of a one-step immunoassay for determination of cadmium in human serum. *Anal Chem* **74**(1):52–58.

Dattelbaum, JD, and Lakowicz, JR (2001). Optical determination of glutamine using a genetically engineered protein. *Anal Biochem* **291**:89–95.

D'Auria, S, Gryczynski, Z, Gryczynski, I, Rossi, M, and Lakowicz, JR (2000). A protein biosensor for lactate. *Anal Biochem* **283**:83–88.

Demas, JN, and DeGraff, BA (2001). Applications of luminescent transition platinum group metal complexes to sensor technology and molecular probes. *Coord Chem Rev* **211**:317–351.

Denk, W, Strickler, JH, and Webb, WW (1990). Two-photon laser scanning fluorescence microscopy. *Science* **248**:73–76.

DiCesare, N, and Lakowicz, JR (2002). Charge transfer fluorescent probes using boronic acids for monosaccharide signaling. *J Biomed Opt* **7**(4):538–545.

Dinely, KE, Malaiyandi, LM, and Reynolds, IJ (2002). A reevaluation of neuronal zinc measurements: Artifacts associated with high intracellular dye concentration. *Mol Pharmacol* **62**(3):618–627.

Dmochowski, IJ, Dmochowski, JE, Oliveri, P, Davidson, EH, and Fraser, SE (2002). Quantitative imaging of cis-regulatory reporters in living embryos. *Proc Nat Acad Sci* **99**(20):12895–12900.

Egner, A, and Hell, SW (2005). Fluorescence microscopy with super-resolved optical sections. *Trends Cell Biol* **15**(4):207–215.

Eis, PS, and Lakowicz, JR (1993). Time-resolved energy transfer measurements of donor–acceptor distance distributions and intramolecular flexibility of a CCHH zinc finger peptide. *Biochemistry* **32**:7981–7993.

Elbaum, D, Nair, SK, Patchan, MW, Thompson, RB, and Christianson, DW (1996). Structure-based design of a sulfonamide probe for fluorescence anisotropy detection of zinc with a carbonic anhydrase-based biosensor. *J Am Chem Soc* **118**(35):8381–8387.

Fierke, CA, and Thompson, RB (2001). Fluorescence-based biosensing of zinc using carbonic anhydrase. *BioMetals* **14**:205–222.

Fisher, G, Ballou, B, Srivastava, M, and Farkas, DL (1996). Far-red fluorescence-based high specificity tumor imaging *in vivo*. *Biophys J* **70**(2):212A.

Forster, T (1948). Intermolecular energy migration and fluorescence in [German]. *Ann Phys* **2**:55–75.

Godwin, HA, and Berg, JM (1996). A fluorescent zinc probe based on metal induced peptide folding. *J Am Chem Soc* **118**:6514–6515.

Goldstein, JL, and Brown, MS (1990). Regulation of the mevalonate pathway. *Nature* **343**:425–430.

Goyet, C, Walt, DR, and Brewer, PG (1992). Development of a fiber optic sensor for measurement of pCO_2 in sea water: Design criteria and sea trials. *Deep-Sea Res* **39**(6):1015–1026.

Grynkiewicz, G, Poenie, M, and Tsien, RY (1985). A new generation of calcium indicators with greatly improved fluorescence properties. *J Biol Chem* **260**(6):3440–3450.

Hanson, GT, Aggeler, R, Oglesbee, D, Cannon, M, Capaldi, RA, Tsien, RY, and Remington, SJ (2004). Investigating mitochondrial redox potential with redox-sensitive green fluorescent protein indicators. *J Biol Chem* **279**:13044–13053.

Haugland, RP (2005). *The Handbook: A Guide to Fluorescent Probes and Labeling Technologies* Invitrogen Corp., Carlsbad, CA.

Hirano, T, Kikuchi, K, and Nagano, T (2004). Zinc fluorescent probes for biological applications. In: C. D. Geddes and J. R. Lakowicz, (eds.), *Reviews in Fluorescence 2004* Kluwer Academic/Plenum Publishers, New York, pp. 55–73.

Huang, C-c, Lesburg, CA, Kiefer, LL, Fierke, CA, and Christianson, DW (1996). Reversal of the hydrogen bond to zinc ligand histidine-119 dramatically diminishes catalysis and enhances metal equilibration kinetics in carbonic anhydrase II. *Biochemistry* **35**(11):3439–3446.

Hummon, AB, Sweedler, JV, and Corbin, RW (2003). Discovering new neuropeptides using single cell mass spectrometry. *Trends Anal Chem* **22**:515–521.

Hunt, JA, Ahmed, M, and Fierke, CA (1999). Metal binding specificity in carbonic anhydrase is influenced by conserved hydrophobic amino acids. *Biochemistry* **38**:9054–9060.

Jiang, P, and Guo, Z (2004). Fluorescent detection of zinc in biological systems: recent development on the design of chemosensors and biosensors. *Coord Chem Rev* **248**:205–229.

Kiefer, LL, Paterno, SA, and Fierke, CA (1995a). Hydrogen bond network in the metal binding site of carbonic anhydrase enhances zinc affinity and catalytic efficiency. *J Am Chem Soc* **117**:6831–6837.

Kiefer, LL, Paterno, SA, and Fierke, CA (1995b). Second shell hydrogen bonds to histidine ligands enhance zinc affinity and catalytic efficiency. *J Am Chem Soc* **117**:6831–6837.

Kikuchi, K, Komatsu, K, and Nagano, T (2004). Zinc sensing for cellular application. *Curr Opin Chem Biol* **8**:182–191.

Lakowicz, JR (2004). Radiative decay engineering 3. Surface plasmon-coupled directional emission. *Anal Biochem* **324**:153–169.

Li, QZ, and Cass, AEG (1991). Periplasmic binding-protein based biosensors. 1. Preliminary study of maltose binding-protein as sensing element for maltose. *Biosens Bioelectron* **6**(5):445–450.

Lippitsch, ME, Pusterhofer, J, Leiner, MJP, and Wolfbeis, OS (1988). Fibre-optic oxygen sensor with the fluorecence decay time as the information carrier. *Anal Chim Acta* **205**:1–6.

Loontiens, FG, and DeBruyne, CK (1965). Beta-xylosidase assay with 4-MU conjugate. *Die Naturwissenschaften*, **52**:661.

Masters, BR, So, PT, and Gratton, E (1997). Multiphoton excitation fluorescence microscopy and spectroscopy of *in vivo* human skin. *Biophys J* **72**:2405–2412.

Milliard, AC, Campagnola, PJ, Mohler, W, Lewis, A, and Loew, LM (2003). Second harmonic imaging microscopy. In: Marriott, G, and Parker, I (eds.), *Methods in Enzymology*. Academic Press, San Diego, pp. 47–69.

Miyawaki, A, Llopis, J, Heim, R, McCaffery, JM, Adams, JA, Ikura, M, and Tsien, RY (1997). Fluorescent indicators for Ca^{2+} based on green fluorescent proteins and calmodulin. *Nature* **388**:882–887.

Moerner, WE, and Kador, L (1989). Optical detection and spectroscopy of single-molecules in a solid. *Phys Rev Lett* **62**:2535.

Monson, E, Brasuel, M, Philbert, MA, and Kopelman, R (2003). PEBBLE nanosensors for *in vitro* bioanalysis. In: Vo-Dinh, T (ed.), *Biomedical Photonics Handbook*. CRC Press, Boca Raton, FL, pp. 1–14.

Orrit, M, and Bernard, J (1990). Single pentacene molecules detected by fluorescence excitation in a *p*-terphenyl crystal. *Phys Rev Lett* **65**:2716–2719.

Outten, CE, and O'Halloran, TV (2001). Femtomolar sensitivity of metalloregulatory proteins controlling zinc homeostasis. *Science* **292**:2488–2492.

Pearce, LL, Gandley, RE, Han, W, Wasserloos, K, Stitt, M, Kanai, AJ, McLaughlin, MK, Pitt, BR, and Levitan, ES (2000). Role of metallothionein in nitric oxide signaling as revealed by a green fluorescent fusion protein. *Proc Nat Acad Sci* **97**(1):477–482.

Peterson, JI, Goldstein, SR, Fitzgerald, RV, and Buckhold, DK (1980). Fiber optic pH probe for physiological use. *Anal Chem* **52**:864–869.

Prasad, AS, and Oberleas, D (1968). Zinc in human serum: Evidence for an amino acid-bound fraction. *J Lab Clini Med* 1006.

Regan, L, and Clarke, ND (1990). A tetrahedral zinc(II)-binding site introduced into a designed protein. *Biochemistry* **29**:10878–10883.

Schwarze, SR, Ho, A, Vocero-Akbani, A, and Dowdy, SF (1999). *In vivo* protein transduction: Delivery of a biologically active protein into the mouse. *Science* **285**:1569–1572.

Szmacinski, H, and Lakowicz, JR (1993). Optical measurements of pH using fluorescence lifetimes and phase-modulation fluorometry. *Anal Chem* **65**:1668–1674.

Thompson, RB (1991). Fluorescence-based fiber optic sensors. In: Lakowicz, JR (ed.), *Topics in Fluorescence Spectroscopy*, Vol. 2: *Principles*. Plenum Press, New York, pp. 345–365.

Thompson, RB (1994). Red and near-infrared fluorometry. In: Lakowicz, JR (ed.), *Topics in Fluorescence Spectroscopy*, Vol. 4: *Probe Design and Chemical Sensing*. Plenum Press, New York, pp. 151–181.

Thompson, RB (2005). Studying zinc biology with fluorescence: Ain't we got fun? *Curr Opin Chem Biol* **9**(5):526–532.

Thompson, RB, and Jones, ER (1993). Enzyme-based fiber optic zinc biosensor. *Anal Chem* **65**:730–734.

Thompson, RB, and Patchan, MW (1995). Lifetime-based fluorescence energy transfer biosensing of zinc. *Anal Biochem* **227**:123–128.

Thompson, RB, Levine, M, and Kondracki, L (1990). Component selection for fiber optic fluorometry. *Appl Spectrosc* **44**(1):117–122.

Thompson, RB, Ge, Z, Patchan, MW, and Fierke, CA (1996a). Performance enhancement of fluorescence energy transfer-based biosensors by site-directed mutagenesis of the transducer. *J Biomed Opt* **1**(1):131–137.

Thompson, RB, Frederickson, CJ, Fierke, CA, Westerberg, NM, Bozym, RA, Cramer, ML, and Hershfinkel, M (2006). Practical aspects of fluorescence analysis of free zinc ion in biological systems: pZn for the biologist. In Thompson, RB (ed.) *Fluorescence Sensors and Biosensors*. CRC Press, Boca Raton, FL pp. 351–376.

Thompson, RB, Maliwal, BP, Feliccia, VL, Fierke, CA, and McCall, K (1998). Determination of picomolar concentrations of metal ions using fluorescence anisotropy: Biosensing with a "reagentless" enzyme transducer. *Anal Chem* **70**(22):4717–4723.

Thompson, RB, Maliwal, BP, and Fierke, CA (1999a). Selectivity and sensitivity of fluorescence lifetime-based metal ion biosensing using a carbonic anhydrase transducer. *Anal Biochem* **267**:185–195.

Thompson RB, Maliwal BP, Fierke CA (1999b) Fluorescence-based sensing of transition metal ions by a carbonic anhydrase transducer with a tethnered fluorophore. In Lakowicz JR, Soper SA, Thompson RB (eds.) *Proceeding of SPIE Conference on Advances in Fluorescence Sensing Technology IV*, SPIE, Bellingham, WA pp. 85–92.

Thompson, RB, Whetsell, WO, Maliwal, BP, Fierke, CA, and Frederickson, CJ (2000). Fluorescence microscopy of stimulated Zn(II) release from organotypic cultures of mammalian hippocampus using a carbonic anhydrase-based biosensor system. *J Neurosci Methods* **96**(1):35–45.

Thompson, RB, Cramer, ML, Bozym, R, and Fierke, CA (2002a). Excitation ratiometric fluorescent biosensor for zinc ion at picomolar levels. *J Biomed Opt* **7**(4):555–560.

Thompson, RB, Peterson, D, Mahoney, W, Cramer, M, Maliwal, BP, Suh, SW, and Frederickson, CJ (2002b). Fluorescent zinc indicators for neurobiology. *J Neurosci Methods* **118**:63–75.

Trulson, MO, Walton, ID, Suseno, A, Matsuzaki, H, and Stern, D (1998). Light scattering by metal sol labels on high density DNA probe arrays. *Systems and Technologies for Clinical Diagnostics and Drug Discovery* San Jose, CA.

Tsien, RY (1989). Fluorescent probes of cell signaling. *Annual Review of Neuroscience*. Annual Reviews, Inc, Palo Alto, CA, p. 27.

Tuchin, VV, Izatt, JA, and Fujimoto, JG (2005). *Coherence Domain Optical Methods and Optical Coherence Tomography in Biomedicine IX*. Society of Photooptical Instrumentation Engineers, 590 pp.

Turro, NJ (1978). *Modern Molecular Photochemistry*, Benjamin/Cummings Publishing Co., Menlo Park, CA.

Ullman, EF, and Schwarzberg, M (1981). Fluorescence quenching with immunological pairs in immunoassays. Syva Company, USA.

van den Berg, C (1984). Determining the copper complexing capacity and conditional stability constants of complexes of copper (II) with natural organic ligands in seawater by cathodic stripping voltammetry of copper-catechol complex ions. *Marine Chem* **15**:1–18.

Walkup, GK, and Imperiali, B (1997). Fluorescent chemosensors for divalent zinc based on zinc finger domains. Enhanced oxidative stability, metal binding affinity, and structural and functional characterization. *J Am Chem Soc* **119**:3443–3450.

Wolfbeis, OS (1991). *Fiber Optic Chemical Sensors and Biosensors*. CRC Press, Boca Raton, FL.

Yang, L, and Ellington, AD (2005). Prospects for the *de novo* design of nucleic acid biosensors. In: Thompson, RB (ed.), *Fluorescence Sensors and Biosensors*. CRC Press, Boca Raton, FL.

Yao, G, Fang, X, Yokota, H, Yanagida, T, and Tan, W (2003). Monitoring molecular beacon DNA probe hybridization at the single-molecule level. *Chem-Eur J* **9**(22):5686–5692.

Zeng, H-H, Bozym, RA, Rosenthal, RE, Fiskum, G, Cotto-Cumba, C, Westerberg, N, Fierke, CA, Stoddard, A, Cramer, ML, Frederickson, CJ, and Thompson, RB (2005). *In situ* measurement of free zinc in an ischemia model and cell culture using a ratiometric fluorescence-based biosensor. *SPIE Conference on Advanced Biomedical and Clinical Diagnostic Systems III* San Jose, CA, pp. 51–59.

Zeng, HH, Thompson, RB, Maliwal, BP, Fones, GR, Moffett, JW, and Fierke, CA (2003). Real-time determination of picomolar free Cu(II) in seawater using a fluorescence-based fiber optic biosensor. *Anal Chem* **75**(24):6807–6812.

CHAPTER 10

ELECTROCHEMISTRY INSIDE AND OUTSIDE SINGLE NERVE CELLS

DANIEL J. EVES AND ANDREW G. EWING

10.1. ELECTROCHEMISTRY INSIDE AND OUTSIDE SINGLE NERVE CELLS

Single-cell analysis can be a useful tool to understand neuronal communication. Specifically, the use of electrochemical detection has led to an increased understanding of the interactions of cellular compartments as they relate to neurotransmitter release. Various types of electrochemical detection are employed in single-cell experiments, and each method provides information about neuronal communication. In this chapter, we describe the development of model neuronal systems and electrochemical techniques for this area.

The brain is a complex network of cells, and this complexity limits our ability to unravel detailed mechanisms of neurotransmission. In order to remove some of this complexity, single cells are frequently studied as a model system. When single cells are examined, the environment can be closely controlled and direct correlations between treatment with a pharmacological agent and its effect can be determined. This closely controlled environment facilitates the study of important cellular processes, like exocytosis, which can be monitored by electrochemistry.

Exocytosis appears to play a major role in cell-to-cell communication. Cells package neurotransmitters into vesicles that are released at the plasma membrane upon stimulus. There are three stages of exocytosis, which have been classified as docking, fusion pore formation, and complete fusion. After fusion, neurotransmitters cross the synapse and bind to related receptors on the postsynaptic terminal, thus propagating chemical signals. Because of the nature of some transmitters, they are detectable by electrochemistry.

Electrochemistry helps to identify and/or quantify neurotransmitters in single vesicles. The electrochemical techniques frequently used for this identification include amperometry, fast scan cyclic voltammetry, and the electrophysiological technique, patch clamp. In amperometry, electroactive species are oxidized at constant potential, thus generating a current that is measured by a potentiostat (Finnegan et al., 1996; Wightman et al., 1991). Fast-scan cyclic voltammetry can be used to provide information about the identity of the molecule in the solution by examining characteristic oxidation–reduction peaks (Travis and Wightman, 1998). Another important technique for measuring exocytosis is patch clamp. Here a small glass capillary is adhered to the cellular membrane to monitor changes in capacitance. Patch clamp has been coupled with amperometry to correlate amounts of release with capacitance

---

New Frontiers in Ultrasensitive Bioanalysis. Edited by Xiao-Hong Nancy Xu
Copyright © 2007 John Wiley & Sons, Inc.

(Albillos et al., 1997; Alés et al., 1999; Tabares et al., 2003). The remainder of the chapter will discuss these techniques in greater detail.

10.2. AMPEROMETRY

Amperometry is well-suited to measuring secretion from cells because of its ability to quantify release from vesicles on a millisecond time scale (Travis and Wightman, 1998) (Figure 10.1). In amperometry, a small electrode is placed near the cell and held at a potential where oxidation is diffusion-limited. A stimulant is applied with a small pipette (Figure 10.1A), resulting in current transients for each exocytosis event. The general oxidation reaction for catechols is shown in Figure 10.1B, and a typical current–time trace is Figure 10.1C. The number of molecules detected from a release event can be calculated using Faraday's law (Figure 10.1D). Experimental determination of charge allows calculation of the number of moles of molecules detected by rearranging the Faraday equation.

For amperometry in biological systems, a carbon fiber electrode is used. In most cases the electrode is constructed in-house; this construction is carried out by aspirating a carbon fiber into a glass capillary, heating and pulling the capillary to a small tip around the fiber, sealing

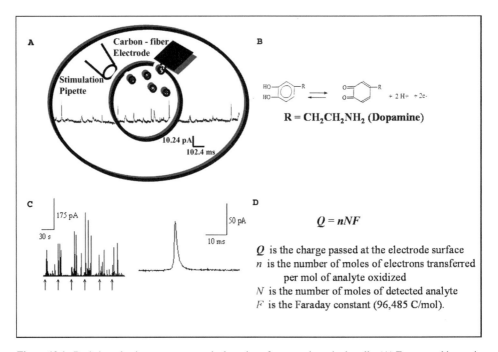

Figure 10.1. Basic introduction to amperometric detection of exocytosis at single cells. (A) Foreground image is the typical setup for amperometry for single cell. Exocytosis is stimulated by a pipette containing a stimulant and the release is monitored by a carbon fiber electrode. The back ground image shows typical amperometric data. See insert for color representation of this figure. (B) The oxidation reaction for catecholamines. The catecholamine is oxidized to the ortho-quinone form, losing two electrons. (C) Left spectra shows a series of stimulations, denoted by the arrows, and the electrochemical responses detected after each stimulation. The right-hand spectra show a single amperometric spike. (D) Faraday's equation, which is used to determine the amount of material released during exocytosis. (Courtesy of Dr. Leslie A. Sombers.)

the fiber in the capillary with epoxy, cutting the electrode and beveling at a 45° angle. The electrode is typically tested by a single sweep in dopamine and used for analysis (Wightman et al., 1991). The electrode is positioned near a cell where exuded neurotransmitter can be detected (Figure 10.1A). When the electrode is placed flush against the cell, there is a small amount of space, about 300 nm, between the cell and the electrode (Anderson et al., 1999; Cans et al., 2003a).

The characteristic shape of amperometric peaks aids in understanding the various aspects of the exocytosis event. The peak shape is dictated by the flow of electroactive agents to the surface of the electrode. Since the electrode is held at a constant potential, 650 mV (versus SSCE), the oxidation of the molecule generates a current that is measured by a potentiostat (Finnegan et al., 1996; Wightman et al., 1991). This current versus time trace can be analyzed to provide information about exocytosis (Figure 10.1C). Specifically, the half-width of the peak (full width at half-maximum) measures the duration of the exocytotic event. The rise time, or the time it takes to go from 10% to 90% of the peak height, equates to the time it takes the fusion pore to open. Thus, the basic amperometric measurement provides data on the amount of transmitter released, the duration of each event, and the opening of the fusion pore.

Amperometry has been used to monitor the release of vesicle fusion in different types of cells with various neurotransmitters. Adrenal chromaffin cells package epinephrine and norepinephrine (Elhamdani et al., 2002; Pihel et al., 1994), and pheochromocytoma 12 cells release dopamine when stimulated (Colliver et al., 2000; Finnegan et al., 1996; Sombers et al., 2004). Mast cell vesicles contain both serotonin and histamine (Jaffe et al., 2004), while pancreatic β cells release insulin but can be loaded with serotonin (Qian et al., 2000, 2003). In order to present the different possibilities of electrochemical detection in cells, the following section has been arranged by cell type with brief discussions of discoveries made using amperometry.

10.3. EXPERIMENTS WITH ADRENAL CHROMAFFIN CELLS

Adrenal chromaffin cells have been examined from various animals and are known to release epinephrine and norepinephrine. These cells have been the most widely used in amperometric studies to understand exocytosis. One interesting feature of these cells is a dense core at the center of the vesicles. This dense core is composed of, among other things, an anionic polypeptide called chromogranin A. The cationic messenger molecules are thought to associate with this polypeptide. During exocytosis, the vesicle fuses, beginning with the formation of a fusion pore. When the fusion pore forms, an influx of extracellular solution changes the pH of the vesicular solution. As the solution in the vesicle becomes more basic, chromogranin A releases the neurotransmitters associated in the matrix and they move into the synapse. Chromogranin A has been hypothesized to force open the fusion pore during full exocytosis (Amatore et al., 2005). The following two sections briefly describe the use of amperometry of adrenal cells to investigate novel cell biology.

10.3.1. Breaking Up F-Actin Increases Release of Catecholamines

Amperometry has been used to study the role of F-actin in exocytosis. F-actin makes up a significant portion of the cytoskeleton of the cell. The cytoskeleton gives support to the plasma membrane, which plays an integral part in exocytosis. Ñeco et al. (2003)

have investigated the effect of a phospholipase type A2 neurotoxin, taipoxin, on stimulated exocytosis. Following 1-hour incubation, taipoxin accumulates along the plasma membrane. Fluorescence microscopy was used to determine that this aggregation is around the F-actin in the cytoskeleton. Amperometry was employed to detect basal vesicular release from taipoxin treated cells, and no difference was noted when compared to control cells. A comparison of amperometric spikes from the stimulated release from taipoxin cells showed a small increase in the number of exocytotic events, thereby increasing catecholamine release. In order to identify the mechanism by which taipoxin affects exocytosis latrunculin A, an F-actin disassembling agent was also added to cells and the amperometry was compared. An increase in release was observed for latrunculin A-treated cells, but it was not significantly different from the increase observed with taipoxin; also, when treated with both additives, there was no variation in the increased exocytosis (Ñeco et al., 2003). This result suggests that taipoxin disrupts F-actin in the same manner as latrunculin A. In order to further clarify the pathway by which taipoxin increases release events, the chromaffin cells were permeablized with digitonin (Ñeco et al., 2003). Digitonin creates holes in the cellular membrane and allows taipoxin to directly enter into the cell and bypass ion channels, particularly Ca^{2+}, in the membrane. Stimulation of permeablized cells containing taipoxin showed an increase in release events, especially in the first few minutes. The area under the amperometric peaks was cumulatively combined and showed both a faster rise in and greater overall area for taipoxin-treated cells. Taken collectively, increased exocytotic activity and degradation of F-actin indicate that taipoxin functions independently of extracellular Ca^{2+} concentrations. This study also showed that breaking up F-actin caused an increase in catecholamine release.

10.3.2. Effect of Aging on Fight-or-Flight Response

Amperometry has also been used to study the fight-or-flight response in chromaffin cells. The fight-or-flight response happens when an animal perceives danger and a physiological response occurs, causing an increase in epinephrine which allows the animal to fight or flee. The fight-or-flight response has been postulated to be less pronounced as age increases. Previous studies of the effects of aging have been conducted using bovine adrenal cells (Elhamdani et al., 1998). The findings indicate a facilitation of L-type Ca^{2+} channels, which are correlated with release of epinephrine in the cells of younger animals. As the animals aged, the channel responsible for this facilitation was absent and the amount of released epinephrine detected decreased. Elhamdani et al. (2002) investigated the correlation of Ca^{2+} influx into cells with epinephrine release in human cells. In order to test the response in both young and old humans, adrenal cells were taken from a large number of human subjects, ages ranging from 13 to 50. The most outstanding difference was found between cells from 26- and 46-year-old male subjects. When Ca^{2+} levels were monitored in these cells, using fluorescence, a strong signal owing to Ca^{2+} influx was observed when young cells were electrically stimulated and compared with older cells. Amperometry was subsequently employed to quantify the amount of release observed from the two classes of cells, young and old. The young human cells had twice as many events as the older cells. In order to correlate these findings with the work done in bovine adrenal cells, the human cells were compared to a calf adrenal cell. In this case, the unstimulated amperometric signals were similar, but upon stimulation the calf cell had twice as many events as the young human cell. The conclusion drawn was that younger human cells also have a good correlation between Ca^{2+} efflux and epinephrine release and therefore had the same Ca^{2+}

channel. Lack of the channel facilitating this correlation in the aging samples explains the gradual loss of the fight-or-flight response.

10.4. EXPERIMENTS AT PC12 CELLS

Pheochromocytoma (PC) 12 is another type of cell that has been used for amperometric studies. One aspect of these cells that is of particular interest is that they function similarly to sympathetic ganglion neurons (Greene and Tischler, 1976). PC12 cells are an immortalized cell line derived from a rat tumor. These cells have small vesicles, around 150 nm in diameter, and generally release the neurotransmitter dopamine (Schubert and Klier, 1977). Chen et al. (1994) measured release from individual exocytosis events at PC12 cells with amperometry, and this model cell line has been widely used since then for studies of exocytosis. The following examples illustrate the power of amperometry to examine exocytotic events at PC12 cells.

10.4.1. The Volume of Vesicles Change Following Increase or Decrease of Dopamine

In 2000, Colliver et al. showed that increasing dopamine levels in PC12 cells by incubation with L-dopa or decreasing dopamine by incubation with reserpine increased or decreased, respectively, average vesicle size as well as dopamine content in vesicles (Colliver et al., 2000). More recently, Sombers et al. (2004) have used the ability to vary vesicle size to examine the effect of vesicle size on release via the fusion pore. During the initial phase of exocytosis, the vesicle fuses with the cell membrane, creating a narrow pore structure between the two. In some release events, this structure is transiently stable and amperometric detection traces exhibit a small pre-spike feature termed the foot of the amperometric signal (Alvarez de Toledo et al., 1993; Chow et al., 1992). This is thought to result from messenger compound diffusing through the open pore.

Although L-dopa and reserpine increase and decrease vesicle size, respectively, the dense core of the vesicle is mostly unchanged (Colliver et al., 2000). This dense core is also made of primarily the protein chromogranin A and is thought to expand during exocytosis, pushing open the fusion pore (Amatore et al., 2005). Although this is a likely outcome with adrenal chromaffin cell vesicles, it appears that the fusion pore in PC12 cell vesicles can pass through a transition state involving a lipid nanotube. Under conditions where the dense core expands, placing tension on the vesicular membrane, the nanotube is stabilized and constricts. This mechanism has been proposed to explain a higher occurrence of events with a foot for the smaller vesicles with larger ratio of dense core to vesicle volume, following treatment with reserpine (Sombers et al., 2004).

In some cases the fusion pore opens and closes without a full exocytosis even occurring. This phenomenon, termed kiss and run, has been observed with amperometry (Staal et al., 2004). If these events represent a significant communication pathway, then the cell's ability to modify vesicle size could be a mechanism of neuronal plasticity (learning and memory).

10.4.2. Investigation of the Role of Munc18 Protein in Exocytosis

Amperometry on PC12 cells has been used to elucidate protein interactions that regulate cellular machinery. The part of the cellular machinery that recruits vesicles to the plasma membrane is called the SNARE (SNAP receptor, where SNAP is soluble N-ethylmaleimide-sensitive factor-attachment protein) complex (Burgoyne et al., 2001). A syntaxin-binding protein, Munc18, has been shown to associate with the SNARE complex and is thought to play a role in regulating exocytosis (Graham et al., 2002). Previous models have shown that Munc18 dissociates from syntaxin 1 prior to SNARE-complex formation and therefore cannot exert any control on the later stages of exocytosis.

Based on data presented regarding a point mutation in a Munc18 homologue in Drosophila (Wu et al., 1998), Burgoyne et al. transfected PC12 cells with a mutated Munc18 protein (Burgoyne et al., 2001). Munc 18 was believed to dissociate from syntaxin, and arginine residue 39 was identified to interact directly with syntaxin, the arginine was changed to a cysteine to reduce this interaction (Misura et al., 2000). Cells transfected with the mutation in Munc18 displayed faster vesicle fusion during exocytosis. This was noted through a decrease in both amplitude and halfwidth of the amperometric signals. The authors suggested that the reduction in amperometric signal resulted from kiss and run fusion. The conclusion drawn from this work is that Munc18 interacts with the SNARE complex and augments the later stages of exocytosis (Burgoyne et al., 2001). It is possible that opening of the fusion pore is kinetically hindered by the SNARE complex needed for its formation. Munc 18 might facilitate disassembly of the SNARE complex once the pore is formed, thereby leading to faster opening.

10.5. EXPERIMENTS AT MAST CELLS

Another cell type that has been extensively studied using amperometry is mast cells. Mast cells contain the neurotransmitter serotonin and the neurohormone histamine and exocytotic release is similar to that thought to take place in neurons and endocrine cells (Pihel et al., 1998). An interesting aspect of mast cells is that the granules (vesicles) are large enough to be visualized optically, sometimes greater than 1 μm in diameter (Jaffe et al., 2004). Due to the large size of the vesicles, mast cells are also a good system to combine video microscopy with amperometry. The following experiments have been carried out to examine the use of amperometry to clarify the role of Ca^{2+} in exocytosis.

10.5.1. Blocking Ryanodine Channels Affects Exocytosis

Amperometry has been used to investigate the role of Ca^{2+} and ryanodine receptors in mast cells. Ryanodine receptors have been identified to modulate exocytosis by triggering a release of internal Ca^{2+} stores. Previous studies of mast cells reported that histamine release was inhibited by the presence of ryanodine, which would suggest that exocytosis was modulated by ryanodine, receptors (Takei et al., 1992). Preliminary data showed that small concentrations of ryanodine and caffeine increased intracellular Ca^{2+} levels in mast cells with no external Ca^{2+}, suggesting an increase in exocytosis. Jaffe et al. (2004) employed amperometry to investigate the role of ryanodine and caffeine in regulating serotonin release from mast cells. A control group of mast cells was stimulated with a mast cell activator, 48/80, which stimulates exocytosis. These control cells were compared to mast cells treated with both a low (1 μM) and high (50 μM) dose of ryanodine. The low dose showed an increase in both amplitude and frequency when compared to control, while the higher dose greatly reduced the half-with and spike area but not peak amplitude. Furthermore, addition of caffeine increased exocytosis. Caffeine (20 mM) was added to mast cells and compared to control conditions, and the caffeine addition caused an increase in the number of events. Further experiments were conducted, and it was determined that caffeine increases exocytosis by increasing the release of intracellular Ca^{2+} stores. These data suggest that the combination of 1 µM ryanodine and 20 mM caffeine work by ryanodine opening ryanodine receptor channels, and caffeine enters the cell through the open receptor channels and facilitates the release of Ca^{2+} from intracellular stores. This study shows that there is a direct correlation between the ryanodine receptor and exocytosis (Jaffe et al., 2004).

10.5.2. Correlation of Ca^{2+} Levels and Vesicle Secretion

Amperometry can be coupled with fluorescent microscopy to give information about vesicle release in mast cells. It has been shown that an increase in Ca^{2+} plays a role in secretion from excitable cells, but there is little known about the effect of Ca^{2+} on nonexcitable cells. While monitoring the intracellular Ca^{2+} levels in mucosal mast cells, serotonin secretion was monitored (Figure 10.2). Increased Ca^{2+} levels correlated to higher serotonin release. When cells were depolarized by K^+, serotonin secretion dropped off, but when repolarized with Na^+, release returned to normal levels. In addition, Ca^{2+} levels were decreased and increased, respectively, and the majority of release events were clustered around

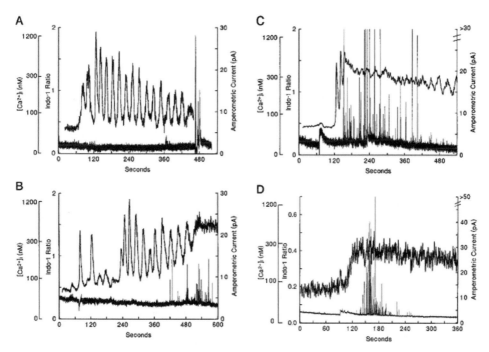

Figure 10.2. Upper traces are calcium level as determined by Indo-1 Ca^{2+}-sensitive dye. Lower traces are amperometric data correlating to the calcium imaging. (A) Shows a decreasing oscillation in calcium levels and very little exocytosis until the end of the spectrum. (B) An increasing oscillation of calcium and an increase of amperometric peaks that correlate with the rise in calcium. (C) A direct increase in calcium that causes exocytosis. (D) A smaller increase in calcium which yields a burst of amperometric spikes. [Reprinted from Kim et al. (1997), with permission.]

the peak Ca^{2+} concentration (Figure 10.2A). When Ca^{2+} levels were steadily increased, exocytosis was persistent (Figure 10.2B). High calcium levels lead to a long prolonged release, and small increases in calcium lead to limited release when the initial value of calcium was varied (Figure 10.2C, D). Coupling calcium imaging with amperometry, the connection between intracellular calcium and exocytotic release was clarified for these nonexcitable mast cells (Kim et al., 1997).

10.6. EXPERIMENTS IN PANCREATIC β-CELLS

Another cell model that has been used to study exocytosis with amperometric detection is the pancreatic β-cell. β-cells naturally release the peptide insulin, but insulin is not easily oxidized with an untreated electrode making amperometry difficult. In order to detect insulin, Huang et al. (1995) coated electrodes with ruthenium oxide and cyanoruthenate. This coating made it possible to catalyze the oxidation of a disulfide bond on the insulin molecule and generate an oxidizing current. A drawback to this method is that the coating degrades over time, so another coating has been devised with greater stability. This latter coating was applied by placing the clean carbon fiber electrode in a solution of RuCl$_3$ and HClO$_4$ and cycling the potential between -0.85 and 0.65 V (versus SSCE) at 100 V/s for 15 minutes. This process produced a coating that could be used for up to 3 days (Gorski et al., 1997). Another method that has been employed to detect release from β-cells is to incubate them with serotonin, which is electroactive and coreleases with insulin, and to monitor its release with conventional amperometry. The following experiments discuss the uses of amperometry in monitoring exocytotic regulatory processes in β-cells.

10.6.1. Differentiation of Fast and Slow Exocytosis in Pancreatic β-Cells

One type of research conducted with pancreatic β-cells was that of Takahashi et al. (1997), were interested in examining both fast and slow exocytosis. This reportedly takes place in two different types of vesicles: small clear and large dense-core vesicles. Small clear vesicles have been reported to undergo fast release, whereas large dense-core-containing vesicles tend toward slower release. Dense-core vesicles contain, among other things, a polyanionic peptide called chromogranin A. It is hypothesized that neurotransmitters in these vesicles are bound inside the dense core. Upon initial fusion of the vesicle to the plasma membrane, the dense core, which is in the low pH environment in the vesicle, is exposed to the more basic environment outside the cell and dissociates and releases the neurotransmitters. This hypothesis assumes that unraveling of the dense core takes time and accounts for the increased time for exocytosis. In order to differentiate between the two types of vesicles, amperomeric spikes were compared. The average time difference between fast and slow release was 1.5 s, an easily differentiated time difference as amperometry has submillisecond resolution. Ca^{2+} plays a large role in cell signaling for exocytosis and in order to control the Ca^{2+} influx in this experiment, photo releasable caged Ca^{2+} was used. Slow and fast releases were also monitored using patch clamp techniques. This method of analysis will be discussed in more detail later in the chapter.

10.6.2. Measuring the Quantal Release of Serotonin

Amperometry was used to measure quantal release of serotonin from β-cells. Zhou and Misler (1996) were able to detect the Ca^{2+} induced release inside of the β-cells with amperometry by bathing the cells in a solution of serotonin to load into vesicles before the experiment. These cells were also incubated with forskolin to increase the Ca^{2+} dependence for exocytotic release. Secretagogues, tolubutamide, a pharmacological agent, and KCl were used to stimulate the cell to release. In the absence of external Ca^{2+}, both tolubutamine and KCl failed to stimulate release. By causing the β-cells to be dependent upon Ca^{2+}, strict control of exocytosis could be maintained. This was necessary to determine the quantal size of the serotonin-loaded vesicles. Using Faraday's law, the quantal size for serotonin loaded β-cells was calculated to range from 0.88 to 1.76×10^5 molecules. This range is smaller than that calculated for the amount of catecholamine released during individual events from chromaffin cells, but is it larger than that calculated for release from events involving synaptic vesicles released from an axon varicosity.

10.7. ARTIFICIAL LIPOSOMAL CELL MODELS OF EXOCYTOSIS

Thus far, cellular model systems have been examined with amperometry. Each of these cell systems affords a chance to examine how various proteins regulate exocytosis. In order to more fully explore the forces that regulate exocytosis, another model system has been devised to explore the role of membranes in neurotransmitter release. This system utilizes liposomes to mimic exocytosis and is illustrated in this section.

A liposome is a spherical configuration for a lipid bilayer that can approximate the environment inside a cell or cellular compartment (Chiu et al., 1999). Zare's laboratory and later Orwar's laboratory started using liposomes as microcontainers for reactions (Davidson et al., 2003; Karlsson et al., 2001, 2003; Moscho et al., 1996). To introduce reagents into the liposomes, a micropipette system can be used with electroporation of the membrane. Proof of injection was verified optically by injecting fluorescein into a giant unilamellar liposome (Karlsson et al., 2000). These liposomes have been manipulated into forming networks of vesicles all connected by lipid nanotubes (Karlsson et al., 2001).

A liposome model for exocytosis has been devised that examines the final stages of vesicle fusion. This model mimics the transition from fusion pore formation to full fusion (Figure 10.3F–I). Release of molecules loaded into the vesicle can be monitored by amperometry, and fusion kinetics can be extrapolated from these measurements. In the liposome model, all protein is removed to eliminate its role in the fusion event. Fusion proteins have been identified to play a central role in fusion pore formation; thus, formation of the fusion pore is not measured here.

In order to create the liposome model for exocytosis, a unilamellar liposome attached to a multilamellar liposome is used. The multilamellar liposome acts as a source of lipid for the unilamellar liposome when it is manipulated. Manipulation of the liposomal system including inflation, nanotube extrusion, and deflation is accomplished with a pulled glass pipette having a tip diameter that is less than 1 μm. A potential is applied across the liposome to disrupt the membrane and allow the pipette to enter into the liposome (Figure 10.3A). To create a model of exocytosis, the pipette is pushed through the opposite side of the liposome (Figure 10.3B) and then drawn back inside. The lipid material of the bilayer adheres to the tip and continues inside the liposome in the form of a lipid nanotube (Figure 10.3C). Inflation

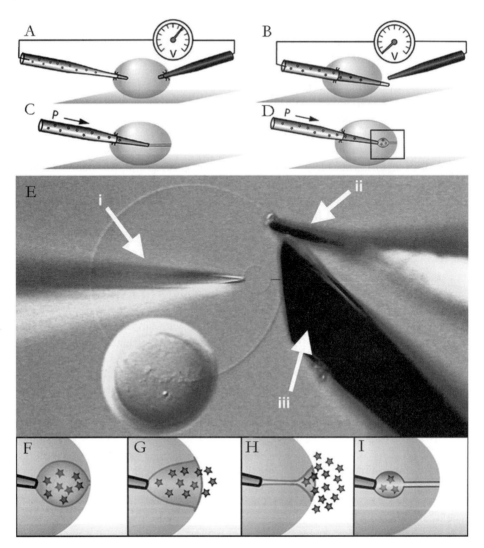

Figure 10.3. (A) A potential is applied between the injection pipette and counter electrode which allows the pipette to enter into the liposome. (B) The pipette continues through the liposome and exits on the other side, and a voltage was applied for this exit as well. (C) The pipette is drawn back inside of the liposome and part of the lipid bilayer adheres to the tip and forms a nanotube. (D) The pressure exuding from the pipette causes a vesicle to form on the end. (E) An experimental view of an artificial cell; the smaller liposome (vesicle) is attached to the pipette (i) on end and to the larger liposome by a nanotube. The nanotube location is marked by a black line since it is not visible by microscopy. Also pictured are the counterelectrode for electroporation (ii) and the detection electrode beveled at 45° (iii). (F–I) Illustration of constant flow from an injection pipette. As the vesicle grows, the nanotube shortens and then spontaneous fusion takes place. After full fusion a vesicle forms again and continues the process. [Reprinted from Cans et al. (2003b), with permission.] See insert for color representation of this figure.

of an inner liposome at the end of the nanotube (Figure 10.3D,E) eventually leads to an unstable pore-like structure between the inner and outer liposomes. This resembles a cell about to undergo the final stage of exocytosis, and opening results in release into the outer solution. The larger liposome acts as a plasma membrane to which the smaller liposome (vesicle) is fused (Figure 10.3F–H). Because they are released easily, oxidized substances are detected at an electrode placed right outside the liposome. A useful aspect of this system is that the distention repeats almost indefinitely (Figure 10.3F–I). Thus it is straightforward to vary conditions such as vesicle size, temperature, and lipid composition with consistent controls without the variability inherent in biological systems (Cans et al., 2003b).

Data from the liposome model have been compared with that obtained from PC12 cells. The liposome model exhibited similar amperometric peak attributes. The pre-spike foot was visible and though the liposomes were larger in size than the PC12 cell vesicles, the release was proportional allowing direct comparison. The added advantage of the liposome model is that it allows analysis of neurotransmitters released via the nanotube, which approximates the fusion pore in cells at its shortest length. With the absence of fusion proteins, it was shown that lipids could drive the opening of the vesicle during exocytosis. Cans et al. (2003b) were able to show that the absence of proteins does not hinder the expansion of a fusion pore in exocytosis. It was shown that for vesicles of like sizes, the time required for full fusion was faster in the liposome model than in a cell. This demonstrates that lipid mechanics is sufficient to drive the opening of the vesicle during exocytosis and leads to speculation that proteins involved in creating the fusion pore might actually limit opening during the final stage of exocytosis.

10.8. FAST-SCAN CYCLIC VOLTAMMETRY

Cyclic voltammetry has also been adopted in order to measure release from single cells. In contrast to amperometry where the signal is proportional to the total mass of messenger released, in the cyclic voltammetry the signal is proportional to the local concentration of electroactive molecules and cyclic voltammetry has been shown to have nanomolar detection limits. The advent of microelctrodes has led to the ability to scan at much higher scan rates than with conventional electrodes. High scan rates provide a means to obtain voltammograms in 10 ms or less, and thus this procedure gives time resolution necessary to monitor exocytosis events. To detect catecholamines, the voltage sweep typically begins at -0.4 V (versus a sodium-saturated calomel reference electrode), ramps to $+1.1$ V, and then goes back to -0.4 V. A typical scan rate is in the range of 800 V/s (Travis and Wightman, 1998). The rapid sweep rate creates a high background current, and therefore background becomes very important to identify what is typically a very small signal in a large background (Figure 10.4).

10.8.1. Single-Cell Experiments with Fast-Scan Cyclic Voltammetry

Fast-scan cyclic voltammetry (FSCV) has been used to identify electroactive species release during exocytosis. Wightman and co-workers used 10-μm carbon fiber electrodes to detect corelease of histamine and serotonin from mast cells (Travis and Wightman, 1998). FSCV has been used to distinguish between two different catecholamines thought to be present in bovine adrenal cells, namely, epinephrine and norepinephrine. When scanned over the potential range from -0.4 to 1.1 V, they appear to be identical. When the scan range was adjusted to 0.0 to $+1.425$ V, epinephrine registers an oxidation peak due to a cyclization reaction of the oxidized product and an added re-reduction peak that can be quantified. This reaction is much less prevalent when norepinephrine is oxidized. When measuring exocytosis from adrenal cells, 20% of the cells analyzed had some events where there was a corelease of both catecholamines, but each event contained at least one of oxidizable

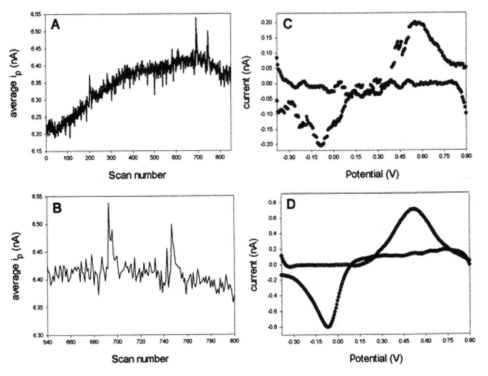

Figure 10.4. (A) Average current within a 60-mV window of the peak oxidation potential of dopamine. Individual current spikes represent faradaic current. (B) Enlargement of the region between scans 600 and 800, showing two current transients used to generate voltammograms. (C) Background-subtracted voltammogram generated by subtracting the average signal of scans 687–690 from the signal of scan 693. (D) Background-subtracted voltammograms of a standard 25 μM dopamine solution. [Reprinted from Kozminski et al. (1998), with permission.]

catecholamine (Pihel et al., 1994). FSCV makes it possible to differentiate between multiple neurotransmitters in the same cell.

10.9. PATCH AMPEROMETRY

Amperometry and patch clamp techniques have been combined to provide a powerful method for the study of exocytosis in single cells is a combination of two different techniques called patch amperometry (Figure 10.5). The patch clamp techniqie records capacitance of the cell as a function of time. Developed by Erwin Neher and Bert Sakmann, it can be used to measure changes in the plasma membrane, including the opening of ion channels and changes in total membrane area such as those related to vesicle fusion and recapture (Sakmann and Neher, 1984). During exocytosis, the capacitance of the cell increases because the membrane of fusing vesicles becomes incorporated in the plasma membrane, thereby increasing the surface area and thus the membrane capacitance. This can be measured by placing a glass capillary with a small opening, typically about 1 μm on the membrane, forming a seal with greater than gigaohm resistance. Loss of membrane area during endocytosis leads to a decrease in membrane capacitance, and this can be measured with the patch technique as well (Sakmann and Neher, 1984).

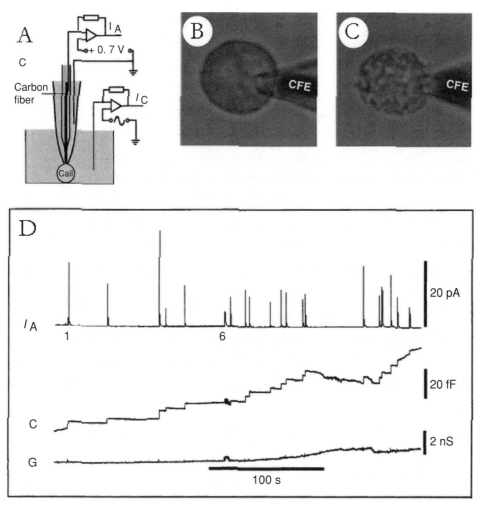

Figure 10.5. (A) Electrical configuration of a patch amperometry electrode/pipette. I_A is the amperometric current (0.7 V versus a Ag/AgCl reference electrode), and I_C is the sine wave current used to measure capacitance. (B) A pipette attached to a chromaffin cell before the experiment. (C) A chromaffin cell after an experiment. (D) The top current trace (I_A) corresponds to the amperometric data. The middle trace (C) is the capacitance trace. The bottom trace (G) is the conductance trace. [Reprinted from Albillos et al. (1997), with permission.]

Combining this technique with amperometry provides a method to correlate the changes in membrane area with release measured at a microelectrode. This is done by placing a carbon fiber inside the patch pipette. Thus capacitance and amperometric data are simultaneously recorded in the same location. This has an added advantage in that the patch pipette affords a more controlled environment for the patched portion of the membrane. Variation of the pipette solution can be used to apply specific conditions to the section of patched membrane. Single cells have been analyzed separately by patch clamp and amperometry, but by placing a carbon fiber electrode in the patch pipette, it makes it possible to study the pore opening and closing at the same area of membrane. Amperometric spikes can be correlated to capacitance, so the amount of messenger released is synchronized to the opening of the pore and can be used to simultaneously determine the vesicular concentration of neurotransmitters (Albillos et al., 1997; Gong et al., 2003). Patch amperometry has also been used to measure the cytoplasmic concentration of messenger molecules (Mosharov et al., 2003).

10.9.1. Initial Use of Patch Amperometry

Patch amperometry was first used to monitor exocytosis at chromaffin cells. In this initial study, Albillos et al. (1997) used adrenal cells as a model system for simultaneous recording of amperometry and capacitance at the cell surface. It was thought that the tightening of the membrane, which occurs when the patch is applied to the cell, would cause release, as in frog neuromuscular junctions, but this was not observed. Stimulation of the cell was accomplished, therefore, by including Ca^{2+} in the patch pipette solution. The vesicular concentration of catecholamine was estimated to be 0.7 M by estimating vesicle size from individual capacitance steps and comparing this to the amount of catecholamine released measured by amperometry. This estimate assumed total neurotransmitter release (Albillos et al., 1997). Cells that were treated with phorbol myristate acetate, an agent that increases the number of vesicles in the readily releasable pool by inhibiting the calcium-dependent protein kinase, had a slightly higher average concentration of 0.96 M. Also in this experiment, feet were visible in the amperometric data. In some instances, the fusion pore was stable enough to allow a significant amount of catechol through before the full fusion event. The ability to combine capacitance measurements and amperometric data should help to better understand the relationship of fusion pore opening and release of messenger (Albillos et al., 1997).

10.9.2. Kiss and Run Exocytosis Is Aided by High Ca^{2+} Concentration

Patch amperometry affords a unique opportunity to easily control the extracellular environment for a small portion of a cell. Using chromaffin cells, Alés et al. (1999) placed increasingly higher concentrations of Ca^{2+} (5 and 90 mM) in the patch pipette. At 5 mM Ca^{2+} the majority of release events showed an increase in capacitance, an earmark of full fusion, but when 90 mM Ca^{2+} was administered, the bulk of the events did not have a lasting increase on the capacitance. These events flickered up and then back down. An analysis of the amperometric peaks of flickering capacitance versus lasting capacitance showed that the amperometric spikes had a similarly normal distribution for both types of fusion. They did not observe a significant difference in either capacitance or amperometry between the two types of events other than the difference in permanent versus flickering conductance. This experiment suggests that a local increase in Ca^{2+} concentration sparks an increase in kiss-and-run exocytosis, but a significant difference was not observed in the amount of neurotransmitter released from kiss and run when compared to normal fusion for chromaffin cells. It is not clear if these results were truly kiss-and-run or rapid recycling following full fusion. The authors suggest that kiss-and-run fusion is a good way for the cell to rapidly recycle vesicles (Alés et al., 1999).

The three major types of electrochemistry used for single cell measurements of exocytosis have been described, and examples of the application of each method have been illustrated. The following section discusses improvements that have been made to aid in better detection in single-cell analysis.

10.10. SINGLE-CELL MEASUREMENTS IN MICROVIALS

One of the challenges in electrochemical detection of exocytosis is that the vesicular concentration of neurotransmitters is greatly reduced upon release and, if this release is not trapped under a closely spaced electrode, the neurotransmitters diffuse into solution at a rapid rate. This dilution of released messenger decreases the likelihood of being able to detect complete release from an entire single cell. Attempts are being made to reduce the amount of dilution that takes place before detection and to detect the total release from a single cell. One method employed is to reduce the size of the dish holding the cell. By reducing the overall volume of the work area, the dilution of sample is also reduced. Electrochemical detection in small volumes allows for study of dynamic events at a single cell (Troyer and Wightman, 2002). With the emergence of micro- and nanofabrication techniques, it has become possible to construct very small compartments to contain cells for analysis.

10.10.1. Microvials and Amperometry

Clark and Ewing carried out electrochemistry in microvials that varied from 1-nL to 9-pL on a silicon wafer (Clark et al., 1997). To provide transparent microvials, Clark et al. constructed vials by imprinting a silicon mold into polystyrene (Figure 10.6) with vial sizes ranging from 310 to 0.4 pL. These vials were used with capillary separation techniques and for electrochemistry.

Use of smaller microvials results in solutions with a high surface-area-to-volume ratio and thus high evaporation rate. Complete evaporation occurs in only a few seconds for 16-pL vial. In order to reduce the amount of evaporation, glycerol was added to lower the vapor pressure of the sample. Using the microvial technology, a single bovine adrenal cell has been placed in a microvial and catecholamine release was measured by cyclic voltammetry giving a concentration of ~60 μMg (Clark et al., 1997).

10.10.2. Microvials and Fast-Scan Cyclic Voltammetry

Troyer and Wightman (2002) developed a technique that measures the active transport of electroactive messengers into single isolated cells immobilized in a microvial. In order to characterize transport, both fast-scan cyclic voltammetry and amperometry were employed. Known concentrations of dopamine were serially injected into microvials, ranging in size from 100 to 200 pL, and a concentration profile was obtained using cyclic voltammetry. Amperometric control experiments were carried out to measure the extent of catecholamine to the surface of the microvial. Injection of 550 amol of dopamine resulted in detection of 480 amol. Thus electrolysis of the sample is nearly complete. Subsequently, single HEK-293 cells were added to the vials, a known dopamine solution added, and uptake by individual cells was monitored. Using the Michaelis–Menten model of saturable kinetics, V_{max}, was calculated and provided a value within a factor of three of the literature value measured for a population of approximately one million cells with rotating disk voltammetry (Earles and Schenk, 1999). This difference was explained by the difference in the cell-volume-to-solution-volume ratio, which was 0.02 for the single-cell experiment and 0.006 for a rotating disk electrode experiment. Thus, the extracellular concentration of catecholamine appears to affect the equilibrium of the uptake transporter. The use of microvial technology provides a high-efficiency method to measure single-cell uptake transporters,

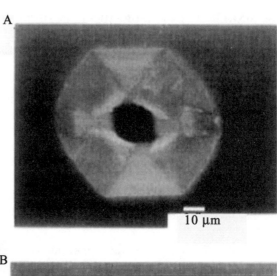
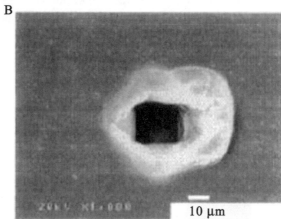
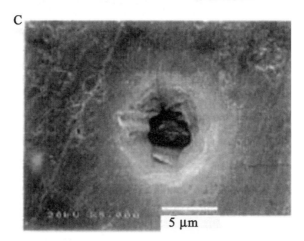

Figure 10.6. Picoliter microvials made by pressing a lithographically constructed mold into hot polystyrene. (A) 75, (B) 20, and (C) 0.4. [Reprinted from Clark et al. (1997), with permission.]

to examine single cell-to-cell heterogeneity, and to examine concentration effects on the transporter.

10.11. SINGLE-CELL DISCOVERIES MADE USING ELECTROCHEMISTRY

Having discussed possible improvements to aid single-cell detection, the following examples outline some of the important discoveries that have been made by use of electrochemical methods in single-cell analysis.

Single cells have been shown to release electroactive messengers from different size vesicles within the same cell type. Bruns et al. (2000) and Chen and Ewing (1995) were able to identify at least two separate vesicle sizes in single-cell experiments. Two distinct size distributions of vesicles were shown to be present in large neurons of the leech and pond snail. These experiments were performed at the cell body of neurons in culture and *in vivo*, respectively.

Cells can be loaded to provide larger signals by bathing with a solution, containing additional messenger (Colliver et al., 2000; Kozminski et al., 1998). In addition, Parkinson's disease symptoms are routinely treated with the dopamine precursor, L-dopa, presumably loading the cells in the brain with dopamine. Adding messenger or precursor had been observed to increase the signal in single-cell experiments, but it was unclear by what means this enhancement was taking place. Kozminski et al. (1998) showed that PC12 cells bathed with L-dopa released quanta with a larger amount of dopamine. Colliver et al. (2000) used amperometry and transmission electron microscopy to determine that the vesicle size in PC12 increased in the presence of excess L-dopa and decreased in the presence of reserpine. This was later corroborated for adrenal cells by Lindau and co-workers using patch amperometry (Gong et al., 2003).

Single synaptic fusion events observed in midbrain cultures have been shown to consist of both a simple and complex fusion (Staal et al., 2004). Complex fusion events were classified by the multiple fusion pore openings or what appears to be flickering fusion. This complex fusion has been hypothesized to be due to kiss-and-run-type fusion of a vesicle with release only via the open fusion pore. Amperometry makes it possible to identify this type of fusion because it records the events with microsecond temporal resolution and great sensitivity. Various pharmacological agents were added to enhance release from these vesicles. Staurosporine, for example, increased the frequency of complex events. Sequential amperometric spikes that would be expected for kiss-and-run events are expected to be two peaks in succession, with the second peak having a reduced amplitude and area. This is shown in Figure 10.7 and represents a milestone in measuring release via the kiss-and-run mechanism.

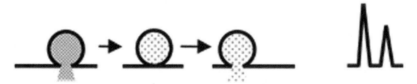

Figure 10.7. Representative amperometric spikes for kiss-and-run fusion along with a visual representation of what happens at the plasma membrane. [Reprinted from Staal et al. (2004), with permission.]

10.12. CONCLUSIONS

In this chapter we have outlined the use of electrochemical methods as they are applied to single-cell model systems. The focus of this work is on correlating these measurements to events occurring in and at nerve cells. The reductionist single-cell analysis approach provides a powerful tool to examine cell-to-cell heterogeneity, examine subcellular-based events, and examine temporal events that would be convoluted in a complex multicelled system. In total, this technology should help us to better understand neuronal communication. Although only applicable to molecules that can be easily oxidized at naked or modified electrodes, electrochemistry allows sensitive measurement of messenger secretion and transport with a high degree of temporal and spatial resolution. Amperometry has been used to monitor changes in fusion kinetics, quantal size, and cellular structure as it relates to exocytosis. Fast-scan cyclic voltammetry has helped to identify multiple neurotransmitter species released from the same cell and in characterizing new systems. Patch amperometry has increased our understanding of pore opening and release from vesicles. The use of microvial technology allows electrochemical detection of total release and the monitoring of active transport processes at singles cells.

ACKNOWLEDGMENTS

The authors gratefully acknowledge the work of many earlier co-workers in this area and the support of the National Institutes of Health and the National Science Foundation.

REFERENCES

Albillos, A, Dernick, G, Horstmann, H, Almers, W, de Toledo, GA, and Lindau, M (1997). The exocytotic event in chromaffin cells revealed by patch amperometry. *Nature* **389**(6650):509–512.

Alés, E, Tabares, L, Poyato, JM, Valero, V, Lindau, M, Alvarez de Toledo, G (1999). High calcium concentrations shift the mode of exocytosis to the kiss-and-run mechanism. *Nat Cell Biol* **1**:40–44.

Alvarez de Toledo, G, Fernandez-Chacon, R, and Fernandez, JM (1993). Release of secretory products during transient vesicle fusion. *Nature* **363**(6429):554–548.

Amatore, C, Arbault, S, Bonifas, I, Bouret, Y, Erard, M, Ewing, AG, and Sombers, LA (2005). Correlation between vesicle quantal size and fusion pore release in chromaffin cell exocytosis. *Biophys J* **88**:4411–4420.

Anderson, BB, Chen, G, Gutman, DA, and Ewing, AG (1999). Demonstration of two distributions of vesicle radius in the dopamine neuron of *Planorbis corneus* from electrochemical data. *J Neurosci Methods* **88**:153–161.

Bruns, D, Riedel, D, Klingauf, J, and Jahn, R (2000). Quantal release of serotonin. *Neuron* **28**:205–220.

Burgoyne, RD, Fisher, RJ, Graham, ME, Haynes, LP, and Morgan, A (2001). Control of membrane fusion dynamics during regulated exocytosis. *Biochem Soc Trans* **29**:467–472.

Cans, A-S, Wittenberg, N, Eves, D, Karlsson, R, Karlsson, A, Orwar, O, and Ewing, A (2003a). Amperometric detection of exocytosis in an artificial synapse. *Anal Chem* **75**:4168–4175.

Cans, A-S, Wittenberg, N, Karlsson, R, Sombers, L, Karlsson, M, Orwar, O, Ewing, A (2003b). Artificial cells: Unique insights into exocytosis using liposomes and lipid nanotubes. *Proc Natl Acad Sci USA* **100**:400–404.

Chen, G, and Ewing, AG (1995). Multiple classes of catecholamine vesicles observed during exocytosis from the *Planorbis* cell body. *Brain Res* **701**:167–174.

Chen, TK, Luo, G, and Ewing, AG (1994). Amperometric monitoring of stimulated catecholamine release from rat pheochromocytoma (PC12) cells at the zeptomole level. *Anal Chem* **66**:3031–3035.

Chiu, DT, Wilson, CF, Karlsson, A, Danielsson, A, Lundqvist, A, Stromberg, A, Ryttsen, F, Davidson, M, Nordholm, S, Orwar, O, and others (1999). Manipulating the biochemical nanoenvironment around single molecules contained within vesicles. *Chem Phys* **247**:133–139.

Chow, RH, von Ruden, L, and Neher, E (1992). Delay in vesicle fusion revealed by electrochemical monitoring of single secretory events in adrenal chromaffin cells. *Nature* **356**(6364):60–63.

Clark, RA, Hietpas, PB, and Ewing, AG (1997). Electrochemical analysis in picoliter microvials. *Anal Chem* **69**:259–263.

Colliver, TL, Pyott, SJ, Achalabun, M, and Ewing, AG (2000). VMAT-mediated changes in quantal size and vesicular volume. *J Neurosci* **20**:5276–5282.

Davidson, M, Karlsson, M, Sinclair, J, Sott, K, and Orwar, O (2003). Nanotube-vesicle networks with functionalized membranes and interiors. *J Am Chem Soc* **125**:374–378.

Earles, C, and Schenk, JO (1999). Multisubtrate mechanism for the inward transport of dopamine by the human dopamine transporter expressed in HEK cells and its inhibition by cocaine. *Synapse* **33**:230–238.

Elhamdani, A, Palfrey, CH, and Artalejo, CR (2002). Ageing changes the cellular basis of the "fight-or-flight" response in human adrenal chromaffin cells. *Neurobiol Aging* **23**:287–293.

Elhamdani, A, Zhou, Z, and Artalejo, CR (1998). Timing of dense-core vesicle exocytosis depends on the facilitation L-type Ca channel in adrenal chromaffin cells. *J Neurosci* **18**:6230–6240.

Finnegan, JM, Pihel, K, Cahill, PS, Huang, L, Zerby, SE, Ewing, AG, Kennedy, RT, and Wightman, RM (1996). Vesicular quantal size measured by amperometry at chromaffin, mast, pheochromocytoma, and pancreatic β-cells. *J Neurochem* **66**:1914–1923.

Gong, L-W, Hafez, I, Alvarez de Toledo, G, and Lindau, M (2003). Secretory vesicles membrane area is regulated in tandem with quantal size in chromaffin cells. *J Neurosci* **23**:7917–7921.

Gorski, W, Aspinwall, CA, Lakey, JRT, and Kennedy, RT (1997). Ruthenium catalyst for amperometric determination of insulin at physiological pH. *J Electroanal Chem* **425**:191–199.

Graham, ME, Washbourne, P, Wilson, MC, and Burgoyne, RD (2002). Molecular analysis of SNAP-25 function in exocytosis. *Ann NY Acad Sci* **971**:210–221.

Greene, LA, Tischler, AS (1976). Establishment of a noradrenergic clonal line of rat adrenal pheochromocytoma cells which respond to nerve growth factor. *Proc Natl Acad Sci USA* **73**:2424–2428.

Huang, L, Shen, H, Atkinson, M, and Kennedy, R (1995). Detection of exocytosis at individual pancreatic {beta} cells by amperometry at a chemistry modified microelectrode. *Proc Natl Acad Sci USA* **92**:9608–9612.

Jaffe, E, Bolanos, P, Galvis, G, and Caputo, C (2004). Ryanodine receptors in peritoneal mast cells: Possible role in the modulation of exocytotic activity. *Pflugers Archiv* **447**:377–386.

Karlsson, M, Nolkrantz, K, Davidson, MJ, Strömberg, A, Ryttsén, F, Åkerman, B, and Orwar, O (2000). Electroinjection of colloid particles and biopolymers into single unilamellar liposomes and cells for bioanalytical applications. *Anal Chem* **72**:5857–5862.

Karlsson, A, Karlsson, R, Karlsson, M, Cans, A-S, Stromberg, A, Ryttsen, F, and Orwar, O (2001). Molecular engineering: Networks of nanotubes and containers. *Nature* **409**(6817):150–152.

Karlsson, A, Karlsson, M, Karlsson, R, Sott, K, Lundqvist, A, Tokarz, M, and Orwar, O (2003). Nanofluidic networks based on surfactant membrane technology. *Anal Chem* **75**:2529–2537.

Kim, TD, Eddlestone, GT, Mahmoud, SF, Kuchtey, J, and Fewtrell, C (1997). Correlating Ca^{2+} responses and secretion in individual RBL-2H3 mucosal mast cells. *J Biol Chem* **272**:31225–31229.

Kozminski, KD, Gutman, DA, Davila, V, Sulzer, D, and Ewing, AG (1998). Voltammetric and pharmacological characterization of dopamine release from single exocytotic events at rat pheochromocytoma (PC12) cells. *Anal Chem* **70**:3123–3130.

Misura, KMS, Scheller, RH, and Weis, WI (2000). Three-dimensional structure of the neuronal-Sec1-syntaxin 1a complex. *Nature* **404**(6776):355–362.

Moscho, A, Orwar, O, Chiu, DT, Modi, BP, and Zare, RN (1996). Rapid preparation of giant unilamellar vesicles. *Proc Natl Acad Sci USA* **93**:11443–11447.

Mosharov, EV, Gong, L-W, Khanna, B, Sulzer, D, and Lindau, M (2003). Intracellular patch electrochemistry: Regulation of cytosolic catecholamines in chromaffin cells. *J Neurosci* **23**:5835–5845.

Ñeco, P, Rossetto, O, Gil, A, Montecucco, C, and Gutiérrez, LM (2003). Taipoxin induces F-actin fragmentation and enhances release of catecholamines in bovine chromaffin cells. *J Neurochem* **85**:329–337.

Pihel, K, Schroeder, TJ, and Wightman, RM (1994). Rapid and selective cyclic voltammetric measurements of epinephrine and norepinephrine as a method to measure secretion from single bovine adrenal medullary cells. *Anal Chem* **66**:4532–4537.

Pihel, K, Hsieh, S, Jorgenson, JW, and Wightman, RM (1998). Quantal corelease of histamine and 5-hydroxytryptamine from mast cells and the effects of prior incubation. *Biochemistry* **37**:1046–1052.

Qian, W-J, Aspinwall, CA, Battiste, MA, and Kennedy, RT (2000). Detection of secretion from single pancreatic β-cells using extracellular fluorogenic reactions and confocal fluorescence microscopy. *Anal Chem* **72**:711–717.

Qian, W-J, Gee, KR, and Kennedy, RT (2003). Imaging of Zn^{2+} Release from pancreatic β-cells at the level of single exocytotic events. *Anal Chem* **75**:3468–3475.

Sakmann, B, and Neher, E (1984). Patch clamp techniques for studying ionic channels in excitable membranes. *Annu Rev Physiol* **46**:455–472.

Schubert, D, and Klier, FG (1977). Storage and release of acetylcholine by a clonal cell line. *Proc Natl Acad Sci USA* **74**:5184–5188.

Sombers, LA, Hanchar, HJ, Colliver, TL, Wittenberg, N, Cans, A, Arbault, S, Amatore, C, and Ewing, AG (2004). The effects of vesicular volume on secretion through the fusion pore in exocytotic release from PC12 cells. *J Neurosci* **24**:303–309.

Staal, RGW, Mosharov, EV, and Sulzer, D (2004). Dopamine neurons release transmitter via a flickering fusion pore. *Nat Neurosci* **7**:341–346.

Tabares, L, Lindau, M, and Alvarez de Toledo, G (2003). Relationship between fusion pore opening and release during mast cell exocytosis studied with patch amperometry. *Biochem Soc Trans* **31**:837—841.

Takahashi, N, Kadowaki, T, Yazaki, Y, Miyashita, Y, and Kasai, H (1997). Multiple exocytotic pathways in pancreatic beta cells. *J Cell Biol* **138**:55–64.

Takei, M, Ueno, M, and Endo, K (1992). Effect of ryanodine on histamine release from rat peritoneal mast cells induced by Anti-IgE. *J Pharm Pharmacol* **44**:523–525.

Travis, ER, and Wightman, RM (1998). Spatio-temporal resolution of exocytosis from individual cells. *Annu Rev Biophys Biomol Struct* **27**:77–103.

Troyer, KP, and Wightman, RM (2002). Dopamine transport into a single cell in a picoliter vial. *Anal Chem* **74**:5370–5375.

Wightman, R, Jankowski, J, Kennedy, R, Kawagoe, K, Schroeder, T, Leszczyszyn, D, Near, J, Diliberto, E, Jr, and Viveros, O (1991). Temporally resolved catecholamine spikes correspond to single vesicle release from individual chromaffin cells. *Proc Natl Acad Sci USA* **88**:10754–10758.

Wu, MN, Littleton, JT, Bhat, Manzoor, A, Prokop, A, and Bellen Hugo, J (1998). ROP, the *Drosophila* Sec1 homolog, interacts with syntaxin and regulates neurotransmitter release in a dosage-dependent manner. *EMBO J* **17**:127–139.

Zhou, Z, and Misler, S (1996). Amperometric detection of quantal secretion from patch-clamped rat pancreatic beta-cells. *J Biol Chem* **271**:270–277.

CHAPTER

11

ELECTROCHEMILUMINESCENCE DETECTION IN BIOANALYSIS

XIAO-HONG NANCY XU AND YANBING ZU

11.1. INTRODUCTION

Electrochemiluminescence (ECL) is defined as the generation of luminescence electrochemically. ECL technique has been used to investigate the photochemical and electrochemical properties of new compounds, complexes and clusters, and mechanisms of organic reactions involving radical intermediates. The early study aimed to pursue the possible application of ECL in displays and electro-optics devices, which still remains a fascinating topic of today's ECL research. ECL research has eventually led to the successful ultrasensitive analysis of biomolecules (e.g., DNA and proteins) in solution and commercialization of several generations of ECL analyzers, which have jointed the quest for the possible early diagnosis of diseases in clinical applications. ECL has been studied extensively since the middle 1960s, and its research progress has been comprehensively reviewed (Faulkner, 1978; Faulkner and Bard, 1977; Greenway, 1990; Knight and Greenway, 1994; Lee, 1997; Gerardi et al., 1999; Bard et al., 2000; Fahnrich et al., 2001; Knight, 2001; Kulmala and Suomi, 2003; Bard, 2004a,b; Richter, 2004).

In this chapter, we will introduce the fundamentals of ECL, provide an overview of historic development of ECL technologies and instrumentation, and focus on describing its applications, its unique features, and future opportunities in ultrasensitive bioanalysis.

11.1.1. Fundamentals

Annihilation ECL Mechanism. ECL has been produced by an energetic electron-transfer reaction between electrogenerated redox species, represented by A^- and D^+ (A and D could be the same species), often radical ions, to form an excited state A^* (or D^*), leading to the releasing of photons by returning the excite state to the ground state (Faulkner and Bard, 1977; Bard, 2004a,b):

$$A + e^- \rightarrow A^{-\bullet} \text{ (electrochemical reduction reaction)} \quad (11.1)$$

$$D - e^- \rightarrow D^{+\bullet} \text{ (electrochemical oxidation reaction)} \quad (11.2)$$

$$A^{-\bullet} + D^{+\bullet} \rightarrow A^* + D \text{ (forming excited state by annihilation)} \quad (11.3)$$

$$A^* \rightarrow A + h\nu \text{ (releasing photon)} \quad (11.4)$$

New Frontiers in Ultrasensitive Bioanalysis. Edited by Xiao-Hong Nancy Xu
Copyright © 2007 John Wiley & Sons, Inc.

The annihilation ECL has been produced in a solution containing A and D at a single electrode, in which the electrode potential is pulsed rapidly to generate reduced and oxidized species at the same electrode. The annihilation ECL has also been generated by rapidly applying oxidation and reduction potentials to two separate electrodes that are located in very close proximity. Typically, "annihilation ECL" involves polyaromatic hydrocarbons, such as 9,10-diphenylanthracene (DPA). DPA can be reversibly oxidized or reduced in organic solvents to produce relatively stable radical cations and anions, respectively. The annihilation of radical ions releases sufficient energy to directly populate both the DPA excited singlet state and the excited triplet state (Beideman and Hercules, 1979). DPA has unit fluorescence efficiency and its ECL efficiency is approximate 25%. Note that ECL efficiency is defined as the fraction of radical ions that are annihilated to generate excited states, multiplied by the fluorescence quantum efficiency (Maness and Wightman, 1994).

Because the annihilation reactions are very energetic (typical free energy change ~2–3 eV), most annihilation ECL systems require the use of organic solvents, because organic solvents offer the wider electrochemical potential window, which allows the generation of radical ions for ECL annihilation reactions. In other words, the potential windows in aqueous solutions are generally too narrow to generate both sufficiently stable oxidized and reduced ECL precursors, radial ions, for annihilation reactions. Thus, it is impossible to carry out annihilation ECL reactions in aqueous solution. Consequently, its application in bioanalysis is quite limited.

The generation of the excited state, rather than producing more energetically favorable ground state in the annihilation ECL reactions, has been interpreted by Marcus electron-transfer (ET) theory, providing the first evidence of the "inverted region" (Marcus, 1965). Basically, in the highly exoergic ET reaction [Eq. (11.3)], the intersection of the potential energy surface of the reactants with that of the electronic ground-state products generates a larger energetic barrier than the intersection of reactants with that of the excited-state products. Thus, the formation rate of the ground-state products is slower than that of the formation of the excited states.

Coreactant ECL mechanim. An alternative means of production of ECL utilizes a coreactant species (such $C_2O_4^{2-}$, $S_2O_8^{2-}$, or amines) in a solution containing luminophore species (typically metal chelates). Upon electrochemical oxidation or reduction, the coreactants immediately undergo chemical decomposition to form a strong reducing or oxidizing intermediate that react with the oxidized or reduced ECL luminophore to generate excited states. The coreactant ECL has been produced using a single potential step or one directional potential scanning at an electrode. The coreactant mechanism allows the ECL to be observed in aqueous solution.

For a typical coreactant ECL system, tris(2,2′-bipyridinyl)ruthenium(II) [Ru(bpy)$_3^{2+}$]/oxalate, the proposed ECL reactions are given below (Chang et al., 1977):

$$Ru(bpy)_3^{2+} - e^- \rightarrow Ru(bpy)_3^{3+} \qquad (11.5)$$

$$Ru(bpy)_3^{3+} + C_2O_4^{2-} \rightarrow Ru(bpy)_3^{2+} + C_2O_4^{-\bullet} \qquad (11.6)$$

$$C_2O_4^{-\bullet} \rightarrow CO_2^{-\bullet} + CO_2 \qquad (11.7)$$

$$Ru(bpy)_3^{3+} + CO_2^{-\bullet} \rightarrow Ru(bpy)_3^{2+*} + CO_2^- \qquad (11.8)$$

$$\text{Ru(bpy)}_3^{2+} + \text{CO}_2^{-\bullet} \rightarrow \text{Ru(bpy)}_3^{+} + \text{CO}_2^{-} \tag{11.9}$$

$$\text{Ru(bpy)}_3^{3+} + \text{Ru(bpy)}_3^{+} \rightarrow \text{Ru(bpy)}_3^{2+*} + \text{Ru(bpy)}_3^{2+} \tag{11.10}$$

$$\text{Ru(bpy)}_3^{2+*} \rightarrow \text{Ru(bpy)}_3^{2+} + h\nu \tag{11.11}$$

In this coreactant ECL process, the oxalate is catalytically oxidized by the electro-generated Ru(bpy)_3^{3+} in the diffusion layer close to the electrode surface. The oxidation product, oxalate radical anion ($\text{C}_2\text{O}_4^{-\bullet}$), breaks down to form a highly reducing radical anion ($\text{CO}_2^{-\bullet}$) ($E^0 \sim -1.9$ V versus NHE) (Butler and Henglein, 1980). The reaction between Ru(bpy)_3^{3+} and $\text{CO}_2^{-\bullet}$ leads to the formation of either the excited-state Ru(bpy)_3^{2+*} or Ru(bpy)_3^{+}. Note that the excited state may also be produced by the reaction between Ru(bpy)_3^{+} and Ru(bpy)_3^{3+}. Because the highly reducing intermediate species are generated upon the coreactant oxidation, this kind of ECL reaction is often referred to as "oxidative-reduction" ECL.

Several coreactant species have been reduced to generate highly oxidizing intermediates that react with the reduced luminophore to produce ECL. Typical example of this "reductive-oxidation" ECL is $\text{Ru(bpy)}_3^{2+}/\text{S}_2\text{O}_8^{2-}$ system (White and Bard, 1982), in which the decomposition of the reduction product of peroxydisulfate results in the formation of a strongly oxidizing species $\text{SO}_4^{-\bullet}$ ($E^0 \geq 3.15$ V versus SCE) (Memming, 1969).

11.1.2. Applications and Unique Features of ECL Technique in Bioanalysis

Since coreactant ECL systems work well in both organic and aqueous solvents, and the emission intensity of coreactant ECL is proportional to the concentration of either the coreactant or the emitter under certain experimental conditions, coreactant ECL has been employed in a wide range of analytical applications, especially untrasensitive bioanalysis.

Many biologically, pharmaceutically, and environmentally important analytes have served, directly or after derivatization, as ECL coreactants, which have been analyzed using ECL detector in a flow-injection separation system, such as liquid chromatography and capillary electrophoresis (Danielson, 2004). Nevertheless, the most popular ECL methods for ultrasensitive bioanalysis use ECL active emitter as a tag to label biomolecules such as protein and nucleic acids, and they analyze these labeled biomolecules in the presence of excess coreactant using commercially available ECL analyzer. This approach has been used in life sciences, pharmaceutical industry, food sciences, and environmental monitoring (Debad et al., 2004).

The most commonly used ECL emitters in bioanalysis are Ru(bpy)_3^{2+} and its derivatives. These molecules show excellent stability and high solubility in aqueous solution, undergo reversible one-electron transfer reactions at easily achievable potentials, and produce ECL emission with a good quantum yield in neutral pH aqueous medium in the presence of oxygen. By attaching suitable groups (e.g., amine) of biomolecular to the bipyridine moieties, Ru(bpy)_3^{2+} has been linked to biological molecules, serving as a tag for bioanalysis using ECL technique in a manner analogous to that of radioactive or fluorescent labels. For example, proteins are most commonly labeled via reaction of lysine amino groups with Ru(bpy)_3^{2+} NHS ester derivatives (Keller and Manak, 1993), while nucleic acids have been labeled at the 5′ terminus using a phosphoramidite-containing derivative, a reagent that modifies the probes with Ru(bpy)_3^{2+} during automatic synthesis (Kenten et al., 1992).

The Ru(bpy)$_3^{2+}$/tri-n-propylamine (TPrA) system exhibits the highest ECL efficiency in aqueous solution among all coreactant ECL systems ever studied, and it has been used for the development of commercial assay buffer solution for bioanalysis. The mechanism of this ECL process is quite complex and remains unclear despite extensive investigation over decades (Noffsinger and Danielson, 1987; Leland and Powell, 1990; Zu and Bard, 2000; Gross et al., 2001; Kanoufi et al., 2001; Miao et al., 2002; Honda et al., 2003; Lai and Bard, 2003; Zhou et al., 2003a; Wightman et al., 2004). Nevertheless, it is generally believed that the ECL reaction of the Ru(bpy)$_3^{2+}$/TPrA system follows the "oxidative-reduction" route, where the oxidation of TPrA molecule leads to the formation of a highly reducing intermediate species, TPrA$^\bullet$, which reduces Ru(bpy)$_3^{3+}$ to generate the excited state, Ru(bpy)$_3^{2+*}$. Emission of a photon from the excite state of Ru(bpy)$_3^{2+*}$ regenerates the ground state, and consequently a single label molecule, Ru(bpy)$_3^{2+}$, participates in multiple reaction cycles to produce multiple photons, leading to the amplification of the ECL signal and enhancing the detection sensitivity.

The ECL detection of biomolecules has been performed homogeneously in solution, referred to as solution-phase ECL assay, and by immobilizing the analytes on solid surfaces, known as surface-phase ECL assay.

In the solution-phase assay, the ECL intensity decreases as a result of binding events of interest. This is attributable to the increased mass and size of the binding pair, leading to the decrease of diffusion coefficient (Carter and Bard, 1990; Rodriguez and Bard, 1990; Xu et al., 2001; Kuwabara et al., 2003). Such assays have been used to (a) measure the binding affinity of DNA with metal chelates and antigen with antibody and (b) develop the ECL sensors for screening tumor markers (Xu et al., 2001).

In surface-phase ECL assays, solid-phase supports (e.g., beads) typically are used to immobilize molecules to form immunoadsorbent. The solid-phase support for the immobilization of biomolecules has been carried out on the working electrode surface or paramagnet beads that are magnetically captured in close proximity to an electrode (Xu et al., 1994; Xu and Bard, 1995; Bard et al., 2000; Debad et al., 2004).

A variety of means have been developed to immobilize biomolecules on the solid surfaces. The simplest way is direct adsorption of biomolecules onto the surface via hydrophobic or electrostatic interactions. However, such interaction is less stable and can be easily disrupted by changing of ionic strength and pH. The modification of surface with an adsorbed capture molecules such as streptavidin or avidin, and with self-assembled monolayers (SAM), has been widely used to immobilize biomolecules via specific interactions, offering much more stable and specific modified surfaces for ECL detection (Xu et al., 1994; Xu and Bard, 1995; Miao and Bard, 2003).

The ECL surface-phase immunoassays have adopted those used in fluorescence immunoassay. As illustrated in Figure 11.1, antibody (or receptor) is immobilized on the solid surface. The labeled protein, such as antigen (or ligand), is then detected by specifically interaction of antigen–antibody (or ligand–receptor) interaction via the direct, sandwich and competitive immunoassays, as depicted in Figures 11.1A–C, respectively. Similar approach has been used to detect ss-DNA as shown in Figure 11.1D, indicating that ss-DNA is immobilized on the surface, which captures the target strand of DNA through its hybridization with the tail part of target strand of DNA. Then, the target strand goes on to hybridize the part of ss-DNA that is labeled with ECL probe.

The surface-phase ECL assays have also been used to determine the activity of enzymes (e.g., protease, kinase, and integrase) as illustrated in Figure 11.2. Typically, the enzyme substrates are immobilized on the surface. The enzyme activities are then measured using

INTRODUCTION

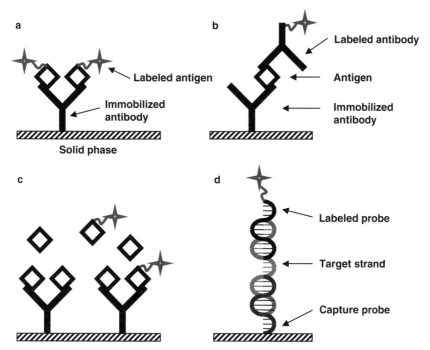

Figure 11.1. Schematics of design of binding assays with ECL detection: (a) Direct immunoassay, which selectively detects antigen tagged with ECL emitter using antibody immobilized on the solid support via specific molecular recognition of antibody with antigen. (b) Sandwich immunoassay, in which antigen is captured by antibody immobilized on the solid support and is detected by specific interaction with another antibody molecule tagged with ECL emitter. (c) Competition assay, showing that unlabeled analytes compete with labeled analytes already bound with the immobilized antibody on the solid support. Decrease in ECL emission is used for analysis of unlabeled analytes. (d) DNA biosensor, illustrating that one terminal of a target strand oligonucleotide is used to specifically interact with the immobilized capture probe on the solid surface, while the other terminal of the target strand specifically recognizes the labeled probe via specific base-pair interaction. This DNA biosensor functions similar to the sandwich assay described in part b. See insert for color representation of this figure.

ECL intensity, which changes in response to removing the part of tagged immobilized substrates from the surface due to cleavage by protease, or binding of the labeled antibody with the reaction product of kinase, as shown in Figures 11.2A and 11.2B, respectively.

Despite the enormous success of the surface-phase assay, it is worthy of noting that the preparation of the immunoadsorbent probes is time-consuming and these immunoassays need to go through separation or washing-step (Xu et al., 2001). Thus, it is very difficult to perform high-throughput analysis and real-time measurement of molecular interactions using a surface-phase assay. Furthermore, the surface-phase immunoassay cannot measure the interactions of free biomolecules (e.g., antigen–antibody) in solution phase where affinity constants are expected to be closer to their affinity *in vivo*.

In comparison with other detection means (e.g., radioisotopes, chemluminescence, and fluorescence), the ECL technique possesses several distinct advantages (Blackburn et al., 1991; Xu et al., 2001; Bard, 2004a,b): (a) ECL is extremely sensitive, because of no needs of light sources, leading to no scattering light and low background. (b) The ECL reactions provide high selectivity and high temporal and spatial resolution, because ECL emission is correlated with the electrochemical potential, which can be easily controlled. (c) The ECL

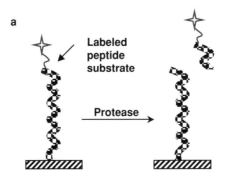

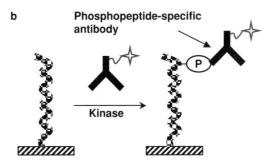

Figure 11.2. Schematics of ECL assays for measuring enzyme activity. (a) Protease activity assay, showing that ECL emission is used to follow the cleavage of the immobilized substrate, leading to the removal of ECL tag from the solid surface and hence change of ECL emission. (b) Kinase activity assay, in which antibody tagged with ECL emitter is used to specifically recognize the phosphorylated product, generated by the reaction of kinase with substrate. See insert for color representation of this figure.

probes, such as Ru(bpy)$_3^{2+}$, offer the minimum perturbation of immune activity, protein solubility, and protein stability, because the ECL probes are small molecules (~1000 Da) and easily conjugated with the target biomolecules. (d) The ECL labels are very stable (over years), which are superior to chemiluminescence and radioisotopic labels (Bard et al., 2000; Blackburn et al., 1991). (e) ECL detection provides a larger dynamic range (six orders of magnitude) for quantification analysis. (f) ECL measurements are simple, rapid (only a few second), and easy for the automation. (g) ECL emission intensity depends on the mass and size of ECL active species, offering one of the best techniques to study the DNA intercalation (Rodriguez and Bard, 1990; Xu et al., 1994; Xu and Bard, 1995), antibody affinities, and protein–protein interactions (Xu et al., 2001). (h) The ECL instrumentation is relatively simple and inexpensive.

11.1.3. Instrumentation

ECL measurements have been carried out in conventional electrochemical apparatus with three-electrode configuration with a photodetector, such as a photomultiplier tube (PMT), a photodiode, or a charge-coupled device (CCD) camera. The annihilation ECL experiments have to be carried out in organic solvents, and its ECL emission is associated with the purity of the solvents, the supporting electrolytes, and the presence of oxygen. Thus, the annihilation ECL experiments are performed inside an oxygen-free dry box or a sealed airtight cell (Hercules, 1964; Visco and Chandross, 1964; Santhanam and Bard, 1965; Tokel and Bard, 1972; Faulkner and Bard, 1977; Xu et al., 1996; Fan, 2004). In contrast, coreactant ECL experiments are typically performed in aqueous solutions containing the high concentration of coreactant and in the presence of oxygen, and its ECL emission has been proved to be insignificantly affected by the presence of oxygen (Bard et al., 2000; Zheng and Zu, 2005a).

The commercial ECL analyzers have achieved full automation by coupling a flow-injection system with an ECL detector, which consists of an electrochemical cell and PMT. Figure 11.3 shows the basic components of such an ECL analyzer originally developed by IGEN International. The ECL analyzer is interfaced with a computer, which controls the entire analysis process, by taking one sample at a time, measuring its ECL emission intensity, refreshing a working electrode, and cleaning up the flow-injection system. Though the ECL analyzer is designed for analysis of biomolecules immobilized on magnetic beads (surface-phase assay), it has been widely used for solution-phase and surface-phase assay and many other ECL studies. Typically, the assay is run off-line in each sample tube, which is then loaded into the instrument's carousel. Each individual sample prepared in the presence of coreactor buffer solution (e.g., TPrA) is drawn into the flow cell, and magnetically responsive beads are captured onto a working electrode (usually platinum or gold) by switching a magnetic field beneath the working electrode; the ECL emission of the sample is recorded using PMT as electrochemical potential is applied to the working electrode (Figure 11.3). Finally, the magnetic field is switched off to release the magnetic beads, and the cleanser solution is used to flush the sample and regenerate the working electrode surface electrochemically. The blank buffer is flowed through the electrochemical cell, and ECL intensity of buffer solution is measured to ensure that the flow-injection system is clean and that the electrode is ready for next measurement.

The other type of commercial ECL analyzer developed by Meso Scale Discovery measures ECL from surface-phase assays carried out directly on disposable electrodes (i.e., screen-printed carbon ink electrodes). This approach allows analysis of an array of ECL samples simultaneously and achieves the high throughput, which is essential for drug screening and genomic and proteomic analysis (Debad et al., 2004).

In addition, a variety of homemade ECL cells have been designed for different applications. For example, ECL detection systems have been developed as detectors for separation techniques, such as liquid chromatography and capillary electrophoresis, in which ECL detectors have met the demands of high speed and low volume required by separation techniques (Knight and Greenway, 1994; Danielson, 2004). Recently, a so-called "wireless ECL" detector has been introduced as a detector for analysis of three amino acids in microfluid devices (Arora et al., 2001). Such an ECL detector consists of a microfabricated "U"-shape floating platinum electrode across a separation channel (Figure 11.4). The electrophoretic separation creates a potential between both ends of U-shape electrode, which is used to generate ECL emission. Thus, the electric field is used to drive both the separation and ECL emission, which is a very clever design and may serve as a model for the future design of ECL detector in microfluid devices.

11.2. RESEARCH OVERVIEW

ECL studies date back to 1964. The initial experiments involved polyaromatic hydrocarbons, such as DPA, anthracence, and rubrene, and were carried out in organic solvents; also,

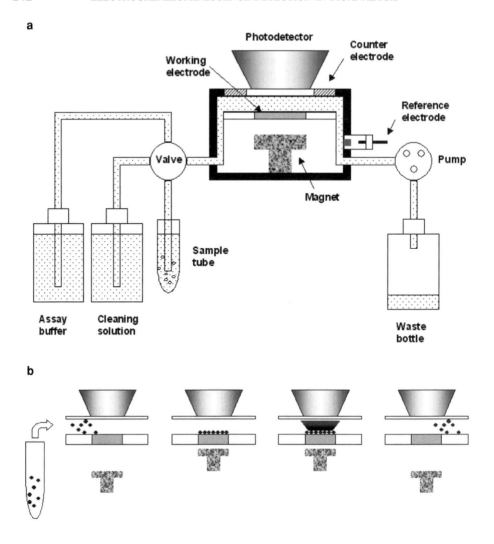

Figure 11.3. Schematics of (a) basic elements of a flow-injection ECL ORIGEN analyzer, include components of flow-injection system (pump, value, sample tube, waste bottle), mobile solutions (ECL assay buffer solution containing TPrA, and cleaning solution), ECL detection system (three-electrode thin-layer electrochemical cell, potentiostat, and photodetector), and magnetic system used to capture magnetic beads onto the electrode surfaces, leading to the pre-concentration of analytes and hence enhancing detection sensitivity. Note that magnetic beads typically serve as a solid support to immobilize antibody for the development of commercial immunoassays (Figures 11.1 and 11.2). (b) Illustration of capturing magnetic beads onto the electrode surface by activating magnetic field, detecting their ECL emission, and then releasing the magnetic field to wash away the beads. [Reprinted from Bard (2004a), *Electrogenerated Chemiluminescence* with permission of Marcel Dekker.]

ECL emission mechanism was suggested to occur via annihilation reactions of radical ions, which were produced as the results of electrochemical reactions (Hercules, 1964; Visco and Chandross, 1964; Santhanam and Bard, 1965). Such exciting new experimental phenomena were quickly explained by Marcus electron-transfer theory in 1965 (Marcus, 1965), while its diffusion and kinetic processes were simulated theoretically by Feldberg in 1966 (Feldberg, 1966a,b). In 1972, the pioneering study of annihilation of metal chelates, $Ru(bpy)_3^{2+}$ was

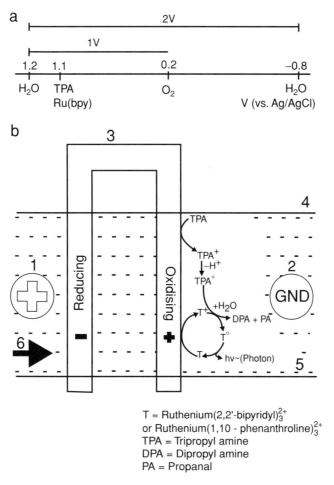

Figure 11.4. Schematics of a "wireless ECL" detector for a microfluid device, showing that both ends of the U-shape floating electrode serve as a working and count electrode, respectively. Potential between both electrodes is provided by electric field created in the separation channel at the position of electrodes, which is used to drive the ECL reactions and separation. Note that "1–6" represents the following: separation (high voltage) anode; separation ground of electric field; U-shaped Pt film electrode bridged a separation channel; separation channel; solution containing Ru(bpy)$_3^{2+}$ or Ru(phen)$_3^{2+}$ and TPA; and electroosmotic flow direction, respectively. [Reprinted from *Anal Chem* **73**:6173–6178 (2001), with permission of American Chemical Society.]

carried out in nonaqueous solution (Tokel and Bard, 1972). The annihilation reaction of Ru(bpy)$_3^{2+}$ produces an excited state with an efficiency approaching 100%, which offers 5% ECL efficiency (Glass and Faulkner, 1981; Luttmer and Bard, 1981). These early studies heavily emphasized the investigation of excited states, ECL mechanisms, and the efficiency of ECL mechanisms.

In 1977, the first coreactant ECL system, Ru(bpy)$_3^{2+}$/oxalate, was reported to produce intense ECL emission simply by oxidizing Ru(bpy)$_3^{2+}$ and oxalate in acetonitrile solution (Chang et al., 1977), opening up the new venue of generating ECL emission. A few years later, ECL emission from this coreactant system was successfully demonstrated in an aqueous solution, showing the maximum emission at pH 6 (Rubinstein and Bard, 1981;

Rubinstein et al., 1983). This study opens up the new era of ECL research and application, which leads to the discovery of many other coreactant ECL systems and makes its application in analytical, biological, clinical and environmental sciences possible.

Further research leads to the discovery of a very efficient coreactant of Ru(bpy)$_3^{2+}$, TPrA (Noffsinger and Danielson, 1987; Leland and Powell, 1990), making it possible to determine Ru(bpy)$_3^{2+}$ at the subpicomolar level in the presence of 0.1 M TPrA at a neutral buffer solution (Blackburn et al., 1991). Thus, Ru(bpy)$_3^{2+}$ has been used to label biomolecules of interest, which are analyzed using ECL detection (Ege et al., 1984; Leland and Powell, 1990; Carter and Bard, 1990; Rodriguez and Bard, 1990; Xu et al., 2001; Kuwabara et al., 2003) and detected by well-designed ECL immunoassays and DNA biosensors (Blackburn et al., 1991; Kenten et al., 1991, 1992; Hsueh et al., 1996, 1998; Yu, 1998; Kijek et al., 2000; Zhang et al., 2001; Xu et al., 1994; Xu and Bard, 1995; Miao and Bard, 2003; Miao and Bard, 2004a,b).

Since the successful immobilization of Ru(bpy)$_3^{2+}$ on polymer-coated electrodes, i.e. Nafion-modified pyrolytic graphite electrode, in 1980 (Rubinstein and Bard., 1980), a variety of approaches have been developed for attaching ECL emitters on electrodes (Abruna and Bard, 1982; Xu and Bard, 1994, 1995; Khramov and Collinson, 2000; Choi et al., 2003; Guo and Dong, 2004; Choi et al., 2005; Ding et al., 2005; Kim et al., 2005a; Zhuang et al., 2005), leading to the development of ECL detectors and sensors for chemical and biochemical analysis. For example, the ECL detector has been developed as a detector for a variety of flow-injection systems (Lee and Nieman, 1995; Shultz et al., 1996), such as liquid chromatography (Lee and Nieman, 1994; Chen and Sato, 1995; Skotty and Nieman, 1995; Ridlen et al., 1997; Forry and Wightman, 2002; Morita and Konishi, 2002, 2003; Ikehara et al., 2005) and capillary electrophoresis (Forbes et al., 1997; Cao et al., 2002; Chiang and Whang, 2002; Liu et al., 2002; Qiu et al., 2003; Gao et al., 2005; Yin and Wang, 2005).

Continuous interest and extensive research in ECL ultimately led to the development of commercial ECL reagents and analyzer, which paves the way for practical application of ECL detection in ultrasensitive bioanalysis, including clinical diagnostics, food sciences, water analysis, and environmental monitoring. The comprehensive list of ECL-based bioassays and their applications have been described in recent review articles (Debad et al., 2004; Richter, 2004). For example, the ECL technique has been used for successful ultrasenstive analysis of tumor markers, such as a-fetoprotein (AFP) (Namba et al., 1999; Yilmaz et al., 2001), Cytokeratin 19 (Sanchez-Carbayo et al., 1999), and prostate-specific antigens (PSA) (Xu et al., 2001; Butch et al., 2002; Haese et al., 2002). These studies demonstrate the possibility of using ECL for ultrasenstive analysis of tumor biomarkers that may lead to the early detection and treatment of cancer. The ECL technique has also been used for analysis of oligonucleotides in solution (Carter and Bard, 1990; Rodriguez and Bard, 1990), development of DNA biosensors (Xu et al., 1994; Xu and Bard, 1995; Miao and Bard, 2004a,b), and study of hybridization of oligonucleotides (Wilson and Johansson, 2003; Spehar et al., 2004).

Furthermore, the ECL technique has been used to detect pharmaceutically and environmentally important molecules, such as phenolic compounds, because these compounds can effectively quench ECL of Ru(bpy)$_3^{2+}$ (McCall et al., 1999; McCall and Richter, 2000; Lin et al., 2004; Kang et al., 2005; Pang et al., 2005; Zheng and Zu, 2005b). By monitoring the decrease of ECL emission, one can effectively detect these molecules of interest, offering a new means of ultrasensitive analysis of these species for environmental monitoring and drug screening.

Note that several crucial biomolecules such as amino acids, peptides (Brune and Bobbitt, 1991, 1992; Lee and Nieman, 1994, 1995), DNA containing guanine residues (Dennany et al., 2003, 2004), and reduced form of nicotinamide adenine dinucleotide (NADH) (Downey and Nieman, 1992) have served as ECL coreactors and been successfully analyzed using ECL. Since NADH is involved in many enzymatic reactions, ECL technique has been developed for ultrasensitive detection of many clinically important enzyme substrates, such as lactate, glucose, and cholesterol (Martin and Nieman, 1993; Yokoyama et al., 1994; Jameison et al., 1996; Martin and Nieman, 1997). Using the similar approach, the ECL technique has also been used for sensitive quantification of the enzyme β-lactamase, offering the opportunity to monitor bacterial resistance to β-lactam antibiotics (Liang et al., 1996a,b). ECL coreactants, 4-(dimethylamino)butyric acid, have been used as a tag to label biomolecules of interest and have been detected in the presence of Ru(bpy)$_3^{2+}$, using the ECL technique (Yin et al., 2005).

With the rapid development of the ECL technique for bioanalysis, study of fundamental mechanisms of ECL reactions and exploring more effective ECL systems have also been carried out extensively. Such fundamental research is mainly motivated by the demand of even higher sensitivity of ECL detection. Ru(bpy)$_3^{2+}$/TPrA system has been the most effective and popular ECL system used for nearly two decades, yet its emission mechanism remains unclear. Extensive research has been dedicated to probe its mechanism (Noffsinger and Danielson, 1987; Leland and Powell, 1990; Zu and Bard, 2000; Gross et al., 2001; Kanoufi et al., 2001; Miao et al., 2002; Honda et al., 2003; Lai and Bard, 2003; Zhou et al., 2003a; Wightman et al., 2004). A new ECL reaction mechanism of the Ru(bpy)$_3^{2+}$/TPrA has been proposed (Miao et al., 2002), suggesting that intense light emission may be produced by the electro-oxidation of TPrA only, in which the oxidation of the Ru(bpy)$_3^{2+}$ is not required. The emission generated by this mechanism has been significantly enhanced at a nonionic fluorosurfactant-modified gold electrode (Li and Zu, 2004).

Several important topics related to the emission efficiency of coreactant ECL have also been investigated. These include the study of the dependence of emission on electrode materials and surface modifications (Zu and Bard, 2000, 2001), effect of presence of oxygen (Marlins et al., 1997; Zheng and Zu, 2005a) and surfactant (Factor et al., 2001; Zu and Bard, 2001; Cole et al., 2003; Li and Zu, 2004; Walworth et al., 2004; Xu et al., 2005).

Several interesting new ECL emitters, such as semiconductor nanoparticles (Ding et al., 2002, Myung et al., 2002, 2003, 2004; Bae et al., 2004; Poznyak et al., 2004; Zou and Ju, 2004; Ren et al., 2005; Zou et al., 2005), cyclometalated iridium (III) complexes (Kim et al., 2005b), ruthenium complexes containing a crown ether moiety (Lai et al., 2002; Muegge and Richter, 2002; Bruce and Richter, 2002a), and multimetallic and dendrimeric ruthenium complexes (Richter et al., 1998; Staffilani et al., 2003; Zhou et al., 2003b), have also been discovered.

In addition to the application of ECL in ultrasensitive bioanalysis, other exciting potential applications of ECL including the development of light source and electro-optics device have been suggested since the discovery of ECL phenomena in 1964. Some of these early suggestions have been well demonstrated. For example, the tiny light source generated at ultramicroelectrodes has been used in scanning optical microscopy, allowing the molecules of interest to be characterized using both electrochemical and spectroscopic methods (Fan et al., 1998; Maus and Wightman, 2001; Zu et al., 2001). ECL emission has also been used as photonic reporters for electrochemically sensing analytes in microfluidic devices (Zhan et al., 2002, 2003a,b).

11.3. RESEARCH HIGHLIGHTS

To illustrate the fundamental and application of ECL techniques in ultrasensitive bioanalysis, several representative research topics are described below in detail.

11.3.1. ECL Assays for Ultrasensitive Bioanalysis

11.3.1.1. Analysis of Tumor Markers Using Solution-Phase ECL Assay

Ultrasensitive analysis of biomarkers offers the possibility of earlier diagnosis of disease. Conventional immunoassays include radioimmunoassay (RIA), fluorescence immunoassay, protein A immunoassay, and bead-format electrochemiluminescence immunoassay, which have been widely used for the analysis of clinically important analytes. However, all these immunoassays require immobilization of antibody or antigen onto a solid surface (e.g., bead) to form an immunoadsorbent. Then, a direct, sandwich or competitive immunoassay is performed to capture analyte (e.g., antigen) onto the solid surface (Figure 11.1). The preparation of the immunoadsorbent probes is time-consuming, and these immunoassays need to be followed by a separation or washing step, which slows down the analysis speed. Thus, these immunoassays are unable to measure real-time kinetics of biomolecular interactions. Furthermore, these immunoassays cannot measure the interactions of biomolecules that are free in solution phase, where affinity constants are expected to be closer to their affinity *in vivo*.

A solution-phase ECL immunoassy has been developed for ultrasensitive detection of tumor markers, such as three molecular forms of prostate-specific antigen (PSA), PSA free, PSA complex, and PSA total (Xu et al., 2001). PSA is typically noncomplexed (or free) or complexed with the serine protease inhibitors alpha 1-antichymotrypsin (PSA-ACT) and alpha 2-macroglobulin (PSA-AMG). Minor forms of PSA include complexes with protein-C inhibitor (PSA-PCI), alpha 1-antitrypsin (PSA-AT), and inter-alpha-trypsin (PSA-IT). Total PSA refers to the sum of all immunodetectable species of PSA, mainly PSA free and PSA-ACT. The concentration of these forms represents the specific pathologic state in the prostate gland. The sensitive and specific detection of these molecular forms of serum PSA has established itself as the most reliable clinical tool for diagnosis and monitoring of prostate cancer (Schellhammer and Wright, 1993; Beduschi et al, 1998).

PSA-free, PSA-total, and PSA-ACT complex are conjugated with an inorganic compound, the *N*-hydroxysuccinimide ester of a ruthenium (II) tris-bipyridine chelate $(Ru(bpy)_3^{2+}–NHS)$, and purified by size-exclusive chromatography. The labeling ratio of PSA with $Ru(bpy)_3^{2+}$ is then carefully characterized using BCA assay and UV-vis spectroscopy, as summarized in Table 11.1.

Table 11.1. Determination of Conjugation Ratio of PSA with Tag, $Ru(bpy)_3^{2+}–NHS$

Protein	Fraction Number	Protein Concentration (μM)	TAG Concentration (μM)	Incorporation Ratio (TAG/Protein)
PSA free	2	1.83	4.16	2.3
PSA total	3	1.91	5.26	2.8
PSA complex	3	0.77	2.48	3.2

Reprinted from *Analyst* **126**:1285–1292 (2001), with permission of the Royal Society of Chemistry.

ECL Reaction Mechinsm

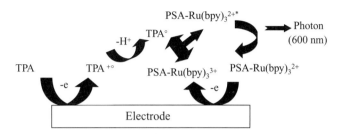

Figure 11.5. Schematic of ECL reaction mechanism of PSA–Ru(bpy)$_3^{2+}$ in the presence of an ECL coreactor, TPrA, showing that the emission of Ru(bpy)$_3^{2+}$ is generated by the energetic electron-transfer reaction between electrochemically generated Ru(bpy)$_3^{3+}$ and an intermediate of oxidized TPrA. [Reprinted from *Analyst* **126**:1285–1292 (2001), with permission of Royal Society of Chemical.]

PSA conjugated with Ru(bpy)$_3^{2+}$ permits PSA be analyzed using ECL, in the presence of coreactor, TPrA. As suggested in Figure 11.5, the luminescent molecule (e.g., Ru(bpy)$_3^{2+}$) conjugated with PSA in solution undergoes an electron-transfer reaction to form a stable species at an electrode (e.g., PSA–Ru(bpy)$_3^{3+}$) that reacts with another species also generated at the electrode [e.g., tripropylamine (TPA$^\bullet$)] through an energetic redox reaction which leads to the formation of an excited state (e.g., Ru(bpy)$_3^{2+*}$), which generates ECL emission by returning to the ground state.

Representative plots of ECL emission intensity versus concentration of labeled PSA free, PSA total, and PSA complex (Figure 11.6) indicate that ECL emission intensity of PSA free (MW 32,000–33,000 daltons), PSA total (34,000 daltons), and PSA complex (95,000 daltons), corrected using the labeled ratio in Table 11.1, are inversely proportional to their molecular weight ($M^{1/2}$) (Xu et al., 2001). The result demonstrates that ECL emission

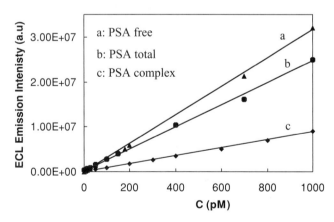

Figure 11.6. ECL calibration curves of three molecular forms of PSA tagged with Ru(bpy)$_3^{2+}$: (a) PSA free (▲), (b) PSA total (●), and (c) PSA-complex (♦) with the slope at 3.3×10^4, 2.5×10^4 and 8.8×10^3 pM^{-1}, respentively. Concentrations of PSA are 0.5–1000 pM. The measurements were taken in assay buffer containing TPrA (pH 7.4) using ORIGIN ECL analyzer. [Reprinted from *Analyst* **126**:1285–1292 (2001), with permission of Royal Society of Chemistry.]

intensity is proportional to the concentration of PSA with a large dynamic range (0.5 pM–1 nM), showing the detection of 1.7 pg/mL PSA in solution using ECL technology. To our knowledge, this is 1000-fold improvement over any conventional PSA immunoassay. In addition, the highest ECL emission intensity is observed from the lowest-molecular-weight PSA free, which is attributable to its highest diffusion coefficient among three molecular forms of PSA.

The ECL immunoassay of detecting these three molecular forms of PSA have been developed by measuring the affinity constact of PSA with its antibody as illustrated below.

$$\text{Ag-Tag} + \text{Ab} \underset{k_{-1}}{\overset{k_1}{\rightleftharpoons}} \text{Tag-Ag-Ab} \quad (11.12)$$

Ag-Tag, Ab, and Ab-Ag represent labeled antigen with Ru(bpy)$_3^{2+}$–NHS (e.g., PSA-Tag), antibody, and antigen–antibody binding complex, respectively. The equilibrium constant (K) of antigen–antibody binding reaction is described in Eq. (11.13).

$$K = \frac{k_1}{k_{-1}} = \frac{[\text{Ab-Ag-Tag}]}{[\text{Ab}][\text{Ag-Tag}]} = \frac{C_b}{(C_{Ab} - C_b)(C_t - C_b)}, \quad (11.13)$$

where C_b, C_{Ab}, and C_t represent the equilibrium concentration of antigen–antibody complex, the total concentration of antibody binding sites, and the total concentration of antigen, respectively.

When the total concentration of antigen (C_t), the molar ratio of antibody binding sites to antigen (R), the mole fractions of binding antigen (X_b), and free antigen (X_f) are used, Eq. (11.13) yields Eq. (11.14):

$$K = \frac{X_b}{C_t(R - X_b)(1 - X_b)}. \quad (11.14)$$

Since ECL emission intensity (I) is proportional to antigen concentration as described by Eq. (11.15), ECL emission (I) from the antigen solution titrated with antibody is described by Eq. (11.16) based upon the mobile equilibrium model (Carter and Bard, 1990; Rodriguez and Bard, 1990).

$$I_{t,0} = B\Phi_f D_f^{1/2} C_t, \quad (11.15)$$

$$I_t = B\Phi_f D_f^{1/2} C_f + B\Phi_b D_b^{1/2} C_b, \quad (11.16)$$

where B is a constant that can be obtained from the calibration curve of ECL emission versus antigen concentration, and Φ_f and Φ_b are the emission efficiency constant of free antigen and bound antigen, respectively. D_f and D_b are the diffusion coefficient of free antigen and bound antigen, respectively. The diffusion coefficient is governed by size and mass of the molecules represented by the Stokes–Einstein equation, Eq. (11.17) (Atkins, 1982; Bard and Faulkner, 1980).

$$D = \frac{kT}{6\pi \eta a}, \quad (11.17)$$

where D, a, k, T, and η represent diffusion coefficient, the size of molecule, Boltzmann constant, temperature and viscosity of solution, respectively. The viscosity (η) is proportional to the square root of molecule weight ($M^{1/2}$) of analytes (Atkins, 1982). Thus, ECL emission intensity will decrease upon the binding of antigen with antibody because of an increase in the mass and size. In contrast, ECL emission intensity will remain constant if no binding occurs. This forms the basis of this solution-phase ECL immunoassay.

The normalized ECL emission ($I_b/I_{t,f}$) is obtained by dividing Eq. (11.16) with Eq. (11.15) as displayed in Eq. (11.18):

$$\frac{I_b}{I_{t,f}} = 1 - \left[1 - \sqrt{\frac{D_b}{D_f}\frac{\Phi_b}{\Phi_f}}\right] X_b. \qquad (11.18)$$

By solving Eq. (11.18) using the normalized ECL emission as a function of affinity constant (K), the mole ratio of antibody binding sites to antigen (R), and the ratio of diffusion coefficient of bound antigen to free antigen, we obtain Eq. (11.19):

$$\frac{I_b}{I_{t,f}} = \left[1 - \sqrt{\frac{D_b}{D_f}\frac{\Phi_b}{\Phi_f}}\right] \frac{(1+R+1/KC_t) - \sqrt{(1+R+1/KC_t)^2 - 4R}}{2}. \qquad (11.19)$$

While all antigen molecules bind with antibody, the mole fraction of bound antigen (X_b) is equal to 1. Thus, Eq. (11.19) is simplified to Eq. (11.20), which indicates that ECL emission intensity decreases to the minimum, forming a baseline unrelated to R, because no antigen molecules are available to bind with further addition of antibody molecules.

$$\frac{I_b}{I_{t,f}} = \sqrt{\frac{D_b}{D_f}\frac{\Phi_b}{\Phi_f}} \qquad (11.20)$$

Addition of PSA-free monoclonal (MAB) results in a decrease of ECL emission intensity, demonstrating that PSA free bound with its MAB because this binding increases size and mass of PSA free and decreases its diffusion coefficient as described in Eqs. (11.16) and (11.17). Emission intensity continues to decrease with addition of MAB until it reaches the minimum, indicating that PSA-free molecules are all bound with MAB. Using the theoretical calculation results [Eq. (11.19)] to match the experimental data, it is found that the binding constant of $K = 1 \times 10^{10}$ M^{-1} with the binding ratio of 1 (Figure 11.7a). Control experiments of PSA-free titrated with PSA-complex MAB shows that ECL emission intensity remains constant as the molar ratio, R, of PSA-complex MAB to PSA-free increases, demonstrating that no binding of PSA-free with PSA-complex MAB occurs (Figure 11.7b). A similar approach has been applied to study the affinity constant of labeled PSA complex with PSA-complex MAB (Xu et al., 2001).

This study demonstrates that an ultrasensitive quantitative solution-phase ECL immunoassay has been developed for detection of different molecular forms of serum PSA (e.g., PSA free and PSA complex) on the basis of their binding affinities with its antibodies. This solution-phase immunoassay has been performed in a single step, which totally eliminates immobilization, separation, and washing steps. Furthermore, this solution-phase immunoassay is able to measure the interactions of free biomolecules (antigen and antibody) in solution. Moreover, this solution-phase immunoassay creates a new platform to measure affinity of protein–protein interactions in solution. One can now extend the application of this solution-phase ECL immunoassay to detect other serum tumor markers, study protein–protein interactions, and measure their affinities in solution. In principle, several samples can be analyzed simultaneously to accomplish high-throughput analysis using the new generation of IGEN Analyzer (M-SERIES).

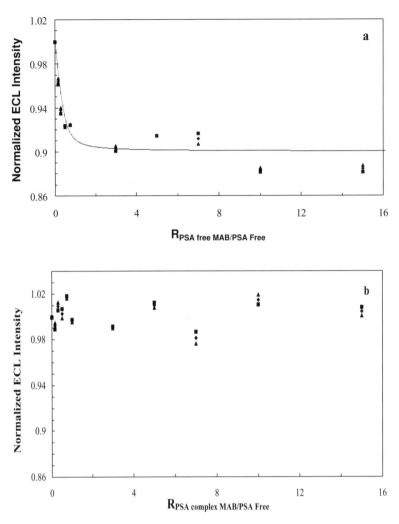

Figure 11.7. Measuring the binding affinity of PSA free. Titration curves of PSA free tagged with Ru(bpy)$_3{}^{2+}$ with (a) monoclonal antibody (MAB) of PSA free and (b) MAB of PSA complex, of a molar ratio (R) of MAB binding sites to PSA at 0.015–15, showing that (a) $K = 1 \times 10^{10} M^{-1}$ and (b) K ∼ 0 (no binding occurs), respectively. ECL emission intensity and its standard deviations are investigated from three independent measurements using ORIGIN ECL analyzer. The solid line is the theoretical calculation results based upon Eq. (11.19). [Reprinted from *Analyst* **126**:1285–1292 (2001), with permission of Royal Society of Chemistry.]

11.3.1.2. Detection of DNA and Protein Using ECL Biosensors

As mentioned above, the study of interaction of metal chelates, such as Os(bpy)$_3{}^{2+}$ and Ru(phen)$_3{}^{2+}$, with DNA in solution using ECL detection has been carried out since 1990 (Carter and Bard, 1990; Rodriguez and Bard, 1990). DNA biosensors, designed for sensing specific sequence of ss-DNA using ECL detection, have also been well demonstrated (Xu et al., 1994; Xu and Bard, 1995). The DNA biosensors are prepared by modifying the gold electrode surface spotted on silicon wafer with self-assembled monolayers, 4-mercaptobutylphosphonic acid (MBPA), Al^{3+}, bisphosphonic, Al^{3+}, and the ss-DNA that is complementary with the target ss-DNA, as shown in Figure 11.8. The metal centers,

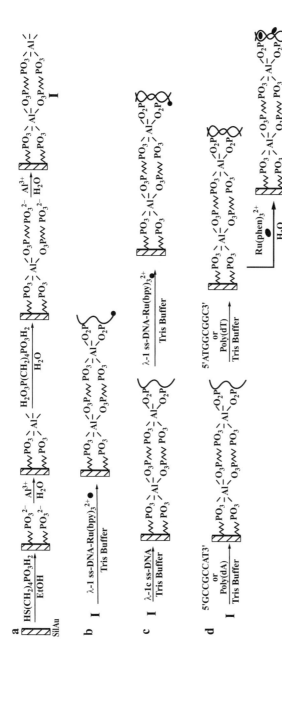

Figure 11.8. Illustration of (a) preparation of Al$_2$(C$_4$BP) film on the Au electrode surface supported on a silicon wafer; (b) immobilization of ss-DNA tagged with Ru(bpy)$_3^{2+}$ on the film; (c) immobilization of ss-DNA on the film, and then hybridization with its complementary strand DNA tagged with Ru(bpy)$_3^{2+}$ via specific base-pair interactions; (d) immobilization of ss-DNA on the film, and then hybridization with its complementary strand DNA to generate ds-DNA, via specific base-pair interactions; finally Ru(phen)$_3^{2+}$ is used to intercalate with ds-DNA for ECL detection. [Reprinted from *J Am Chem Soc* 117:2627–2631 (1995), with permission of American Chemical Society.]

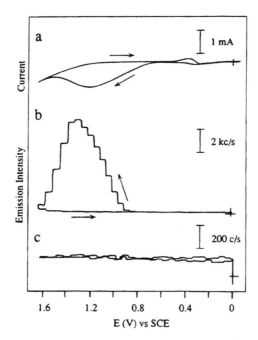

Figure 11.9. (a) Cyclic voltammogram, (b) ECL emission-potential transient of the λ-1 ss-DNA-Ru(bpy)$_3^{2+}$, and (c) λ-1 ss-DNA (nonlabeled) immobilized on the Al$_2$(C$_4$BP) film of the Au electrode in 0.19 M phosphate buffer/0.13 M TPrA (pH 7). The electrodes are prepared as described in the text and in Figure 11.8. The potential is scanned from 0 to 1.60 V at scan rate of 50 mV/s. The staircase shape of the emission curve is the result of integration of the photon flux in the single-photon-counting system (c/s = count per second). [Reprinted from *J Am Chem Soc* **117**:2627–2631 (1995), with permission of American Chemical Society.]

Al (III), interact strongly with the phosphate groups of DNA and phosphonic acids, forming a well-ordered self-assemble monolayer, which is used to effectively immobilize ss-DNA on the electrode surface (Xu et al., 1994; Xu and Bard, 1995).

The immobilized ss-DNA is able to selectively recognize the complementary strand DNA via specific base-pair hybridization reaction, which is then detected via ECL reaction in the presence of TPrA, either using Ru(phen)$_3^{2+}$ to intercalate with the hybridized ds-DNA or using a target strand of ss-DNA tagged with Ru(bpy)$_3^{2+}$, as shown in Figure 11.8. ECL emission is observed as immobilized ss-DNA is hybridized with its complementary stand DNA, whereas control experiment shows that no ECL emission is detected in the presence of noncomplementary strand DNA (Figure 11.9), clearly demonstrating the selectivity of ECL DNA biosensors. The monolayer of ss-DNA has been well-detected using such ECL DNA biosensors.

Similar approaches have also been used to covalently immobilize ss-DNA and antibody of C-reactive protein (CRP) on Au (111) electrodes, which serves as sensors to selectively detect its complementary strand DNA and its antigen (CRP), respectively, as shown in Figure 11.10 (Miao and Bard, 2003). In this study, the Au electrode modified with a SAM of 3-mercaptopropanoic acid (3-MPA) is used to covalently attach a 23-mer synthetic ss-DNA on the electrode surface, by reacting an amino group at the 5′-terminal of ss-DNA with the carboxyl group of 3-MPA, in the presence of 1-ethyl-3-(3-dimethylaminopropyl) carbodiimide hydrochloride (EDAC). The immobilized ss-DNA is then hybridized with a

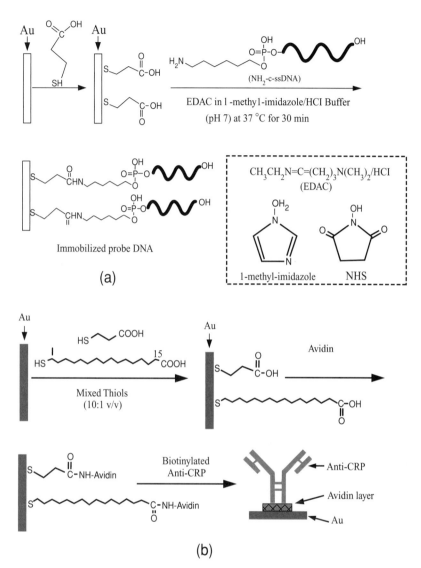

Figure 11.10. Illustration of (a) covalently attaching ss-DNA-tagged with Ru(bpy)$_3^{2+}$ on the Au (111) electrode surface modified with 3-MPA for ECL detection; (b) attaching a biotinylated antibody of protein (CRP) onto the electrode surface, modified with avidin via biotin–avidin interaction, which is then used to capture its antibody tagged with Ru(bpy)$_3^{2+}$, for ECL detection. The assay is similar to sandwich assay described in Figure 11.1b. [Reprinted from *Anal Chem* **75**:5825–5834 (2003), with permission of American Chemical Society.]

target complementary strand DNA tagged with Ru(bpy)$_3^{2+}$, which is then detected, in the presence of TPrA, using ECL technique. Likewise, avidin is covalently attached onto the Au(111) electrode surface modified with 3-MPA, by reacting of amine group of avidin with carboxyl group of 3-MPA. The biotinylated antibody of CRP is then immobilized on the Au (111) surface via the specific interaction of biotin with avidin. Using the sandwich assay approach as illustrated in Figure 11.1, the immobilized antibody of CRP recognizes CRP, and finally captured CRP use molecular recognition to specifically snatch the antibody of CPR tagged with Ru(bpy)$_3^{2+}$, which is used to quantitatively analyze CRP (1–24 μg/mL) using ECL detection, in the presence of TPrA. Great care, such as blocking a free carboxyl group with ethanolamine and filling pinholes with BSA, has been taken to prevent nonspecific interactions of labeled species with the electrode surface, leading to the higher selectivity of sensors.

Taken together, these studies have demonstrated the proof-of-concept that the ECL technique can be used to detect ss-DNA and proteins with high specificity and sensitivity for a variety of applications.

11.3.1.3. Enzyme-Based ECL Sensors

NADH plays a vital role in an array of biological processes, and its concentration is directly regulated by a variety of dehydrogenases. Thus, sensitive and selective assays of NADH have been developed on the basis of its association with dehydrogenases. In general, NAD$^+$ serves as a cofactor, aiding dehydrogenases to digest its substrate into small molecules.

NADH has been used as an ECL coreactor of Ru(bpy)$_3^{2+}$ for producing ECL emission (Martin and Nieman, 1993; Yokoyama et al., 1994; Jameison et al., 1996; Liang et al., 1996a,b; Martin and Nieman, 1997), because the reduced form of NADH containing a tertiary amine group acts as an effective coreactant for Ru(bpy)$_3^{2+}$, whereas its oxidized form containing an aromatic secondary amine is an ineffective ECL coreactor. In these studies, cation-exchange polymer films, such as Eastman AQ and Nafion, are used to immobilize Ru(bpy)$_3^{2+}$ and dehydrogenases, which has served as sensors to detect NADH, and substrates of enzyme (e.g., glucose, L-lactate, and alcohol), as illustrated in Figure 11.11. The Pt electrode modified with two separate polymer films, on top of each other on its surface, embedded with enzyme and Ru(bpy)$_3^{2+}$, is used to detect substrates of enzyme (e.g., glucose, L-lactate, and alcohol), in the presence of its cofactor (NAD$^+$). The immobilized Ru(bpy)$_3^{2+}$ is oxidized, which reacts with NADH generated by the immobilized enzyme accompanying the consuming of substrates, to produce ECL emission (Figure 11.11). Ru(bpy)$_3^{2+}$ and enzyme embedded in the polymer films on the electrode are regenerated after each reaction cycle. The electrode itself is compact and needs a very limited amount of solution. Reactions at the electrode are rapid because reactants on the electrode surface are not limited by the diffusion. These features make it possible to serve as an electrode for thin-layer electrochemical and ECL detectors in flow-injection analyzer.

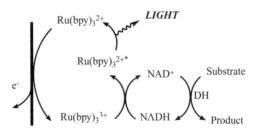

Figure 11.11. Illustration of enzyme-based ECL assay, showing that an enzyme, dehydrogenase (DH), and ECL emitter, Ru(bpy)$_3^{2+}$, embedded in cation exchange polymer films sitting next to each other on the electrode surface, generate ECL emission in the presence of NADH, which produces NAD$^+$, a cofactor of enzyme, which aids the enzyme in breaking down its substrates into small molecules and regenerating NADH simultaneously. [Reprinted from *Biosens Bioelectron* **12**:479–489 (1997), with permission of Elsevier.]

By varying immobilized enzymes (such as glucose dehydrogenase, L-lactate dehydrogenase, and alcohol dehydrogenase), the enzyme substrate (glucose, lactate, and ethanol) is analyzed using ECL detection (Martin and Nieman, 1997). As a unique feature of ECL detection, the sensors show the larger dynamic ranges for detection of glucose, lactate, and ethanol at 0.010–3.0 mM, 0.050–2.5 mM, and 0.025–1.5 mM, respectively.

11.3.2. Study Mechanism of Ru(bpy)$_3^{2+}$/TPrA System

As pointed out above, the success of ECL applications in ultrasensitive bioanalysis is largely attributable to the discovery of the efficient ECL systems, such as Ru(bpy)$_3^{2+}$/TPrA, in aqueous solution. The desire of rational design of effective ECL emitters and coreactors to improve its detection sensitivity has inspired the further investigation of its mechanism. It has been generally believed that ECL reaction of the Ru(bpy)$_3^{2+}$/TPrA system follows the "oxidative-reduction" mechanism as described below:

On the basis of TPrA oxidation, two routes have been proposed:

Catalytic route:

$$\text{Ru(bpy)}_3^{2+} - e \rightarrow \text{Ru(bpy)}_3^{3+} \quad (E^0 \sim 1.02 \text{ V versus SCE}) \quad (11.21)$$

$$\text{Ru(bpy)}_3^{3+} + \text{TPrA} \rightarrow \text{Ru(bpy)}_3^{2+} + \text{TPrA}^{\bullet+} \quad (11.22)$$

Direct oxidation route:

$$\text{TPrA} - e \rightarrow \text{TPrA}^{\bullet+} \quad (E^0 \sim 0.9 \text{ V versus SCE}) \quad (11.23)$$

then

$$\text{TPrA}^{\bullet+} \rightarrow \text{TPrA}^{\bullet} + \text{H}^+ \quad (11.24)$$

followed by either

$$\text{Ru(bpy)}_3^{3+} + \text{TPrA}^{\bullet} \rightarrow \text{Ru(bpy)}_3^{2+*} + \text{Pr}_2\text{NC}^+\text{HCH}_2\text{CH}_3 \quad (11.25)$$

$$\text{Ru(bpy)}_3^{2+} + \text{TPrA}^{\bullet} \rightarrow \text{Ru(bpy)}_3^{+} + \text{Pr}_2\text{NC}^+\text{HCH}_2\text{CH}_3 \quad (11.26)$$

or

$$\text{Ru(bpy)}_3^{3+} + \text{Ru(bpy)}_3^{+} \rightarrow \text{Ru(bpy)}_3^{2+*} + \text{Ru(bpy)}_3^{2+} \quad (11.27)$$

where

$$\text{TPrA}^{\bullet+} = \text{Pr}_3\text{N}^{\bullet+} \quad \text{and} \quad \text{TPrA}^{\bullet} = \text{Pr}_2\text{NC}^{\bullet}\text{HCH}_2\text{CH}_3$$

Several studies suggest that the excited state is produced primarily by reaction of Ru(bpy)$_3^{3+}$ and TPrA$^{\bullet}$ (Leland and Powell, 1990; Gross et al., 2001; Kanoufi et al., 2001; Wightman et al., 2004). The contribution of the catalytic route and the direct oxidation route to overall ECL intensity depend on concentrations of Ru(bpy)$_3^{2+}$ and TPrA as well as TPrA oxidation rate at electrode. For the presence of higher concentration of Ru(bpy)$_3^{2+}$, TPrA oxidation mainly proceeds via the catalytic pathway. In contrast, in the presence of trace amount of Ru(bpy)$_3^{2+}$ and the higher concentration of TPrA, direct oxidation of TPrA at the electrode plays a dominate role. Typically, the trace amount of biomolecules tagged with $Ru(bpy)_3^{2+}$ in the presence of high concentration of TPrA is used in ultrasensitive analysis. Thus, its mechanism involves the direct oxidation of TPrA. The investigation of this mechanism is of great interest in improving the sensitivity of ECL detection.

In addition, it has been demonstrated that the anodic oxidation rate of TPrA is strongly dependent on the electrode material. More facile TPrA oxidation has been observed at glassy carbon (GC), compared to Au and Pt electrodes (Zu and Bard, 2000). More hydrophobic electrode surfaces, such as the presence of surfactant at the electrode, also favor this reaction (Zu and Bard, 2001). A new reaction pathway [Eq. (11.28)] describing the intermediacy of TPrA cation radical, $TPrA^{\bullet+}$, for the formation of $Ru(bpy)_3^{2+}$ excited state has been proposed (Miao et al., 2002).

$$Ru(bpy)_3^+ + TPrA^{\bullet+} \rightarrow Ru(bpy)_3^{2+*} + TPrA \qquad (11.28)$$

The mechanism has been investigated using electrochemical methods, scanning electrochemical microscopy (SECM), and electron spin resonance (ESR) spectra. Compared with the conventional ECL routes (catalytic and direct oxidation routes), a unique feature of this new emission pathway is that oxidation of $Ru(bpy)_3^{2+}$ is not required to produce the excited state of $Ru(bpy)_3^{2+}$. Since the standard oxidation potential of TPrA is lower than that of $Ru(bpy)_3^{2+}$, a low-oxidation-potential (LOP) ECL emission has been observed in the potential region below 0.9 V vs SCE (Zu and Bard, 2000; Miao et al., 2002), showing the conventional and LOP ECL intensity-potential transient at a freshly polished glassy carbon electrode (Figure 11.12). It is believed that the LOP ECL mechanism accounts for the bead-based ECL process, in which $Ru(bpy)_3^{2+}$ immobilized on the surface of the beads located well out of the electron tunneling distance from the electrode still produces ECL emission (Miao et al., 2002). Apparently, LOP ECL intensity primarily depends on TPrA oxidation rate, which is significantly facilitated by the adsorption of nonionic fluorosurfactant species at noble metal electrodes, especially the Au electrode (Li and Zu, 2004). Consequently,

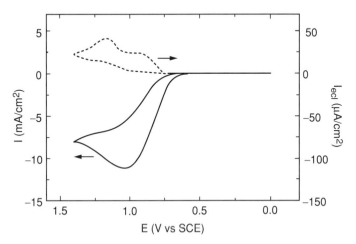

Figure 11.12. Cyclic voltammogram (solid line) and ECL emission-potential transient (dashed line) acquired at a glassy carbon electrode in 0.15 M phosphate buffer solution (pH 7.5) containing 100 mM TPrA and 1 μM $Ru(bpy)_3^{2+}$, showing that ECL emission is generated at the potentials at 0.75 V and 1.1 V, where oxidation of TPrA and $Ru(bpy)_3^{2+}$ takes place, respectively. Potential scan rate, 0.1 V/s. The unit of ECL emission transient is an arbitrary unit. [Reprinted from *Anal Chem* **72**:3223–3232 (2001), with permission of American Chemical Society.]

a more intense LOP ECL emission has been observed at a fluorosurfactant-modified Au electrode. These studies ultimately may lead to the design of more efficient ECL systems for ultrasentive bioanalysis.

Typically, for ECL reactions involving Ru(bpy)$_3^{2+}$/TPrA, the presence of oxygen produces little effect on ECL intensity. However, a recent study (Zheng and Zu, 2005a) indicates that, in the presence of a low concentration of TprA, direct oxidation reaction plays a predominant role in producing ECL. Thus, the ECL intensity is highly sensitive to the presence of O$_2$. In the presence of a high concentration of TPrA, the dissolved oxygen within the ECL reaction layer is rapidly consumed by the higher concentration of TPrA•, leading to the insignificant quenching effect on ECL by oxygen. In contrast, for the low concentration of TPrA, because the lower amount of its radical anion will be generated, the presence of oxygen can intercept and destroy the intermediates before they participate in the ECL reactions, leading to the obvious reduction of the emission intensity. In other words, the study also suggests that, for ECL reactions via the catalytic route, the quenching of ECL emission by the presence of oxygen is mainly attributable to the quenching of the excited state of Ru(bpy)$_3^{2+*}$, because the catalytic route does not involve the consumption of intermediate radical ions by the presence of oxygen. These results suggest that removing environmental oxygen or achieving rapid oxidation of TprA may aid the improvement of ECL detection sensitivity.

11.3.3. New ECL Emitters, Semiconductor Nanocrystals (Quantum Dots)

Even though the synthesis, characterization, and application of nanocrystals (NCs) have been extensively studied over the past two decades (Alivisatos, 1996), the water-soluble NCs have only been developed as biological fluorescence labels in recent years (Chan and Nie, 1998; Bruchez, et al., 1998). Because of confinement of the electrons and holes, atomic-like electronic energy levels are formed in NCs. The spacing of the highest occupied and lowest unoccupied quantum confined orbital is strongly dependent on the size of the NC. Thus, the quantum size effect permits continuous tuning of a wide range of the band gap simply by changing the size of NC, making its possible to prepare multi-colors of probes for multiplexing analysis, as well described in Chapter 4.

In order to generate ECL emission from NCs, it is crucial to produce the stable oxidized and reduced forms of NCs. A recent study shows that monolayer-protected bare (nonpassivated) Si NCs are chemically stable upon electron–hole injection (Ding et al., 2002), the prerequisite of the generation of ECL emission from NCs. In fact, the well-dispersed Si NCs in organic solvent have been oxidized and reduced at a Pt electrode, generating ECL emission via annihilation and coreactant ECL mechanisms. The band gap between the onset of oxidation and reduction is related to the energy between the highest occupied and lowest unoccupied quantum confined orbital.

The relatively higher ECL intensity is observed in the cathodic reactions, indicating that the electrogenerated oxidized forms are more stable in maintaining their charged states prior to the annihilation reaction. Higher ECL intensity has also been produced via coreactant ECL reaction mechanism, suggesting that the coreactants overcome the limited potential window of the solvent or short life of anion or cation, the same as those observed in conventional ECL systems (Bard, 2004a,b).

Unlike those observed in photoluminescence, the peak wavelength of ECL spectra (color) of bare NCs is insensitive to its core size and capping agent, but highly sensitive to the surface properties of NCs. The ECL spectra of bare Si NCs show a peak wavelength of 640 nm, significantly red-shifted from that observed in its photoluminescence spectra (∼420 nm).

Similar results have been observed in the ECL studies of other nonpassivated semiconductor NCs, such as CdSe (Myung et al., 2002), CdTe (Bae et al., 2004), and Ge (Myung et al., 2004). The profound distinction of ECL and photoluminescence spectra of bare Si NCs may be attributable to the different emission mechanisms resulting from photoluminescence and ECL. However, the study of the passivated NCs (CdSe NCs capped with ZnSe) shows the similar photoluminescence and ECL spectra (Myung et al., 2003). Taken together, these studies suggest that the excitation and emission of NC cores is primarily a mechanism of photoluminescence. In contrast, the electron transfer between surface states at NC surface plays a vital role in generating ECL emission.

Recently, ECL investigations of the films of cross-linked CdSe/CdS and CdSe/ZnS core-shell NCs (Poznyak et al., 2004), CdSe hollow spherical (Zou et al., 2005), and CdS spherical assemblies (Ren et al., 2005) have also been carried out, suggesting that other mechanisms may be involved in generating ECL emission of NCs. More extensive research on this topic will certainly be forthcoming, giving new insights into the ECL mechanisms of NCs. Since these NCs exhibit rich optical and electronic properties, its potential application in ECL-based bioassays is very exciting and promising. The new phenomena and discoveries in this topic will surely be seen for years to come.

11.4. SUMMARY AND OUTLOOK

In summary, the studies have demonstrated that ECL is a powerful ultrasensitive analytical tool in bioanalysis. However, its inherent potentials in ultrasenstive detection are not yet fully realized. In this chapter, we overview the historic research progress in ECL, introduce its fundamentals and instrumentation, highlight the unique features and distinct advantages of ECL techniques in ultrasensitive detection, illustrate several representative ECL assays and their applications in bioanalysis, and describe current forefront research in ECL mechanisms, new ECL emitters, and coreactors.

There is no doubt that new progresses, new discoveries, and new breakthroughs in ECL research will be continuously made, which will shed new light on ECL mechanisms and lead to the development of new applications. To further improve the sensitivity of ECL detection, the upcoming research will focus on the following approaches, such as increasing labeling ratio using more effective conjugation methods and micro/nanosphere embedded with multiple molecules of ECL emitters, rational design of more effective ECL emitters and coreactors via mechanism study, and reducing background emission by exploring its mechanism.

The future major advances of ECL detection in ultrasensitive bioanalysis may rely on breakthroughs in the following research topics: (i) new approaches of probing ECL mechanisms in real time at single-molecule level, leading to better understanding of ECL reaction mechanisms, and hence rational design of more efficient ECL emitters and coreactors; (ii) design of new assays that will not require the use of antibody and direct labeling of molecules of interest with ECL emitter; (iii) multiplexing analysis of ECL emitters, aiming high-throughput analysis and study of affinity and binding kinetics of molecular interactions; (iv) exploring the possibility of analysis of single living cells using ECL detection (for example, quantitatively mapping single receptors on single living cells by lighting up the receptors with nanoprobes using SECM); (v) development of new ECL instrumentation that will permit the application of ECL detection to probing intracellular components and mechanisms of living cells.

ACKNOWLEDGMENTS

Zu thanks the support of the Research Grants Council of the Hong Kong Special Administrative Region, China (Projects HKU 7061/04P and HKU 7059/05P). Xu acknowledges the support of NIH (R21-RR15057-01), Old Dominion University, and her research group, especially those (Jeffers, Logan, Viola, Wen, and Yoneyama) who contribute to her ECL research program.

REFERENCES

Abruna, HD, and Bard, AJ (1982). Electrogenerated chemiluminescence. 40. A chemiluminescent polymer based on the tris(4-vinyl-4'-methyl-2,2'-bipyridyl)ruthenium(II) system. *J Am Chem Soc* **104**: 2641–2642.

Alivisatos, A (1996). Perspectives on the physical chemistry of semiconductor nanocrystals. *J Phys Chem* **100**(31):13226–13239 and references therein.

Arora, A, Eijkel, JCT, Morf, WE, and Manz, A (2001). A wireless electrochemiluminescence detector applied to direct and indirect detection for electrophoresis on a microfabricated glass device. *Anal Chem* **73**:3282–3288.

Atkins, PW (1982). *Physical Chemistry*. Freeman, San Francisco, CA, pp. 823–905.

Bae, Y, Myung, N, and Bard, AJ (2004). Electrochemistry and electrogenerated chemiluminescence of CdTe nanoparticles. *Nano Lett* **4**:1153–1161.

Bard, AJ (2004a). *Electrogenerated Chemiluminescence*. Marcel Dekker, New York and references therein.

Bard, AJ (2004b). Miscellaneous topics and conclusions. In: Bard, AJ (ed.), *Electrogenerated Chemiluminescence*. Marcel Dekker, New York, pp. 523–532 and references therein.

Bard, AJ, Debad, JD, Leland, JK, Sigal, GB, Wilbur, JL, and Wohlstadter, JN (2000). Chemiluminescence, electrogenerated. In: Meyers RA (ed.), *Encyclopedia of Analytical Chemistry: Applications, Theory and Instrumentation*, Vol. II, John Wiley & Sons, New York, pp. 9842–9849 and references therein.

Beduschi, R, Beduschi, MB, and Oesterling, J. (1998): Percent free-prostate-specific antigen test: Improving both sensitivity and specificity. *Infect Urol* **11**(5):133–138, and references therein.

Beideman, FE, and Hercules, DM (1979). Electrogenerated chemiluminescence from 9,10-diphenylanthracene cations reacting with radical anions. *J Phys Chem* **83**:2203–2209.

Blackburn, GF, Shah, HP, Kenten, JH, Leland, J, Kamin, RA, Link, J, Peterman, J, Powell, MJ, Shah, A, Talley, DB, Tyagi, SK, Wilkins, E, Wu, TG, and Massey, RJ (1991). Electrochemiluminescence detection for development of immunoassays and DNA probe assays for clinical diagnostics. *Clin Chem* **37**:1534–1539.

Bruce, D, and Richter, MM (2002a). Electrochemiluminescence in aqueous solution of a ruthenium(II) bipyridyl complex containing a crown ether moiety in the presence of metal ions. *Analyst* **127**:1492–1494.

Bruce, D, and Richter, MM (2002b). Green electrochemiluminescence from ortho-metalated Tris(2-phenylpyridine)iridium(III). *Anal Chem* **74**:1340–1342.

Bruchez, M, Jr., Moronne, M, Gin, P, Weiss, S, and Alivisatos, AP (1998). Semiconductor nanocrystals as fluorescent biological labels. *Science* **281**:2013–2016.

Brune, SN, and Bobbitt, DR (1991). Effect of pH on the reaction of tris(2,2'-bipyridyl)ruthenium(III) with amino acids: Implications for their detection. *Talanta* **38**:419–424.

Brune, SN, and Bobbitt, DR (1992). Role of electron-donating/withdrawing character, pH, and stoichiometry on the chemiluminescent reaction of tris(2,2'-bipyridyl)ruthenium(III) with amino acids. *Anal Chem* **64**:166–170.

Butch, AW, Crary, D, and Yee, M (2002). Analytical performance of the Roche total and free PSA assays on the Elecsys 2010 immunoanalyzer. *Clin Biochem* **35**:143–145.

Butler, J, and Henglein, A (1980). Elementary reactions of the reduction of thallium(1+) in aqueous solution. *Rad Phys Chem* **15**:603–612.

Cao, W, Jia, J, Yang, X, Dong, S, and Wang, E (2002). Capillary electrophoresis with solid-state electrochemiluminescence detector. *Electrophoresis* **23**:3683–3698.

Carter, MT, and Bard, AJ (1990). Electrochemical investigations of the interaction of metal chelates with DNA. 3. Electrogenerated chemiluminescent investigation of the interaction of tris(1,10-phenanthroline)ruthenium(II) with DNA. *Bioconjugate Chem* **1**:257–263.

Chan, W, and Nie, S (1998). Quantum dot bioconjugates for ultrasensitive nonisotopic detection. *Science* **281**:2016–2018.

Chang, MM, Saji, T, and Bard, AJ (1977). Electrogenerated chemiluminescence. 30. Electrochemical oxidation of oxalate ion in the presence of luminescers in acetonitrile solutions. *J Am Chem Soc* **99**:5399–5403.

Chen, X, and Sato, M (1995). High-performance liquid chromatographic determination of ascorbic acid in soft drinks and apple juice using Tris(2,2'-bipyridine)ruthenium(II) electrochemiluminescence. *Anal Sci* **11**:749–754.

Chiang, MT, and Whang, CW (2002). Tris(2,2'-bipyridyl)ruthenium(III)-based electrochemiluminescence detector with indium/tin oxide working electrode for capillary electrophoresis. *J Chromatogr A* **934**:59–66.

Choi, HN, Cho, SH, and Lee, WY (2003). Electrogenerated chemiluminescence from tris(2,2'-bipyridyl)ruthenium(II) immobilized in titania-perfluorosulfonated Ionomer composite films. *Anal Chem* **75**:4250–4256.

Choi, HN, Cho, SH, Park, YJ, Lee, DW, and Lee, WY (2005). Sol–gel-immobilized tris(2,2'-bipyridyl)ruthenium(II) electrogenerated chemiluminescence sensor for high-performance liquid chromatography. *Anal Chim Acta* **541**:49–56.

Cole, C, Muegge, BD, and Richter, MM (2003). Effects of poly(ethylene glycol) *tert*-octylphenyl ether on tris(2-phenylpyridine)iridium(III)-tripropylamine electrochemiluminescence. *Anal Chem* **75**:601–604.

Danielson, ND (2004). Analytical applications: Flow injection, liquid chromatography, and capillary electrophoresis. In: Bard, AJ (ed.), *Electrogenerated Chemiluminescence*. Marcel Dekker, New York, pp. 397–444.

Debad, JD, Glezer, EN, Wohlstadter, J, and Sigal, GB (2004). Clinical and biological applications of ECL. In: Bard AJ (ed.), *Electrogenerated Chemiluminescence*. Marcel Dekker, New York, pp. 359–396 and references therein.

Dennany, L, Forster, RJ, and Rusling, JF (2003). Simultaneous direct electrochemiluminescence and catalytic voltammetry detection of DNA in ultrathin films. *J Am Chem Soc* **125**:5213–5218.

Dennany, L, Forster, RJ, White, B, Smyth, M, and Rusling, JF (2004). Direct electrochemiluminescence detection of oxidized DNA in ultrathin films containing $[Os(bpy)_2(PVP)_{10}]^{2+}$. *J Am Chem Soc* **126**:8835–8841.

Ding, ZF, Quinn, BM, Haram, SK, Pell, LE, Korgel, BA, and Bard, AJ (2002). Electrochemistry and electrogenerated chemiluminescence from silicon nanocrystal quantum dots. *Science* **296**:1293–1297.

Ding, SN, Xu, JJ, and Chen, HY (2005). Tris(2,2'-bipyridyl)ruthenium(II)-zirconia-Nafion composite films applied as solid-state electrochemiluminescence detector for capillary electrophoresis. *Electrophoresis* **26**:1737–1744.

Downey, TM, and Nieman, TA (1992). Chemiluminescence detection using regenerable tris(2,2'-bipyridyl)ruthenium(II) immobilized in Nafion. *Anal Chem* **64**:261–268.

Ege, D, Becker, WG, and Bard, AJ (1984). Electrogenerated chemiluminescent determination of tris(2,2'-bipyridine)ruthenium ion (Ru(bpy)32+) at low levels. *Anal Chem* **56**:2413–2417.

Factor, B, Muegge, B, Workman, S, Bolton, E, Bos, J, and Richter, MM (2001). Surfactant chain length effects on the light emission of tris(2,2'-bipyridyl)ruthenium(II)/tripropylamine electrogenerated chemiluminescence. *Anal Chem* **73**:4621–4624.

Fahnrich, KA, Pravda, M, and Guilbault, GG (2001). Recent applications of electrogenerated chemiluminescence in chemical analysis. *Talanta* **54**:531–559.

Fan, FRF (2004). Experimental techniques of electrogenerated chemiluminescence. In: Bard, AJ (ed.), *Electrogenerated Chemiluminescence*. Marcel Dekker, New York, pp. 359–396 and references therein.

Fan, FRF, Cliffel, D, and Bard, AJ (1998). Scanning electrochemical microscopy. 37. Light emission by electrogenerated chemiluminescence at SECM tips and their application to scanning optical microscopy. *Anal Chem* **70**:2941–2948.

Faulkner, LR (1978). Chemiluminescence from electron-transfer processes. *Methods Enzymol* **57**:494–526.

Faulkner, LR, and Bard, AJ (1977). Techniques of electrogenerated chemiluminescence. In: Bard, AJ (ed.), *Electroanalytical Chemistry*, Vol. 10. Marcel Dekker, New York, pp. 1–95.

Feldberg, SW (1966a). A possible method for distinguishing between triplet–triplet annihilation and direct singlet formation in electrogenerated chemiluminescence. *J Phys Chem* **70**:3928–3930.

Feldberg, SW (1966b). Theory of controlled potential electrogeneration of chemiluminescence. *J Am Chem Soc* **88**:390–393.

Forbes, GA, Nieman, TA, and Sweedler, JV (1997). On-line electrogenerated Ru(bpy)$_3^{3+}$ chemiluminescent detection of β-blockers separated with capillary electrophoresis. *Anal Chem Acta* **347**:289–293.

Forry, SP, and Wightman, RM (2002). Electrogenerated chemiluminescence detection in reversed-phase liquid chromatography. *Anal Chem* **74**:528–532.

Gao, Y, Tian, YL, and Wang, EK (2005). Simultaneous determination of two active ingredients in Flos daturae by capillary electrophoresis with electrochemiluminescence detection. *Anal Chim Acta* **545**:137–141.

Gerardi, RD, Barnett, NW, and Lewis, AW (1999). Analytical applications of tris(2,2'-bipyridyl)ruthenium(III) as a chemiluminescent reagent. *Anal Chim Acta* **378**:1–41.

Glass, RS, and Faulkner, LR (1981). Electrogenerated chemiluminescence from the tris(2,2'-bipyridine)ruthenium(II) system. An example of S-route behavior. *J Phys Chem* **85**:1160–1165.

Greenway, GM (1990). Analytical applications of electrogenerated chemiluminescence. *Trends Anal Chem* **9**:200–203.

Gross, EM, Pastore, P, and Wightman, RM (2001). High-frequency electrochemiluminescent investigation of the reaction pathway between tris(2,2'-bipyridyl)ruthenium(II) and tripropylamine using carbon fiber microelectrodes. *J Phys Chem B* **105**:8732–8738.

Guo, ZH, and Dong, SJ (2004). Electrogenerated chemiluminescence from Ru(bpy)$_3^{2+}$ ion-exchanged in carbon nanotube/perfluorosulfonated ionomer composite films. *Anal Chem* **76**:2683–2688.

Haese, A, Dworschack, RT, Piccoli, SP, Sokoll, LJ, Partin, AW, and Chan, DW (2002). Clinical evaluation of the Elecsys total prostate-specific antigen assay on the Elecsys 1010 and 2010 systems. *Clin Chem* **48**:944–947.

Hercules, DM (1964). Chemiluminescence resulting from electrochemically generated species. *Science* **143**:808–809.

Honda, K, Yoshimura, M, Rao, TN, and Fujishima, A (2003). Electrogenerated chemiluminescence of the ruthenium tris(2,2')bipyridyl/amines system on a boron-doped diamond electrode. *J Phys Chem B* **107**:1653–1663.

Hsueh, YT, Smith, RL, and Northrup, MA (1996). A microfabricated, electrochemiluminescence cell for the detection of amplified DNA. *Sens Actuators B* **33**:110–114.

Hsueh, YT, Collins, SD, and Smith, RL (1998). DNA quantification with an electrochemiluminescence microcell. *Sens Actuators B* **49**:1–4.

Ikehara, T, Habu, N, Nishino, I, and Kamimori, H (2005). Determination of hydroxyproline in rat urine by high-performance liquid chromatography with electrogenerated chemiluminescence detection using tris(2,2′-bipyridyl) ruthenium(II). *Anal Chim Acta* **536**:129–133.

Jameison, F, Sanchez, RI, Dong, L, Leland, JK, Yost, D, and Martin, MT (1996). Electrochemiluminescence-based quantitation of classical clinical chemistry analytes. *Anal Chem* **68**:1298–1302.

Kang, JZ, Liu, JF, Yin, XB, Qiu, HB, Yan, JL, Yang, XR, and Wang, EK (2005). Capillary electrophoresis with indirect electrochemiluminescence detection. *Anal Lett* **38**:1179–1191.

Kanoufi, F, Zu, Y, and Bard, AJ (2001). Homogeneous oxidation of trialkylamines by metal complexes and its impact on electrogenerated chemiluminescence in the trialkylamine/Ru(bpy)$_3^{2+}$ system. *J Phys Chem B* **105**:210–216.

Keller, GH, and Manak, MM (1993). *DNA Probes*, 2nd edition. Stockton Press, New York, pp. 173–196.

Kenten, JH, Casadei, J, Link, J, Lupold, S, Willey, J, Powell, M, Rees, A, and Massey, R (1991). Rapid electrochemiluminescence assays of polymerase chain reaction products. *Clin Chem* **37**:1626–1632.

Kenten, JH, Gudibande, S, Link, J, Willey, JJ, Curfman, B, Major, EO, and Massey, RJ (1992). Improved electrochemiluminescent label for DNA probe assays: Rapid quantitative assays of HIV-1 polymerase chain reaction products. *Clin Chem* **38**:873–879.

Khramov, AN, and Collinson, MM (2000). Electrogenerated chemiluminescence of tris(2,2′-bipyridyl)ruthenium(II) ion-exchanged in Nafion-silica composite films. *Anal Chem* **72**:2943–2948.

Kijek, TM, Rossi, CA, Moss, D, Parker, RW, and Henchal, EA (2000). Rapid and sensitive immunomagnetic-electrochemiluminescent detection of staphyloccocal enterotoxin B. *J Immun Methods* **236**:9–17.

Kim, DJ, Lyu, YK, Choi, HN, Min, IH, and Lee, WY (2005a). Nafion-stabilized magnetic nanoparticles (Fe$_3$O$_4$) for [Ru(bpy)(3)](2+) (bpy = bipyridine) electrogenerated chemiluminescence sensor. *Chem Commun* 2966–2968.

Kim, JI, Shin, IS, Kim, H, and Lee, JK (2005b). Efficient electrogenerated chemiluminescence from cyclometalated iridium(III) complexes. *J Am Chem Soc* **127**:1614–1615.

Knight, AW (2001). Electrogenerated chemiluminescence. In: Grarcia-Campana, AM, and Baeyens, WRG (eds.), *Chemiluminescence in Analytical Chemistry*. Marcel Dekker, New York, pp. 211–247.

Knight, AW, and Greenway, GM (1994). Occurrence, mechanisms and analytical applications of electrogenerated chemiluminescence. *Analyst* **119**:879–890.

Kulmala, S, and Suomi, J (2003). Current status of modern analytical luminescence methods. *Anal Chim Acta* **500**:21–69.

Kuwabara, T, Noda, T, Ohtake, H, Ohtake, T, Toyama, S, and Ikariyama, Y (2003). Classification of DNA-binding mode of antitumor and antiviral agents by the electrochemiluminescence of ruthenium complex. *Anal Biochem* **314**:30–37.

Lai, RY, and Bard, AJ (2003). Electrogenerated chemiluminescence. 70. The application of ECL to determine electrode potentials of tri-*n*-propylamine, its radical cation, and intermediate free radical in MeCN/benzene solutions. *J Phys Chem A* **107**:3335–3340.

Lai, RY, Chiba, M, Kitamura, N, and Bard, AJ (2002). Electrogenerated chemiluminescence. 68. Detection of sodium ion with a ruthenium(II) complex with crown ether moiety at the 3,3′-positions on the 2,2′-bipyridine ligand. *Anal Chem* **74**:551–553.

Lee, WY (1997). Tris(2,2′-bipyridyl)ruthenium(II) electrogenerated chemiluminescence in analytical science. *Mikrochim Acta* **127**:19–39.

Lee, WY, and Nieman, TA (1994). Determination of dansyl amino acids using tris(2,2′-bipyridyl)ruthenium(II) chemiluminescence for post-column reaction detection in high-performance liquid chromatography. *J Chromatogr* **659**:111–118.

Lee, WY, and Nieman, TA (1995). Evaluation of use of tris(2,2′-bipyridyl)ruthenium(III) as a chemiluminescent reagent for quantitation in flowing streams. *Anal Chem* **67**:1789–1796.

Leland, JK, and Powell, MJ (1990). Electrogenerated chemiluminescence: An oxidative-reduction type ECL reaction using tripropyl amine. *J Electrochem Soc* **137**:3127–3131.

Li, F, and Zu, Y (2004). Effect of nonionic fluorosurfactant on the electrogenerated chemiluminescence of the tris(2,2′-bipyridine)ruthenium(II)/tri-*n*-propylamine system: Lower oxidation potential and higher emission intensity. *Anal Chem* **76**:1768–1772.

Liang, P, Dong, L, and Martin, MT (1996a). Light emission from ruthenium-labeled penicillins signaling their hydrolysis by β-lactamase. *J Am Chem Soc* **118**:9198–9199.

Liang, P, Sanchez, RI, and Martin, MT (1996b). Electrochemiluminescence-based detection of β-lactam antibiotics and β-lactamases. *Anal Chem* **68**:2426–2431.

Lin, XQ, Li, F, Pang, YQ, and Cui, H (2004). Flow injection analysis of gallic acid with inhibited electrochemiluminescence detection. *Anal Bioanal Chem* **378**:2028–2033.

Liu, J, Cao, W, Qiu, H, Sun, X, Yang, X, and Wang, E (2002). Determination of sulpiride by capillary electrophoresis with end-column electrogenerated chemiluminescence detection. *Clin Chem* **48**:1049–1058.

Luttmer, JD, and Bard, AJ (1981). Electrogenerated chemiluminescence. 38. Emission intensity-time transients in the tris(2,2′-bipyridine)ruthenium(II) system. *J Phys Chem* **85**:1155–1159.

Maness, KM, and Wightman, RM (1994). Electrochemiluminescence in low ionic strength solutions of 1,2-dimethoxyethane. *J Electroanal Chem* **396**:85–95.

Marcus, RA (1965). On the theory of chemiluminescent electron-transfer reactions. *J Chem Phys* **43**:2654–2657.

Marlins, C, Vandeloise, R, Walton, D, and Vander Donckt, E (1997). Ultrasonic modification of light emission from electrochemiluminescence processes. *J Phys Chem A* **101**:5063–5068.

Martin, AF, and Nieman, TA (1993). Glucose quantitation using an immobilized glucose-dehydrogenase enzyme reactor and a tris(2,2′-bipyridyl)ruthenium(II) chemiluminescence sensor. *Anal Chim Acta* **281**:475–481.

Martin, AF, and Nieman, TA (1997). Chemiluminescence biosensors using tris(2,2′-bipyridyl)ruthenium(II) and dehydrogenases immobilized in cation exchange polymers. *Biosens Bioelectron* **12**:479–489.

Maus, RG, and Wightman, RM (2001). Microscopic imaging with electrogenerated chemiluminescence. *Anal Chem* **73**:3993–3998.

McCall, J, and Richter, MM (2000). Phenol substituent effects on electrogenerated chemiluminescence quenching. *Analyst* **125**:545–548.

McCall, J, Alexander, C, and Richter, MM (1999). Quenching of electrogenerated chemiluminescence by phenols, hydroquinones, catechols, and benzoquinones. *Anal Chem* **71**:2523–2527.

Memming, R (1969). Mechanism of the electrochemical reduction of persulfates and hydrogen peroxide. *J Electrochem Soc* **116**:785–790.

Miao, W, and Bard, AJ (2003). Electrogenerated chemiluminescence. 72. Determination of immobilized DNA and C-Reactive protein on Au(111) electrodes using tris(2,2′-bipyridyl)ruthenium(II) labels. *Anal Chem* **75**:5825–5834.

Miao, W, and Bard, AJ (2004a). Electrogenerated chemiluminescence. 80. C-reactive protein determination at high amplification with [Ru(bpy)$_3$]$^{2+}$-containing microspheres. *Anal Chem* **76**:7109–7113.

Miao, W, and Bard, AJ (2004b). Electrogenerated chemiluminescence. 77. DNA hybridization detection at high amplification with [Ru(bpy)$_3$]$^{2+}$-containing microspheres. *Anal Chem* **76**:5379–5386.

Miao, W, and Choi, JP (2004). Coreactants. In: Bard, AJ (ed.) *Electrogenerated Chemiluminescence*. Marcel Dekker, New York, pp. 213–271.

Miao, W, Choi, JP, and Bard, AJ (2002). Electrogenerated chemiluminescence 69: The tris(2,2′-bipyridine)ruthenium(II), (Ru(bpy)$_3{}^{2+}$)/tri-*n*-propylamine (TPrA) system revisited—A new route involving TPrA$^{\bullet+}$ cation radicals. *J Am Chem Soc* **124**:14478–14485.

Morita, H, and Konishi, M (2002). Electrogenerated chemiluminescence derivatization reagents for carboxylic acids and amines in high-performance liquid chromatography using tris(2,2′-bipyridine)ruthenium(II). *Anal Chem* **74**:1584–1589.

Morita, H, and Konishi, M (2003). Electrogenerated chemiluminescence derivatization reagent, 3-isobutyl-9,10-dimethoxy-1,3,4,6,7,11b-hexahydro-2H-pyrido[2,1-a]isoquinolin-2.ylamine, for carboxylic acid in high-performance liquid chromatography using tris(2,2′-bipyridine)ruthenium (II). *Anal Chem* **75**:940–946.

Muegge, BD, and Richter, MM (2002). Electrochemiluminescent detection of metal cations using a ruthenium(II) bipyridyl complex containing a crown ether moiety. *Anal Chem* **74**:547–550.

Myung, N, Ding, ZF, and Bard, AJ (2002). Electrogenerated chemiluminescence of CdSe nanocrystals. *Nano Lett* **2**:1315–1319.

Myung, N, Bae, Y, and Bard, AJ (2003). Effect of surface passivation on the electrogenerated chemiluminescence of CdSe/ZnSe nanocrystals. *Nano Lett* **3**:1053–1055.

Myung, N, Lu, X, Johnston, KP, and Bard, AJ (2004). Electrogenerated chemiluminescence of Ge nanocrystals. *Nano Lett* **4**:183–185.

Namba, Y, Usami, M, and Suzuki, O (1999). Highly sensitive electrochemiluminescence immunoassay using the ruthenium chelate-labeled antibody bound on the magnetic micro beads. *Anal Sci* **15**:1087–1093.

Noffsinger, JB, and Danielson, ND (1987). Generation of chemiluminescence upon reaction of aliphatic amines with tris(2,2′-bipyridine)ruthenium(III). *Anal Chem* **59**:865–868.

Pang, YQ, Cui, H, Zheng, HS, Wan, GH, Liu, LJ, and Yu, XF (2005). Flow injection analysis of tetracyclines using inhibited Ru(bpy)$_3{}^{2+}$/tripropylamine electrochemiluminescence system. *Luminescence* **20**:8–15.

Poznyak, SK, Talapin, DV, Shevchenko, EV, and Weller, H (2004). Quantum dot chemiluminescence. *Nano Lett* **4**:693–698.

Qiu, HB, Yan, JL, Sun, XH, Liu, JF, Cao, WD, Yang, XR, and Wang, EK (2003). Microchip capillary electrophoresis with an integrated indium tin oxide electrode-based electrochemiluminescence detector. *Anal Chem* **75**:5435–5440.

Ren, T, Xu, JZ, Tu, YF, Xu, S, and Zhu, JJ (2005). Electrogenerated chemiluminescence of CdS spherical assemblies. *Electrochem Commun* **7**:5–9.

Richter, MM (2004). Electrochemiluminescence (ECL). *Chem Rev* **104**:3003–3036.

Richter, MM, Bard, AJ, Kim, W, and Schmehl, RH (1998). Electrogenerated chemiluminescence. 62. Enhanced ECL in bimetallic assemblies with ligands that bridge isolated chromophores. *Anal Chem* **70**:310–318.

Ridlen, JS, Skotty, DR, Kissinger, PT, and Nieman, TA (1997). Determination of erythromycin in urine and plasma using microbore liquid chromatography with tris(2,2′-bipyridyl)ruthenium(II) electrogenerated chemiluminescence detection. *J Chromatogr B* **694**:393–400.

Rodriguez, M, and Bard, AJ (1990). Electrochemical studies of the interaction of metal chelates with DNA. 4. Voltammetric and electrogenerated chemiluminescent studies of the interaction of tris(2,2′-bipyridine)osmium(II) with DNA. *Anal Chem* **62**:2658–2662.

Rubinstein, I, and Bard, AJ (1980). Polymer films on electrodes. 4. Nafion-coated electrodes and electrogenerated chemiluminescence of surface-attached tris(2,2′-bipyridine)ruthenium(2+). *J Am Chem Soc* **102**:6641–6642.

Rubinstein, I, and Bard, AJ (1981). Electrogenerated chemiluminescence. 37. Aqueous ecl systems based on tris(2,2′-bipyridine)ruthenium(2+) and oxalate or organic acids. *J Am Chem Soc* **103**:512–516.

Rubinstein, I, Martin, CR, and Bard, AJ (1983). Electrogenerated chemiluminescent determination of oxalate. *Anal Chem* **55**:1580–1582.

Sanchez-Carbayo, M, Espasa, A, Chinchilla, V, Herrero, E, Megias, J, Mira, A, and Soria, F (1999). New electrochemiluminescent immunoassay for the determination of CYFRA 21–1: Analytical evaluation and clinical diagnostic performance in urine samples of patients with bladder cancer. *Clin Chem* **45**:1944–1953.

Santhanam, KSV, and Bard, AJ (1965). Chemiluminescence of electrogenerated 9,10-diphenylanthracene anion radical. *J Am Chem Soc* **87**:139–140.

Schhellhammer, PF, and Wright, GL (1993). Biomolecular and clinical characteristics of PSA and other candidate prostate tumor markers. *Urol Clin North Am* **20**:597–606 and references therein.

Shultz, LL, Stoyanoff, JS, and Nieman, TA (1996). Temporal and spatial analysis of electrogenerated Ru(bpy)$_3^{3+}$ chemiluminescent reactions in flowing streams. *Anal Chem* **68**:349–354.

Skotty, DR, and Nieman, TA (1995). Determination of oxalate in urine and plasma using reversed-phase ion-pair high-performance liquid chromatography with tris(2,2′-bipyridyl)ruthenium(II)-electrogenerated chemiluminescence detection. *J Chromatgr B* **665**:27–36.

Spehar, AM, Koster, S, Kulmala, S, Verpoorte, E, Rooij, N, and Koudelka-Hep, M (2004). The quenching of electrochemiluminescence upon oligonucleotide hybridization. *Luminescence* **19**:287–295.

Staffilani, M, Hoss, E, Giesen, U, Schneider, E, Harti, F, Josel, HP, and De Cola, L (2003). Multimetallic ruthenium(II) complexes as electrochemiluminescent labels. *Inorg Chem* **42**:7789–7798.

Tokel, N, and Bard, AJ (1972). Electrogenerated chemiluminescence. IX. Electrochemistry and emission from systems containing tris(2,2′-bipyridine)ruthenium(II) dichloride. *J Am Chem Soc* **94**:2862–2863.

Visco, RE, and Chandross, EA (1964). Electroluminescence in Solutions of Aromatic Hydrocarbons. *J Am Chem Soc* **86**:5350–5351.

Walworth, J, Brewer, KJ, and Richter, MM (2004). Enhanced electrochemiluminescence from Os(phen)$_2$(dppene)$^{2+}$ (phen = 1,10-phenanthroline and dppene = bis(diphenylphosphino)ethene) in the presence of Triton X-100 (polyethylene glycol *tert*-octylphenyl ether). *Anal Chem Acta* **503**:241–245.

White, HS, and Bard, AJ (1982). Electrogenerated chemiluminescence. 41. Electrogenerated chemiluminescence and chemiluminescence of the Ru(2,2′-bpy)$_3^{2+}$–S$_2$O$_8^{2-}$ system in acetonitrile–water solutions. *J Am Chem Soc* **104**:6891–6895.

Wightman, RM, Forry, SP, Maus, R, Badocco, D, and Pastore, P (2004). Rate-determining step in the electrogenerated chemiluminescence from tertiary amines with tris(2,2′-bipyridyl)ruthenium(II). *J Phys Chem B* **108**:19119–19125.

Wilson, R, and Johansson, MK (2003). Photoluminescence and electrochemiluminescence of a Ru(II)(bpy)$_3$-quencher dual-labeled oligonucleotide probe. *Chem Commun* 2710–2711.

Wilson, R, Akhavan-Tafti, H, DeSilva, R, and Schaap, AP (2001). Comparison between acridan ester, luminol, and ruthenium chelate electrochemiluminescence. *Electroanalysis* **13**:1083–1092.

Xu, G, Pang, HL, Xu, B, Dong, S, and Wong, KY (2005). Enhancing the electrochemiluminescence of tris(2,2′-bipyridyl)ruthenium(II) by ionic surfactants. *Analyst* 541–544.

Xu, XH, and Bard, AJ (1994). Electrogenerated chemiluminescence. 55. Emission from adsorbed Ru(bpy)$_3^{2+}$ on graphite, platinum, and gold. *Langmuir* **10**:2409–2414.

Xu, XH, and Bard, AJ (1995). Immobilization and hybridization of DNA on an aluminum(III) alkanebisphosphonate thin film with electrogenerated chemiluminescent detection. *J Am Chem Soc* **117**:2627–2631.

Xu, XH, Yang, HC, Mallouk, TE, and Bard, AJ (1994). Immobilization of DNA on an aluminum(III) alkanebisphosphonate thin film with electrogenerated chemiluminescent detection. *J Am Chem Soc* **116**:8386–8387.

Xu, XH, Shreder, K, Iverson, B, and Bard, AJ (1996). Generation by electron transfer of an emitting state not observed by photoexcitation in a linked Ru(bpy)$_3^{2+}$-methyl viologen. *J Am Chem Soc* **118**:3656–3660.

Xu, XH, Jeffers, R, Gao, J, and Logan, B (2001). Novel solution-phase immunoassays for molecular analysis of tumor markers. *Analyst* **126**:1285–1292.

Yilmaz, N, Erbagci, AB, and Aynacioglu, AS (2001). Cytochrome P450C9 genotype in southeast Anatolia and possible relation with some serum tumour markers and cytokines. *Acta Biochim Polon* **48**:775–782.

Yin, XB, and Wang, E (2005). Capillary electrophoresis coupling with electrochemilurninescence detection: A review. *Anal Chim Acta* **533**:113–120.

Yin, XB, Qi, B, Sun, X, Yang, X, and Wang, E (2005). 4-(Dimethylamino)butyric acid labeling for electrochemiluminescence detection of biological substances by increasing sensitivity with gold nanoparticle amplification. *Anal Chem* **77**:3525–3530.

Yokoyama, K, Sasaki, S, Ikebukuro, K, Takeuchi, T, Karube, I, Tokitsu, Y, and Masuda, Y (1994). Biosensing based on NADH detection coupled to electrogenerated chemiluminescence from ruthenium tris(2,2'-bipyridine). *Talanta* **41**:1035–1040.

Yu, H (1998). Comparative studies of magnetic particle-based solid phase fluorogenic and electrochemiluminescent immunoassay. *J Immun Methods* **218**:1–8.

Zhan, W, Alvarez, J, and Crooks, RM (2002). Electrochemical sensing in microfluidic mystems using electrogenerated chemiluminescence as a photonic reporter of redox reactions. *J Am Chem Soc* **124**:13265–13270.

Zhan, W, Alvarez, J, and Crooks, RM (2003a). A two-channel microfluidic sensor that uses anodic electrogenerated chemiluminescence as a photonic reporter of cathodic redox reactions. *Anal Chem* **75**:313–318.

Zhan, W, Alvarez, J, Sun, L, and Crooks, RM (2003b). A multichannel microfluidic sensor that detects anodic redox reactions indirectly using anodic electrogenerated chemiluminescence. *Anal Chem* **75**:1233–1238.

Zhang, L, Schwartz, G, O'Donnell, M, and Harrison, RK (2001). Development of a novel helicase assay using electrochemiluminescence. *Anal Biochem* **293**:31–37.

Zheng, H, and Zu, Y (2005a). Emission of tris(2,2'-bipyridine)ruthenium(II) by coreactant electrogenerated chemiluminescence: From O-2-insensitive to highly O-2-sensitive. *J Phys Chem B* **109**:12049–12053.

Zheng, H, and Zu, Y (2005b). Highly efficient quenching of coreactant electrogenerated chemiluminescence by phenolic compounds. **109**:16047–16051.

Zhou, M, Heinze, J, Borgwarth, K, and Grover, CP (2003a). Direct voltammetric evidence for a reducing agent generated from the electrochemical oxidation of tripropylamine for electrochemiluminescence of ruthenium tris(bipyridine) complexes? *ChemPhysChem* **4**:1241–1243.

Zhou, M, Roovers, J, Robertson, GP, and Grover, CP (2003b). Multilabeling biomolecules at a single site. 1. Synthesis and characterization of a dendritic label for electrochemiluminescence assays. *Anal Chem* **75**:6708–6717.

Zhuang, YF, Zhang, DM, and Ju, HX (2005). Sensitive determination of heroin based on electrogenerated chemiluminescence of tris(2,2′-bipyridyl)ruthenium(II) immobilized in zeolite Y modified carbon paste electrode. *Analyst* **130**:534–540.

Zou, GZ, and Ju, HX (2004). Electrogenerated chemiluminescence from a cdse nanocrystal film and its sensing application in aqueous solution. *Anal Chem* **76**:6871–6876.

Zou, GZ, Ju, HX, Ding, WP, and Chen, HY (2005). Electrogenerated chemiluminescence of CdSe hollow spherical assemblies in aqueous system by immobilization in carbon paste. *J Electroanal Chem* **579**:175–180.

Zu, Y, and Bard, AJ (2000). Electrogenerated chemiluminescence. 66. The role of direct coreactant oxidation in the ruthenium tris(2,2′)bipyridyl/tripropylamine system and the effect of halide ions on the emission intensity. *Anal Chem* **72**:3223–3232.

Zu, Y, and Bard, AJ (2001). Electrogenerated chemiluminescence. 67. Dependence of light emission of the tris(2,2′)bipyridyl ruthenium(II)/tripropylamine system on electrode surface hydrophobicity. *Anal Chem* **73**:3960–3964.

Zu, Y, Ding, Z, Zhou, J, Lee, Y, and Bard, AJ (2001). Scanning optical microscopy with an electrogenerated chemiluminescent light source at a nanometer tip. *Anal Chem* **73**:2153–2156.

CHAPTER

12

SINGLE-CELL MEASUREMENTS WITH MASS SPECTROMETRY

ERIC B. MONROE, JOHN C. JURCHEN, STANISLAV S. RUBAKHIN, AND JONATHAN V. SWEEDLER

12.1. INTRODUCTION

Recent advances in qualitative and quantitative chemical measurement techniques, at both single and subcellular levels, have generated considerable interest within the biological, chemical, and medical sciences, promising to further our understanding of the complexities of cellular function. Knowledge gleaned from these single-cell studies has begun to provide answers to a diverse set of biological questions regarding inter- and intracellular communication, cell differentiation, physiological responses to stimuli or disease states, and gene expression. In order to address these questions, ongoing advances in analytical techniques and sampling protocols continue to enhance our ability to monitor the chemical composition of the specific cells associated with dynamic cellular events. As a result of this continued progress, single-cell systems have become attractive models for the study of fundamental *in vivo* processes. However, along with the constraints inherent to the small physical dimensions and limited sample sizes associated with single-cell analyses, they also present a set of vast, complex analytical challenges.

Notwithstanding these challenges, the desire for greater chemical information is a driving force in the continued effort to improve single-cell analytical techniques. This is particularly apparent in the field of mass spectrometry (MS), where recent advances in instrumentation and sample preparation protocols have combined to enable the analysis of biologically relevant molecules within the complex environment of tissues and cellular domains. Ideally suited to the analysis of a range of samples, from unknown to highly complex, MS facilitates the collection of unparalleled amounts of quality chemical information, and unlike many other analytical techniques developed for single-cell analysis, preselection of the analyte of interest is often not required. In addition, MS is inherently capable of multiplexed detection, allowing for the analysis of hundreds or thousands of molecular species within the context of a single experiment. Basically, MS separates molecules on the basis of their molecular mass, which is intrinsically linked to their atomic composition and identity. Molecules may also be selectively fragmented in tandem mass spectrometry (MS/MS) experiments to help elucidate the structure or composition of an unknown analyte or to verify the identity of a known molecule. As a result of these separation/fragmentation features, single-cell MS experiments often present tens to hundreds of signals that are unknown, in large part due to

---

New Frontiers in Ultrasensitive Bioanalysis. Edited by Xiao-Hong Nancy Xu
Copyright © 2007 John Wiley & Sons, Inc.

the sheer complexity of single cells and their accompanying molecular diversity (Chaurand et al., 2004a; Hummon et al., 2005; Rubakhin et al., 2005; Yanagisawa et al., 2003). Efforts are underway to add proteomic and metabolomic measurement capabilities to single-cell analyses; however, accomplishing these goals will also require additional improvements in instrument sensitivity and sample handling.

Also contributing to the appeal of single-cell analysis is the desire to minimize sample complexity; individual cell samples are often more simple chemically than homogenate samples comprised of large numbers of cells. Sample simplicity is particularly beneficial in MS analysis, because often the effective dynamic range of a particular measurement is insufficient to monitor the diverse dynamic range of compounds found within tissues. Moving to single cells in order to simplify samples is particularly relevant in studies of the nervous system and certain secretory organs, primarily due to the heterogeneity of the cells within these systems.

While reducing the sample size from that of an entire animal or tissue to a single cell decreases sample complexity, issues relating to detection thresholds become important. Current MS instrumentation is highly sensitive, with detection limits in the attomole range or below, making the analysis of biological materials in a cell more feasible. It is this decreased complexity of samples that allows for analysis of analytes that are often present at relatively high *local* concentrations at the single-cell level, despite being present at low *global* concentrations in the tissue as a whole.

This chapter provides an overview of single-cell analysis via mass spectrometry and includes discussions regarding several MS methodologies used in this application: matrix-assisted laser desorption ionization MS (MALDI MS), secondary ion mass spectrometry (SIMS), and electrospray ionization MS (ESI MS). Single-cell MS analyses encompass a wide range of analytes, from simple atomic ions to complex proteins, and utilize biological models ranging from bacteria to mammalian cells. Although current results are impressive, continuing advances in MS promise further improvements to the sensitivity and selectivity of this measurement technique, thereby enhancing the quality and quantity of information obtained from single-cell analyses of biological materials.

12.2. MS FUNDAMENTALS

Mass spectrometry measures the mass of molecular or atomic ions for analyte detection and characterization, setting this method apart from the optical and radiographic techniques covered elsewhere in this book. In fact, MS is often able to detect molecules and atomic ions that would otherwise be inaccessible to the aforementioned approaches. Generally speaking, MS requires first that the analyte be transferred from the condensed phase to the gas phase by desorption or sputtering. Next, the analyte is ionized, followed by its separation in vacuum; and lastly, it is detected. These events occur within a mass spectrometer, an instrument consisting of three primary parts: an ion source, a mass analyzer, and a detector. Analytes may be ionized by using a variety of ion sources (discussed further in following sections); the resulting ions are then separated in a vacuum on the basis of their mass-to-charge ratio (m/z). As the name suggests, the m/z is based on the mass of the atomic or molecular ion, as determined by the chemical composition of the analyte, and the charge on the ionized species, which depends on the method of ionization, the properties of the analyte, and the surrounding matrix.

In the case of single-cell analysis, the method of analyte desorption/ionization used is critical to the experiment and often determines the classes of compounds detected. MALDI,

SIMS, and ESI have been used predominately as ionization methods for this application and will be discussed in the following sections. Each ionization method performs particularly well under a given set of sample conditions. MALDI performs well with solid samples, while SIMS is better suited to surface analyses, and ESI is ideal for liquid samples. Information on additional aspects of the instrumentation, such as mass analyzers and detectors, is not presented in this chapter, but can be found in a number of recent reviews and textbooks (Gross and Caprioli, 2004; Honour, 2003; Rubakhin et al., 2005; Sickmann et al., 2003; Skoog et al., 1997).

12.2.1. Matrix-Assisted Laser Desorption Ionization (MALDI)

MALDI MS has many features that make this technique optimal for single-cell analysis. The femto- to attomole detection limits are ideal for measuring compounds in biological materials. First, the method is surprisingly salt-tolerant. While adducts of sodium $[M + Na]^+$ and potassium $[M + K]^+$ may appear, they do not deter the measurement process; their signal intensities in the mass spectra can often be reduced by washing the sample or reducing the rate of matrix crystal formation during sample preparation (Garden et al., 1996; Monroe et al., 2006a). Secondly, during cell isolation, although one may observe the degradation of peptides and proteins due to the presence of enzymes in the cellular samples, this enzymatic degradation can be quenched by the application of an acidic MALDI matrix. Last, but not least, MALDI MS experiments are often carried out in an array format ideal for high-throughput analysis, allowing for the individual pretreatment of isolated cells prior to analysis. These pretreatments include procedures such as labeling, enzymatic treatments, electrophysiological recording or stimulation, or application of pharmacological agents.

Developed by Karas and Hillenkamp (Hillenkamp et al., 1991; Karas and Hillenkamp, 1988) and Tanaka et al. (1988) in the second half of the 1980s, MALDI MS has become one of the most powerful methods available to ionize intact biomolecules. Samples are first prepared by incorporating analytes into a low-molecular-weight organic matrix. A wide variety of matrices (crystalline and liquid) are available to analyze a broad spectrum of materials; these matrices include derivatives of benzoic and cinnamic acids, or glycerol. Not surprisingly, some matrices are more effective in the ionization of peptides and proteins, while others are better suited for analysis of oligonucleotides or polymers. Following sample preparation with the appropriate matrix, radiation is then projected at the sample in the form of nanosecond ultraviolet or infrared laser pulses. The radiation is strongly absorbed by the matrix, causing rapid, localized heating and subsequent ejection of both neutral and charged analyte molecules, matrix molecules, and matrix–analyte clusters, as illustrated in Figure 12.1. This process may occur in a vacuum ($<10^{-6}$ torr) or at atmospheric pressure. As a result of these particle interactions, analyte ionization occurs both on the target in the matrix crystals and within the MALDI plume following the laser pulse. In either positive or negative ionization modes, analyte molecules tend to form monoprotonated $[M + H]^+$ or deprotonated $[M - H]^-$ ions, respectively. As a result of the rapid expansion and cooling of the of the ejection plume, there is little fragmentation of analyte molecules during the MALDI desorption/ionization process; therefore, MALDI MS is considered a "soft" ionization technique.

There are a number of experimental parameters to be considered when using MALDI MS to analyze single cells. Various matrices perform differently in regard to the ionization of specific classes of biomolecules; therefore, care should be taken to select an appropriate matrix, because this decision can greatly affect the results of an experiment. For instance, dihydroxybenzoic acid (DHB) and α-cyano-4-hydroxycinnamic acid (CHCA) are

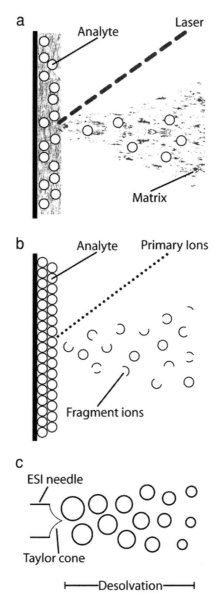

Figure 12.1. Methods of ionization in single-cell MS. (a) MALDI uses a pulsed laser source to irradiate a sample containing an absorbing matrix and the analyte to vaporize and ionize the sample. (b) SIMS sputters the sample surface with primary ions, which leads to significant analyte fragmentation during the ionization process. (c) In ESI, a high potential is applied between the ESI needle emitter and the entrance to the mass spectrometer, causing the analyte to be sprayed from the tip.

appropriate for peptide and protein analysis, while 3-hydroxypicolinic acid (HPA) assists in the detection of nucleic acids. In order to detect an analyte, it must first be incorporated into the matrix, making solvent selection an important consideration. One must also determine whether the matrix selected and analyte being studied are adequately soluble together. Cell lysis and sample salt concentrations should also be controlled during sample preparation.

The addition of other chemicals, such as trifluoroacetic acid (TFA), to the matrix solution may also increase the intensity of analyte signals. Although MALDI MS has proven highly successful in the detection of larger analytes such as peptides and proteins in single cells, both the qualitative and quantitative analyses of smaller molecules (<1000 Da) is not as straightforward. The measurement of these metabolites is complicated by the suppressive effects of the highly abundant matrix ions present in the low mass region, as well as by the presence of lipid signals. However, the discrimination between matrix and analyte ion signals can be greatly facilitated by the addition of MS/MS capabilities to a MALDI MS experiment.

12.2.2. Secondary Ion Mass Spectrometry (SIMS)

Secondary ion mass spectrometry (SIMS) was reported as early as 1910 by Sir Thomson (Honig, 1985) with additional advancements made by Milligan (Arnot and Milligan, 1936) in 1936. SIMS imaging was first demonstrated in the early 1960s by Castaing and Slodzian (1962). In order to ionize an analyte, a primary ion, generated by an ion source, impacts the surface of the sample, thereby causing the ejection of atoms, molecules, and clusters of different species (Figure 12.1) (Briggs et al., 1989). Ejected species largely form singly charged secondary ions due either to the addition of hydrogen, the loss of an electron in positive ion modes of operation, or the loss of hydrogen or gain of an electron in negative modes. Unlike MALDI, this sputtering of ions from the surface is a relatively destructive process and often results in a sizable amount of analyte fragmentation. It is only recently, with the advent of gold and other cluster ion sources (McArthur et al., 2004; Touboul et al., 2005; Winograd, 2005), that this fragmentation has been decreased to allow for the analysis of larger molecules. The addition of matrix similar to that used in MALDI MS analyses and coating the sample in a thin (several nm) layer of gold or silver has also been shown to increase the mass range available for SIMS analysis by limiting molecular fragmentation and/or increasing charge transfer. As a result of these advances, SIMS has become a useful method to study molecules up to approximately 1000 m/z (and occasionally, even higher).

Traditionally, SIMS has been used as a surface analysis technique, or to analyze atomic ions in cellular samples. Recent advances in sampling protocols and the development of new primary ion sources have made the analysis of biological samples even more attractive (McArthur et al., 2004; Touboul et al., 2005; Winograd, 2005). Why use SIMS in single-cell analysis? One of the greatest benefits of single-cell SIMS analysis is the high spatial resolution afforded by the ion sputtering process. Primary ion beams are regularly focused to a spot with a diameter of less than a micron, enabling the high-resolution, subcellular chemical imaging of isolated cells. The method is ideally suited for the detection of small molecules and is often applied to the analysis of cellular membrane lipids and small lipophilic molecules. SIMS has also proven to be highly sensitive, with parts per billion detection limits.

It is important to understand that not all molecules ionize well in SIMS. For instance, in studies of the lipid membrane, detected signals consist largely of lipid headgroups and small lipophilic molecules, such as cholesterol and vitamin E, in addition to atomic ions. Extensive fragmentation of molecules often occurs and is largely the cause for the limited number of biologically relevant substances available for analysis. This fragmentation can, however, aid in the identification of unknown signals in instances where several fragmentation products can be traced to a parent molecule (Monroe et al., 2005). To stabilize the distribution of analytes in single cell samples, chemical fixation or rapid freezing can be used. Freeze-fracture techniques enable the interrogation of the interior of cells (Chandra and Morrison, 1992). It is important to note that SIMS is inherently surface-sensitive, requiring that any contaminants be removed from the sample surface. Also, special care has to be taken to limit sample charging during analysis.

12.2.3. Electrospray Ionization

ESI is one of the most common ionization methods used for biological mass spectrometry. Thus, it may be surprising that of the three ionization techniques discussed, ESI MS has thus far been the least used in single cell analyses. ESI MS does, however, have several advantages over the other techniques, particularly when interfaced with an ion trap or Fourier transform mass spectrometry mass analyzer. ESI is often hyphenated to an initial separation method such as liquid chromatography (LC) or capillary electrophoresis (CE). This allows for complex mixtures to be separated prior to introduction to the mass spectrometer. The separation of various components can be based on any number of molecular characteristics, as determined by the separation parameters applied to the given experiment.

Developed by Fenn in the 1980s, ESI uses an electric field to create an aerosol of highly charged droplets composed of volatile solvents and analytes (Fenn et al., 1989). These droplets subsequently undergo size reduction, using a combination of solvent evaporation and coulombic explosions, until both charged and neutral analyte molecules are ultimately introduced into the gas phase (Figure 12.1). In practice, a high electric potential (1–4 kV) is applied between an emitter, typically a metal or metal-coated pulled glass capillary filled with analyte, and the entrance to a mass analyzer. ESI produces primarily multiply charged ions as well as sodium and potassium adducts. This complicates spectral interpretation because a single analyte may be represented by multiple charge states and adducts. In general, available mass analyzers are capable of analyzing ions below m/z 3000. The analysis of higher charge states permits the detection of higher mass compounds and eases fragmentation studies, because multiply charged molecules tend to fragment more completely. Similar to MALDI, ESI is a soft ionization method that is characterized by little or no analyte fragmentation during ionization.

Single-cell analysis presents several challenges with ESI MS. It is difficult to scale down traditional desalting procedures to the volume of single cells and ESI performance suffers in highly salty conditions. In addition, multiple charge states and salt adducts reduce sensitivity as the response for each analyte is spread among multiple peaks in the mass spectrum. ESI is a solution-based ionization technique requiring that the cell be placed into solution. This often results in significant dilution of the femtoliter-to-picoliter cellular contents. Current research is, however, addressing many of these issues and shows promise for future studies.

12.2.4. Subsidiary (Non-MS) Techniques

Mass spectrometry provides a wealth of information about individual cellular content. However, in many cases, additional methods have to be used in conjunction with MS to solve particular experimental problems. For example, separation techniques such as capillary electrophoresis (CE), solid-phase extraction (SPE), and microfluidics may desalt samples, preconcentrate analytes, mix analytes with MS friendly buffers, including MALDI matrix solutions, and fractionate analytes. The goal is to simplify samples and improve the dynamic range of analyses, thereby enhancing the number of observed, and ultimately identified, substances (Caprioli et al., 1987; Hofstadler et al., 1995). Chemical information such as hydrophobicity and charge is also obtainable from these techniques and can be used for analyte characterization and identification.

Barriers to sensitivity may also be overcome, not only by sample preconcentration and desalting, but also by increasing the quantity of the analyte of interest in a cell using a variety of different strategies. These include molecular biological approaches to induce overexpression, pharmacological methods that block the transport, release, degradation, uptake, and/or processing of analytes, and adjustment of the diet, or stimulation of a particular behavior, in the entire organism being studied. The identification of specific cells for analysis is also highly important as tissues are often laden with heterogeneous cells. Tissue staining, including immunostaining, is one of the most powerful ways to identify cells of interest and may aid in the identification or verification of a particular analyte (Chaurand et al., 2004b). Unfortunately, many staining protocols require some aspect of tissue fixation which may not be compatible with MS analysis. Electrophysiological techniques may also be used to identify individual cells.

12.2.5. Quantitative Analysis with MS

Despite the almost unmatched performance of MS methods in the qualitative characterization of biological samples, quantitative characterization remains a challenge; repeated MS experiments on the same sample can yield peaks of varying intensities. The use of a variety of isotopic approaches is a rapidly developing field that allows both semiquantitative and quantitative measurements (Cannon et al., 2000). However, quantitative measurements from single cells and other small-volume samples are particularly problematic and have not been well-addressed. The difficulties in extracting peptides from a single cell without excessive sample dilution, preventing adsorption on the surface of vials, and performing the labeling reactions on a nanoliter or smaller scale have all slowed the successful application of isotopic measurements to single-cell samples. In SIMS, however, semiquantitative measurements are possible when ionization defects are accounted for by normalizing signals to a ubiquitous ion (Monroe et al., 2005; Ostrowski et al., 2004). Quantitative and semiquantitative MS investigations of single cells currently represent a great opportunity for further research.

12.2.6. Analyte Identification with MS

A major strength of MS is its capability to detect and identify multiple analytes within an individual biological sample, down to the single-cell level, without the need for analyte preselection or specific labeling. Current MS methodology and instrumentation allow for the detection of the mass of atomic and molecular ions with high mass resolution. Moreover, the majority of mass spectrometers are capable of isolating and fragmenting specific analytes, often creating a characteristic set of ions which assist in analyte identification. For example, in MALDI MS, an excessive increase in laser irradiation of a sample may lead to fragmentation and/or metastable decay of investigated compounds. Using these phenomena, two techniques, in-source decay (ISD) and post-source decay (PSD), have been developed and successfully applied to single-cell analyses (Marvin et al., 2003). Collision cells have been also added to some MALDI instruments, resulting in collisionally induced dissociation (CID) of analyte molecules via impacts with injected gas molecules (Harvey, 1999). Fragmentation may be induced by other methods, including electron capture dissociation (ECD), electron transfer dissociation (ETD), blackbody infrared radiative dissociation (BIRD), infrared multiphoton dissociation (IRMPD), and sustained off-resonance irradiation (SORI)

(Mirgorodskaya et al., 2002; Sleno and Volmer, 2004; Wysocki et al., 2005). Each fragmentation method is well-characterized and can be used with different ionization methods. As a result, the fragmentation patterns of parent ions are often predictable, permitting parent ion identification. In addition, multiple stages of mass analysis (MS^n) and fragmentation can be performed on a single instrument, significantly improving the identification of unknown signals. Analyte identification with MS has proven beneficial in cases of protein and peptide analysis where the amino acid sequence of a peptide or protein may be deduced from the observed fragment ions.

12.3. CURRENT RESEARCH TOPICS AND HIGHLIGHTS

Much of the current research in single-cell analysis by MS is divided between method/instrument development and the analysis of single cells. Method development includes improvements in sample handling and preparation, as well as improvements in instrumental sensitivity, mass range, and physical resolution, with the aim of enabling subcellular analyses. Biological studies have covered nearly every class of substance involved in cell function, from inorganic ions, such as sodium and potassium, to large biological molecules, although the main focus has been directed toward characterizing lipids, proteins, peptides, and cell-to-cell signaling molecules. Since the information obtained is distinct and dependent upon the ionization approach used to obtain the data, MALDI MS, SIMS, and ESI MS experiments are described separately.

12.3.1. Overview

The initial development of MS for single cell analysis may have been the application of laser microprobe mass spectrometry (LMMS) to the study of atomic ions in the late 1970s (Schmidt et al., 1980). These early studies involved the analysis of atomic ions on the basis of the harsh nature of ionization by the direct ablation of samples with a laser. Using this technique, Seydel presented the quantitative measurements of Na^+ and K^+ from single bacteria, as well as the metabolism of isonicotinic acid hydrazine in treated cells (Seydel and Linder, 1981). The changes in the K^+ and Na^+ ratio in *Escherichia coli* induced by drug application were studied as a criterion of the physiological state of a cell and of its viability (Lindner and Seydel, 1983). Similar studies with the bacteria responsible for leprosy, *Mycobacterium leprae*, and for *in vitro* drug screening and *in vivo* therapy control have also been reported (Seydel and Lindner, 1988). Observation of aluminum neurotoxicity in cultured rat hippocampal neurons indicated that Al potentiation of Fe-induced oxidative stress might contribute to the facilitation of oxidative injury (Xie et al., 1996).

More recent work evolved from LMMS analysis, primarily due to the addition of various matrixes that permitted, for the first time, the ionization of intact larger molecules, such as peptides and proteins, with MALDI MS. The first MALDI MS single-cell experiments began shortly after the first demonstration of the technique, suggesting that the instrumentation is particularly well-suited to the analysis of single cells.

Another important and growing area is the analysis of intact microorganisms and viruses. Most studies have concentrated on identifying a virus based on the peak pattern of mass spectra and not the identification of the chemical species, such that limited chemical information regarding the cells under analysis has been reported. Although these studies are beyond the scope of this chapter, an analysis of the chemical components of these systems are presented below. For more information, see the excellent reviews by Siuzdak (Fuerstenau et al., 2001; Siuzdak, 1998) and Fenselau and Demirev (2001).

12.3.2. MALDI MS

Performing single-cell MALDI MS experiments is relatively straightforward. Typically, following animal dissection, the tissue of interest is removed and treated with the appropriate chemicals (enzymes, glycerol, low calcium solution, MALDI matrix, etc.) and temperature (elevated or decreased) to allow for single-cell isolation and manipulation. Single cells are typically isolated in the presence of physiological saline or a solution of MALDI matrix. Samples are then moved to a MALDI sample plate and rinsed to remove salts using deionized water or matrix solution (Garden et al., 1996). For single cells, a matrix can be directly deposited on fresh tissue either dropwise (Garden et al., 1996; Jimenez et al., 1994) or by spraying the matrix over a tissue section for high-resolution chemical imaging (Caprioli et al., 1997). A small amount (0.1% v/v) of TFA is often added to the matrix to enhance analyte extraction and ionization, and cells are lysed to enhance analyte–matrix incorporation. Following matrix application, samples are allowed to dry prior to insertion into the mass spectrometer. The particular matrix used, the extraction protocols, and the matrix-to-analyte ratio should be optimized for different sample types. Therefore, before initiating a new MALDI MS experiment, reports of previous experiments with similar samples should be consulted.

The first single-cell MALDI MS experiments were performed by van Veelan and coworkers in 1993 (van Veelan et al., 1993). Large neurons from the freshwater invertebrate, *Lymnaea stagnalis*, were isolated and transferred to a 0.5 to 1.0-μL droplet of matrix placed on a sample plate. Three types of neurons were examined and known peptides from different prohormones in each cell were detected, as well as several unknown peptides. For example, the analysis of a single neuroendocrine cerebral cell from *Lymnaea* revealed a complex set of peptides encoded by the egg-laying prohormone, the caudodorsal cell hormone (Jimenez et al., 1994). In another instance, Worster et al. (1998) reported the detection of alternative RNA splicing of the FMRFamide gene; a mutually exclusive expression of distinct sets of peptides was observed, revealing the pattern of post-translational processing for two peptide precursors.

Many single-cell MALDI MS experiments have been performed using another common neurobiological model, the opisthobranch mollusk, *Aplysia californica*. As an early example, novel processing of the egg laying hormone (ELH) and acidic peptide was uncovered (Garden et al., 1998). The intracellular processing of the *Aplysia* insulin prohormone has also been reported (Floyd et al., 1999). Using a combination of molecular techniques and MALDI MS, several hundred previously unreported *Aplysia* peptides have been characterized. Many of these peptides are a result of the novel processing of previously identified peptide precursors (Garden et al., 1998, 1999; Li et al., 1998). Using a dual-matrix approach, in which both CHCA and DHB matrices were applied to a single neuron, the complete *de novo* sequence of an unknown peptide was obtained (Li et al., 1999). In an impressive feat of sample preparation and subcellular MS, Rubakhin et al. (2000) were able to show that peptide products from the processing of several prohormones are colocalized into individual secretory vesicles, ~1 μm in diameter, within the *Aplysia* atrial gland.

One of the most difficult challenges in conventional peptide analysis is the identification of post-translational modifications (PTMs). In many instances, both modified and unmodified peptides may be detected in cellular samples. As PTMs almost always result in a mass shift detectible by MS, a residue-specific enzymatic reaction may be applied to verify the presence of a PTM by removing the PTM from the peptide. This change may then be detected via single-cell mass spectrometry. For example, single-cell MALDI MS spectra of R3-14 neurons treated with pyroglutamate (pGlu) aminopeptidase have been used to verify the presence of unusual pGlu modified peptides from N-terminal glutamate (Glu) in *Aplysia* (Garden et al., 1999).

In addition to mollusk cells, single-cell MALDI MS has been applied to a wide variety of cells originating from species ranging from crayfish to insects to vertebrates. One particularly interesting experiment combined both ELISA and single-cell MALDI MS to examine the expression of hyperglycemic hormone (cHH) precursors in the crayfish *Orconectes limosus* (Redeker et al., 1998). Following single-cell MALDI MS analysis, the samples could be recovered from the sample plate by redissolving the sample in acetonitrile and transferring this solution to a standard ELISA immunosorbant assay, while maintaining ~70% of the reactivity of controls that did not undergo MALDI preparations. In another example, neurons identified via retrograde labeling with dextran-tetramethylrhodamine in the American cockroach, *Periplaneta americana*, were selectively dissected and analyzed with single-cell MALDI MS, resulting in the detection of 21 FMRFamide-related peptides in each cell, as illustrated in Figure 12.2 (Neupert and Predel, 2005). Single-cell MS has also been used to verify immunohistochemical staining. Evidence of this was seen in the processing of pheromone biosynthesis activating neuropeptide (PBAN) in the corn earworm moth, *Helicoverpa zea*, where MS was able to detect several previously unknown peptides (Ma et al., 2000); one peptide indicated by staining was not detected via MS, rather, an

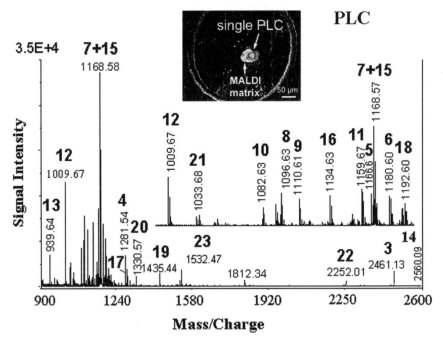

Figure 12.2. A representative MALDI mass spectrum of a morphologically identified PLC neuron from the metathoracic ganglion from *Periplaneta americana*. Signals corresponding to FMRFamide-related peptides are labeled in accordance with their position within the precursor molecule. The inset shows the single PLC with applied matrix on the MALDI sample plate. Scale bar represents 50 μm. [Reprinted from Neupert and Predel (2005), with permission from Elsevier.]

N-terminally extended form of the peptide was detected. As one last example, the processing products of proopiomelanocortin (POMC) were examined in melanotrope cells from the pituitary intermediate lobe of the frog, *Xenopus laevis*, by single-cell MALDI MS; mass signals were detected from both known and novel peptides (van Strien et al., 1996).

Because of their smaller size, fewer single-cell MS studies have been performed using mammalian model systems in comparison to those described above. As an example, Li et al. (1996) were able to detect both subunits of hemoglobin from a single, lysed human erythrocyte, without sample cleanup, further illustrating the sensitivity and salt tolerance of MALDI MS analyses. In cultured rat pheochromocytoma cells (PC12) grown on target, numerous different, but unidentified, signals were observed when treated with nerve growth factor (NGF) as compared to undifferentiated PC12 cells (Bergquist, 1999). In mouse, the increase of histamine in mouse bone-marrow-derived mast cells grown on monolayers of murine myelomonocytic leukemia cells during maturation has been observed (Shimizu et al., 2003); also, individual cells from mouse pituitary have been profiled, and numerous peptides have been detected that are encoded by the POMC prohormone (Rubakhin et al., 2006).

Traditional application of single-cell MALDI MS entails the analysis of single, isolated cells. There is an alternative approach, termed MALDI MS imaging (MSI), that allows chemical images to be obtained with a spatial resolution approaching the size of single cells. In MALDI MSI, a tissue slice is coated in matrix and mass spectra are acquired by rastering the laser across the sample in a stepwise manner to generate ion images of one or more m/z values, which can then be transformed and displayed as a spatial map of the analyte(s) of interest (Stoeckli et al., 1999). The approach is relatively new and is undergoing rapid development, especially in improving methods to incorporate matrix while minimizing analyte redistribution and preserving spatial resolution (Jurchen et al., 2005; Kruse and Sweedler, 2003; Monroe et al., 2006b).

Although initial spatial resolutions in MALDI MSI were insufficient to analyze single cells, the technique has produced quite impressive results from tissue analyses. Molecular ion images of mouse brain (Stoeckli et al., 2001), human glioma xenografts (Schwartz et al., 2004), rat pituitary (Caprioli et al., 1997), and mouse epididymis (Chaurand et al., 2003) have been obtained. Typically, a few hundred signals are observed in any given 25- to 100-μm tissue slice region. MALDI MSI also has the potential to be a powerful tool for high-throughput screening and discovery of biochemical markers of different diseases, including cancer (Chaurand et al., 2004c; Stoeckli et al., 2001), Alzheimer's (Stoeckli et al., 2002), and Parkinson's (Pierson et al., 2004). This methodology is also being expanded to include practical clinical use; development of histological staining protocols compatible with MALDI MS analyses will facilitate both the optical examination and chemical imaging of normal and cancerous tissues (Chaurand et al., 2004b). Serving as a diagnostic tool, this technique has been used to examine the relationships between sample morphology, preparation strategy, and spectral quality (Garden and Sweedler, 2000). Peptide distribution within the exocrine secretory gland (Kruse and Sweedler, 2003) and individual cultured neurons (Rubakhin et al., 2003) from *Aplysia* have also been successfully examined with MALDI MSI (Figure 12.3).

12.3.3. SIMS

Unlike MALDI MS, SIMS single-cell research uses chemical imaging extensively, in large part, because of the high resolution afforded by the primary ion beam, reduced surface damage, and the relative ease with which ion beams may be controlled. Several sample preparation techniques have been developed to retain the native distribution of analytes

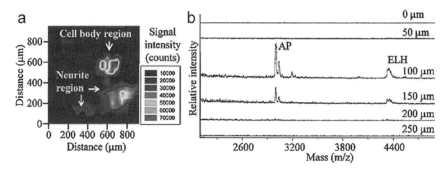

Figure 12.3. The fast application and drying of MALDI matrix preserves the neuropeptide spatial distribution and allows imaging of cells in culture. (a) MS image of the spatial distribution of the peptide with m/z 4617 in a single cultured cerebral ganglion neuron. (b) Mass spectra were taken from spots located on a line with their centers separated by 50 μm. Each spot on the MS image represents an average of 25 laser pulses. [Reprinted with permission from Rubakhin et al. (2003). Copyright 2003, American Chemical Society.]

and/or increase the ionization yield of an analyte. The surface sensitivity of SIMS analyses requires that the sample surface be smooth and flat in order to limit ionization and mass-measurement defects due to sample topography and charging. Although complicated sample topography can be a sizable obstacle in that single cells are not flat by nature, placing cells on a smooth surface appears adequate for most analyses. It is also beneficial for the surface to be conductive in order to limit the effects of sample charging.

A number of methods have been developed to prepare samples containing volatile compounds (such as hydrated biological specimens) for SIMS imaging. Cryofixation by flash-freezing in liquid-nitrogen-slushed liquid propane preserves analyte spatial localization, even for small ions such as sodium and potassium (Clerc et al., 1997). Frozen samples may be inserted directly into the mass spectrometer using a specialized cold transfer stage and analyzed in a hydrated form at low temperatures (Colliver et al., 1997). As a result, the frozen water matrix may also increase ion yields and limit analyte fragmentation. The deposition of a thin layer of metal onto the sample surface has been demonstrated to increase ion yields from cellular samples (Nygren et al., 2004a). Alternatively, samples may be freeze-fractured using a sandwich technique that exposes different cellular regions, including the internal space of the plasma membrane (Colliver et al., 1997; Roddy et al., 2002). Chemical fixation with materials such as glutaraldehyde (Clerc et al., 1997) and blotting of freeze-dried cells on a metal surface to form a molecular imprint (Sjovall et al., 2003) have also been demonstrated.

As stressed throughout this book, the localization of biomolecules within individual cells is an important target for biological and medical research. Many intracellular processes are directional and/or localized to specific regions of a cell, making SIMS nearly ideal for the spatial analysis of biological systems; the high spatial resolution (commonly 100 nm to 1 μm) afforded by this technique enables isolated cell and tissue analysis at the single, and even subcellular, levels.

In addition to imaging larger molecules, SIMS can also provide imaging atomic species, allowing the technique to be used to diagnose cellular physiological states, as well as to track labeled compounds. Studies using freeze-fractured renal epithelial LLC-PK1 cells revealed a well-preserved, intracellular ionic composition of potassium and calcium (Chandra, 2001). An increased ^{39}K signal was observed in the interior of healthy cells, whereas both ^{40}Ca and ^{23}Na signals were relatively higher outside of these cells. Further research has shown that damaged cells may be identified by a reduced potassium signal and the abnormal presence of calcium within the cell (Chandra, 2001). A technique that uses SIMS to map the destination of amino acids and their metabolites has been presented by Chandra (2004). L-Arginine and phenylalanine were labeled with stable ^{13}C and ^{15}N isotopes and then loaded into cells, and the corresponding ^{13}C^{15}N signals were detected from different subcellular regions, demonstrating a surprisingly specific localization of elevated amounts of L-arginine-related compounds in nuclei.

There have also been several reports that have addressed the study of malignant cancers using single-cell SIMS. The accumulation of copper ions in the cytoplasm of carcinoma PC3 cells has been detected and linked to an increased metastatic potential (Gazi et al., 2004); low calcium levels have been found in the spindle region of T98G malignant glioma cells during metaphase (Chandra, 2003); and the known mutagen, PhIP, has been observed to be primarily located on the surface of the outer leaflet membrane in breast cancer cells (Quong et al., 2004).

Not only do disease states often result in gross morphological and/or chemical changes to a biological system, so can the application of pharmacological agents. Examination of the effect of mitosis-arresting drugs (e.g., taxol, monastrol, nocodazole) on M-phase arrested cells revealed a reduced level of calcium heterogeneity in the arrested cell and resulted in the imaging of chromosome alignment (Chandra, 2001). In the yeast, *Candida glabrata* (Cliff et al., 2003), SIMS imaging detected the retention of the antibiotic Clofazamine (whose mechanism of action is not understood) on the cell surface, but not in the interior, indicating that it does not penetrate through the cell wall. The distribution of the boron neutron capture therapy drug, *p*-boronophenylalanine, in cultured mouse melanoma cells (B16) was observed to be confined to the periphery of the cells and nuclei (Oyedepo et al., 2004). This preferential deposition of ^{10}B in the tumor cell nucleus is thought to increase the chance of destroying the cell following neutron irradiation, as desired. The homogeneous integration of cocaine into the single-cell organism, *Paramecium*, has also been observed (Colliver et al., 1997).

Analyzing the mass spectra obtained from various cells and microorganisms has made the identification and classification of specific samples possible. Two strains each of *Saccharomyces cerevisiae* and *Candida glabrata* were identified on the basis of their lipid signature present in their mass spectrum. A blind study resulted in the successful assignment of each spectrum to its species or strain of origin (Jungnickel et al., 2005). Research on a series of 28 bacterial strains using principal component analysis (PCA) enabled Ingram et al. (2003) to differentiate gram-positive from gram-negative bacteria and, in some cases, differentiate individual species. The spores and vegetative cells of the pathogenic bacteria, *Bacillus megaterium*, have been identified by their surface phospholipid fingerprints using polyatomic ion sources (Au_n^+ and C_{60}^+) and an *in situ* freeze-fracture stage (Thompson et al., 2004).

Molecular imaging of the lipid membrane on a single-cell scale aims to provide an understanding of the functional role of specific lipids in the function of cellular membranes. Ewing and co-workers have presented the lipid imaging of very small (15 μm) rat pheochromocytoma cells with a frozen, hydrated freeze-fracture approach (Roddy et al., 2002). Using molecular imprint-imaging, Sjövall et al. (2003) demonstrated that phosphocholine is predominantly located in the nuclear membrane, and cholesterol is most abundant in the plasma membrane, of blood cells. The mechanisms of membrane fusion of conjugating (mating) protozoan *Tetrahymena* cells have been investigated, and an elevated concentration of the high-curvature lipid 2-aminoethylphosphonolipid in the fusion region between two cells (Figure 12.4) was detected (Ostrowski et al., 2004). This particular finding may lead to a

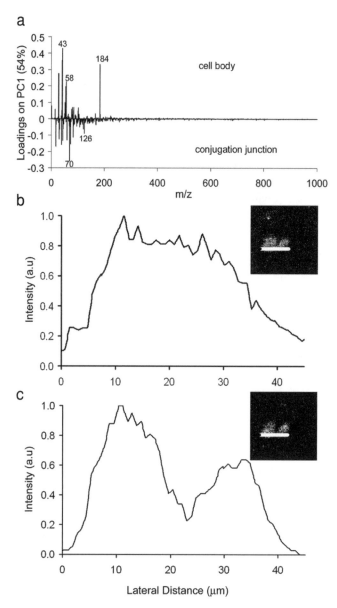

Figure 12.4. SIMS study of two mating cells from *Tetrahymena* showing that the conjugation junction has a different lipid composition than the rest of the cell membranes, and specifically contains elevated amounts of the lipid 2-aminoethylphosphonolipid. (a) Loadings plot from principal component analysis comparing the mass spectra of the cell bodies and the conjugation junction. (b) Line scan for m/z 69 through the conjugation junction, illustrating that the total lipid content is relatively constant across the mating cells. The inset shows the SIMS image for m/z 69, highlighting the pixels used for the line scan. (c) Line scan for m/z 184 through the junction, demonstrating a sharp decrease in signal at the conjugation junction. The inset shows the SIMS image for m/z 184, highlighting the pixels used for the line scan. [Reprinted with permission from Ostrowski et al. (2004). Copyright 2004, AAAS.]

better understanding of the mechanisms of many other cellular events, such as endocytosis and exocytosis, where highly curved membrane surfaces are typically formed. Recently, vitamin E has been observed to be localized at the junction between the cell soma and the extending neurite in the lipid membrane of neurons isolated from *Aplysia* (Monroe et al., 2005), suggesting an important biological role for vitamin E in neuronal function.

Cholesterol ions and phospholipid ions were detected by SIMS imaging in distinct areas of brain slices rich in cell bodies with particularly strong cholesterol signals coming from morphological structures containing myelinated axons, including the corpus callosum, the anterior commissure, the nucleus triangularis septi, and the caudate putamen (Todd et al., 2001; Touboul et al., 2004). In rat kidney sections, intense cholesterol signals originated from nuclear areas of epithelial cells and the basal lumina of the renal tubes (Nygren et al., 2004b). Application of a MALDI matrix to a tissue slice from *Lymnaea* allowed detection of molecules with molecular masses > 2500 Da, although some loss in spatial resolution was observed (McDonnell et al., 2005).

12.3.4. Electrospray MS

As mentioned previously, of the three ionization methods, ESI MS has been less commonly applied to single-cell MS analysis; however, advances in instrumentation and sample handling techniques suggest that the future of single-cell investigation with ESI MS is promising. Ewing and Smith have applied capillary electrophoresis–electrospray ionization Fourier transform ion cyclotron resonance MS, or CE–ESI–FTICR MS, to the study of a single human erythrocyte (Hofstadler et al., 1995, 1996). In this study, a single red blood cell was inserted into a capillary column and lysed prior to spraying into the mass spectrometer. Both the α- and β-chains of hemoglobin were detected, with an average mass resolution of 45,000 (FWHM) (Figure 12.5). In another study, a solvent-extracted aliquot of human blood, equivalent to the volume of a single erythrocyte, allowed for the detection of carbonic anhydrase ($\sim$7 amol) (Valaskovic et al., 1996). Capillary electrophoresis–electrospray ionization–time-of-flight MS, or CE–ESI–TOF MS, has also been demonstrated in the separation and analysis of the α- and β-chains of hemoglobin from a single erythrocyte (Cao and Moini, 1999).

12.3.5. Alternative Sampling Methods

In addition to manual cell isolation and imaging, several alternate sampling protocols show promise with regard to single-cell MS measurements. Laser capture microdissection uses an infrared laser to transiently melt an ethylene vinyl acetate (EVA) thermoplastic film applied to a tissue section (Caldwell and Caprioli, 2005). During the infrared laser pulse, the EVA transfer film fuses with the underlying cells. When the film is separated from the tissue section, the cells remain adhered to the film, thereby allowing selective removal of single or small groups of cells. Application of a MALDI matrix, in a manner similar to single-cell preparation, has been shown in the analysis of cancerous and normal human epithelial cells from human breast tissue (Palmer-Toy et al., 2000; Xu et al., 2002) and mouse prostate (Todd et al., 2001).

Another research area receiving considerable attention is the analysis of cellular release, especially in neurons. As might be expected, analysis of material released by a cell presents a different set of considerations; the releasate is rapidly diluted after release, and only a small fraction of the material within the cell is released. Using solid-phase extraction (SPE)

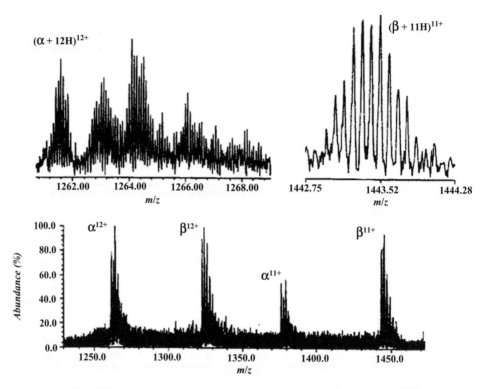

Figure 12.5. ESI-FTICR mass spectrum acquired following the injection, in-column lysing, and CE separation of a single human erythrocyte. Multiply charged species corresponding to the α- and β-chain of human hemoglobin are evident. [Reprinted from Hofstadler et al. (1996), copyright John Wiley and Sons Ltd. Reproduced with permission.]

beads, Hatcher et al. (2005) have developed a technique to collect cellular releasate by placing SPE beads directly on tissue, or isolated neurons. In this procedure, an SPE bead is placed on a slice of tissue or an isolated cell, and the sample is stimulated to initiate release. The bead is then removed, and it is rinsed to wash away physiological salts; then the trapped peptides are eluted from the bead with traditional MALDI matrices. By selectively placing the bead on the neurite of a cultured *Aplysia* bag cell neuron, activity-dependent ELH, acidic peptide, and β-bag cell peptide release was detected, while a bead placed 200 μm from the cell presented only chemical noise, similar to beads placed on the cell both pre- and post-stimulation (Figure 12.6).

12.4. OUTLOOK

The continued development of instrumentation and sampling protocols tailored to specific instrumental requirements will allow for the expansion of single-cell MS analyses into new realms. Innovative studies will involve systems of increasing complexity, not only by examining cells from higher animals, but also by observing molecules present in smaller quantities and their interactions with other biological compounds.

While past development of MALDI TOF mass spectrometers focused predominantly on increasing mass resolution, manufacturers have also recognized the need to design more sensitive instruments. This increased sensitivity will enable the analysis of molecules present

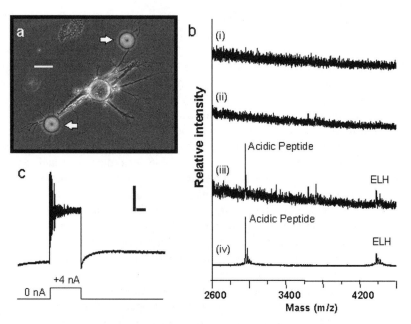

Figure 12.6. Single-bead collection and MALDI MS detection of peptide release from an individual bag cell neuron. (a) Single SPE beads (arrows) were placed on the same and distinct neuronal processes of a cultured bag cell neuron before and during stimulation. Scale bar represents 50 μm. (b) Action potentials in cultured bag cell neurons were electrically stimulated. (c) Mass spectra obtained from single SPE beads before (i) and during (ii and iii) stimulation; (ii) presents the mass spectrum obtained from an SPE bead placed ∼200 m from cellular processes; (iii) shows mass spectrum obtained from an SPE bead placed directly on a neurite. Peaks with masses indicative of bag cell peptides present only in the release spectrum from (iii) confirm activity-dependent exocytosis of acidic peptide and ELH. The identity of the cultured bag cell neuron is verified by its characteristic single cell mass spectrum (iv). [Reprinted with permission from Hatcher et al. (2005). Copyright 2005, American Chemical Society.]

in lower quantities within cells than previously possible, and, in turn, smaller and more complex cells may be analyzed. Development of improved commercial chemical imaging packages will advance the study of tissues on a single-cell level, addressing issues derived from single-cell isolation while increasing experimental throughput. The addition of tandem MS capabilities to MALDI MS will allow for the further detection and characterization of molecules. Unknown peptide signals, for instance, may be at least partially sequenced within the mass spectrometer. By mapping the transition from a molecular ion to a particular fragment ion or ions, the identity of a molecule may be identified and/or separated from ions of similar mass. This may be most beneficial in the study of smaller molecules such as drugs (Reyzer et al., 2003) or lipids, which have traditionally presented a problem in MALDI MS analyses.

With recent advances in cluster ion sources, single-cell molecular analysis via SIMS is currently seeing substantial growth. Further advances in ion source technology and applications, including source longevity, beam size, and secondary ion yield, will further assist in this growth. Realizing the prospect of achieving molecular depth profiling (Cheng and Winograd, 2005; Wucher et al., 2004) of a single cell to create a three-dimensional chemical reconstruction may be of great benefit to understanding cellular processes and the distribution of small molecules on submicron scales. Expanding the range of molecules available for analysis via new cluster ion sources (Brunelle et al., 2005; Touboul et al., 2005), larger clusters (Nagy et al., 2005; Novikov et al., 2005), or other experimental parameters will be required, although currently available sources have proven to work remarkably well in experiments regarding the composition and distribution of lipids and small molecules in the cellular membrane. Further development of freeze-fracture techniques and/or depth profiling to access the interior of single cells will expand analyses beyond lipids and allow for the analysis of inter- and intracellular signaling molecules. The analysis of membrane dynamics as a result of either external stimulation (such as drug application) or intracellular communication events appears to be a promising avenue of future research that will push the limits of instrument capability and drive further advancements.

There are, however, some improvements needed before ESI MS assumes a prominent role in the field—specifically, development of ESI MS instrumentation and sample handling protocols that are better-optimized for single-cell analysis. Recent advances in ion trap technology with regard to loading capacities, resolution, and sensitivity enable ESI MS to be applied to a greater range of applications in single-cell MS analysis. However, further advances are needed with regard to speed of analysis, dynamic range, and sample introduction/ionization. Finally, to address many of the sampling problems observed with single-cell analysis, there has been promising recent research into electrospray sources integrated into microfluidic devices (Mogensen et al., 2004; Sung et al., 2005).

Advances in sample preparation for single-cell MS should focus on the development of protocols that allow for an increase in sample throughput, cellular isolation, sensitivity, and the range of analytes available for analysis. Sampling protocols that retain the spatial information of single cells with respect to native tissue may ease functional determinations. Chemical imaging and mapping experiments on a single-cell scale have thus far proven promising, although the trade-off between analyte extraction (sensitivity) and analyte redistribution (spatial resolution) must be addressed. Efforts to improve cell preparation for SIMS analysis via matrix addition, MALDI-like (McArthur et al., 2004) or frozen water (Roddy et al., 2002), may increase the feasibility of molecular and subcellular analyses of readily diffusible molecules via chemical or cryo-preservation and/or freeze fracture. Protocols designed to isolate and analyze ever smaller cells will enable the routine examination of mammalian cells from tissues and cultures at the single and subcellular levels. Increasing the throughput of single-cell analyses will also be beneficial in developing an understanding of complex biological systems. Intriguing developments in the novel preparation and collection of inter- and intracellular signaling molecules may also serve to put single-cell MS into routine use when studying biological systems. As is common with methodology development, the specific problem to be addressed serves to drive the development process. The application of single-cell MS to a wide range of intriguing biological problems, such as systems biology, disease states, addiction, and stimulus response, may guide future developments in the chemical, biological, and medical fields.

12.5. QUESTIONS DRIVING FUTURE RESEARCH

The future of single-cell analysis appears bright, and we expect that the major advances will occur as driven by the issued raised by the following key questions:

Instrumentation/sampling: Is true quantitation from single-cell samples within our grasp using current technologies? What can be done to improve cellular analyte extraction? Quantitation with mass spectrometry has involved the addition of selective labels. However, when assaying single-cell samples, where the amount of material is close to the detection limit and extra sample steps often cause sample losses that prevent detection, quantitation is especially difficult. Further work will involve the development of new sample handling procedures as well as an increased use of normalizing factors to account for variations in samples and sample preparation strategies. New and improved extraction and preparation protocols are required to increase detectability and to extend the mass range available for analysis.

Whole organ and organism studies: In light of the considerable success of single-cell MS, how can we prepare samples so that cells in entire three-dimensional regions of interest can be characterized at the cellular level? Through high-throughput analyses and the development of advanced molecular imaging, high-resolution chemical reconstructions can advance both medical and biological science. Using serial tissue slices, several three-dimensional chemical images have been constructed using MALDI MSI (Crecelius et al., 2005). By increasing the resolution and speed of data acquisition, analyses of entire organs, as well as small regions of interest on a cellular level, should be obtainable. Using SIMS, the application of depth profiling with cluster ion sources, particularly with C_{60} sources (Wucher et al., 2004), may also prove beneficial and allow for the three-dimensional chemical imaging of single cells on a submicron scale (Chandra, 2005).

Cellular classification in the CNS of "higher" animals: Is it possible to classify cells in vertebrate CNS (subtypes of neurons, glia, etc.) based on their mass fingerprints? Will such classifications change for specific neurons during learning or behavioral plasticity? The classification and analysis of single mammalian cells from tissues presents numerous challenges with respect to both instrument sensitivity and preparation protocols, but also represents the next step in single-cell analyses.

Response to applied stressors: How do the distributions of chemicals in an organism's central nervous system change after the application of a stressor such as a drug of abuse? In which cells are these changes manifested? This problem appears ideally suited for the combination of MSI and single-cell analysis. An understanding of chemical changes within a biological system may lead to advances in disease or disorder treatment and prevention, is the driving force behind many of the projects mentioned within this chapter, and presents an intriguing avenue for further research.

Biomarker assay: Can single-cell mass spectrometry be used to detect chemical signatures indicative of disease states? Will such detections be of any use in clinical settings for the virtual real-time diagnosis of these disease states (as in single-cell-sized biopsies)? Further information is required to address these questions. The capability to classify cancerous cells on the basis of mass spectra (Schwartz et al., 2004) indicates that such capabilities are feasible. In order to gain acceptance in the clinical community, reproducibility and throughput must be increased. This does, however, present an intriguing application of single-cell analysis because chemical changes often occur well before morphological changes may be observed; obviously, the earlier that one may detect changes in individual cells resulting from disease, the earlier the treatment may begin.

While the studies that have been performed using single-cell MS have certainly been exciting, they are merely the beginning. As the throughput, information content, and ease of use continue to improve, a greater range of experiments will highlight the complexities of biology with a combination of temporal, spatial, and chemical information provided by the next generation MS-based single-cell experiments.

ACKNOWLEDGMENTS

The assistance of Stephanie Baker in manuscript preparation and editing is greatly appreciated. The support of NIH through DA018310 and DK070285 is gratefully acknowledged.

REFERENCES

Arnot, FL, and Milligan, JC (1936). A new process of negative ion formation. *Proc R Soc Lond, A, Math Phys Sci* **156**:538–560.

Bergquist, J (1999). Cells on the target matrix-assisted laser-desorption/ionization time-of-flight mass-spectrometric analysis of mammalian cells grown on the target. *Chromatographia* **49**:S41–S48.

Briggs, D, Brown, A, and Vickerman, JC (1989). *Handbook of Static Secondary Ion Mass Spectrometry*. John Wiley & Sons, Chichester.

Brunelle, A, Touboul, D, and Laprevote, O (2005). Biological tissue imaging with time-of-flight secondary ion mass spectrometry and cluster ion sources. *J Mass Spectrom* **40**:985–999.

Caldwell, RL, and Caprioli, RM (2005). Tissue profiling by mass spectrometry: A review of methodology and applications. *Mol Cell Proteomics* **4**:394–401.

Cannon, DM, Jr, Winograd, N, and Ewing, AG (2000). Quantitative chemical analysis of single cells. *Annu Rev Biophys Biomol Struct* **29**:239–263.

Cao, P, and Moini, M (1999). Separation and detection of the alpha- and beta-chains of hemoglobin of single intact red blood cells using capillary electrophoresis/electrospray ionization time-of-flight mass spectrometry. *J Am Soc Mass Spectrom* **10**:184–186.

Caprioli, RM, DaGue, B, Fan, T, and Moore, WT (1987). Microbore HPLC/mass spectrometry for the analysis of peptide mixtures using a continuous flow interface. *Biochem Biophys Res Commun* **146**:291–299.

Caprioli, RM, Farmer, TB, and Gile, J (1997). Molecular imaging of biological samples: Localization of peptides and proteins using MALDI-TOF MS. *Anal Chem* **69**:4751–4760.

Castaing, R, and Slodzian, G (1962). Microanalyse par emission ionique secondaire. *Microscopie* **1**:395–410.

Chandra, S (2001). Studies of cell division (mitosis and cytokinesis) by dynamic secondary ion mass spectrometry ion microscopy: LLC-PK1 epithelial cells as a model for subcellular isotopic imaging. *J Microsc* **204**:150–165.

Chandra, S (2003). SIMS ion microscopy as a novel, practical tool for subcellular chemical imaging in cancer research. *Appl Surf Sci* **203**:679–683.

Chandra, S (2004). Subcellular SIMS imaging of isotopically labeled amino acids in cryogenically prepared cells. *Appl Surf Sci* **231–232**:462–466.

Chandra, S (2005). Quantitative imaging of subcellular calcium stores in mammalian LLC-PK1 epithelial cells undergoing mitosis by SIMS ion microscopy. *Eur J Cell Biol* **84**:783–797.

Chandra, S, and Morrison, GH (1992). Sample preparation of animal tissues and cell cultures for secondary ion mass spectrometry (SIMS) microscopy. *Biol Cell* **74**:31–42.

Chaurand, P, Fouchecourt, S, DaGue, BB, Xu, BJ, Reyzer, ML, Orgebin-Crist, MC, and Caprioli, RM (2003). Profiling and imaging proteins in the mouse epididymis by imaging mass spectrometry. *Proteomics* **3**:2221–2239.

Chaurand, P, Sanders, ME, Jensen, RA, and Caprioli, RM (2004a). Proteomics in diagnostic pathology—Profiling and imaging proteins directly in tissue sections. *Am J Pathol* **165**:1057–1068.

Chaurand, P, Schwartz, SA, Billheimer, D, Xu, BGJ, Crecelius, A, and Caprioli, RM (2004b). Integrating histology and imaging mass spectrometry. *Anal Chem* **76**:1145–1155.

Chaurand, P, Schwartz, SA, and Caprioli, RM (2004c). Assessing protein patterns in disease using imaging mass spectrometry. *J Proteome Res* **3**:245–252.

Cheng, J, and Winograd, N (2005). Depth profiling of peptide films with TOF-SIMS and a C60 probe. *Anal Chem* **77**:3651–3659.

Clerc, J, Fourre, C, and Fragu, P (1997). SIMS microscopy: Methodology, problems and perspectives in mapping drugs and nuclear medicine compounds. *Cell Biol Int* **21**:619–633.

REFERENCES

Cliff, B, Lockyer, N, Jungnickel, H, Stephens, G, and Vickerman, JC (2003). Probing cell chemistry with time-of-flight secondary ion mass spectrometry: Development and exploitation of instrumentation for studies of frozen-hydrated biological material. *Rapid Commun Mass Spectrom* **17**:2163–2167.

Colliver, TL, Brummel, CL, Pacholski, ML, Swanek, FD, Ewing, AG, and Winograd, N (1997). Atomic and molecular imaging at the single-cell level with TOF-SIMS. *Anal Chem* **69**:2225–2231.

Crecelius, AC, Cornett, DS, Caprioli, RM, Williams, B, Dawant, BM, and Bodenheimer, B (2005). Three-dimensional visualization of protein expression in mouse brain structures using imaging mass spectrometry. *J Am Soc Mass Spectrom* **16**:1093–1099.

Fenn, JB, Mann, M, Meng, CK, Wong, SF, and Whitehouse, CM (1989). Electrospray ionization for mass spectrometry of large biomolecules. *Science* **246**:64–71.

Fenselau, C, and Demirev, PA (2001). Characterization of intact microorganisms by MALDI mass spectrometry. *Mass Spectrom Rev* **20**:157–171.

Floyd, PD, Li, L, Rubakhin, SS, Sweedler, JV, Horn, CC, Kupfermann, I, Alexeeva, V, Ellis, TA, Dembrow, NC, Weiss, KR, and Vilim, FS (1999). *Aplysia californica* insulin prohormone processing, distribution, and relation to metabolism. *J Neurosci* **19**:7732–7741.

Fuerstenau, SD, Benner, WH, Thomas, JJ, Brugidou, C, Bothner, B, and Siuzdak, G (2001). Mass spectrometry of an intact virus. *Angew Chem Int Ed Engl* **40**:541–544.

Garden, RW, Moroz, LL, Moroz, TP, Shippy, SA, and Sweedler, JV (1996). Excess salt removal with matrix rinsing: Direct peptide profiling of neurons from marine invertebrates using matrix-assisted laser desorption/ionization time-of-flight mass spectrometry. *J Mass Spectrom* **31**:1126–1130.

Garden, RW, Shippy, SA, Li, L, Moroz, TP, and Sweedler, JV (1998). Proteolytic processing of the *Aplysia* egg-laying hormone prohormone. *Proc Natl Acad Sci USA* **95**:3972–3977.

Garden, RW, Moroz, TP, Gleeson, JM, Floyd, PD, Li, L, Rubakhin, SS, and Sweedler, JV (1999). Formation of N-pyroglutamyl peptides from *N*-Glu and *N*-Gln precursors in *Aplysia* neurons. *J Neurochem* **72**:676–681.

Garden, RW, and Sweedler, JV (2000). Heterogeneity within MALDI samples as revealed by mass spectrometric imaging. *Anal Chem* **72**:30–36.

Gazi, E, Lockyer, NP, Vickerman, JC, Gardner, P, Dwyer, J, Hart, CA, Brown, MD, Clarke, NW, and Miyan, J (2004). Imaging ToF-SIMS and synchrotron-based FT-IR micro spectroscopic studies of prostate cancer cell lines. *Appl Surf Sci* **231–232**:452–456.

Gross, ML, and Caprioli, RM (2004). Mass analysis and associated instrumentation. In: Gross ML, and Caprioli RM (eds.), *The Encyclopedia of Mass Spectrometry*. Elsevier Science, Oxford.

Harvey, DJ (1999). Matrix-assisted laser desorption/ionization mass spectrometry of carbohydrates. *Mass Spectrom Rev* **18**:349–450.

Hatcher, NG, Richmond, TA, Rubakhin, SS, and Sweedler, JV (2005). Monitoring activity-dependent peptide release from the CNS using single-bead solid-phase extraction and MALDI TOF MS detection. *Anal Chem* **77**:1580–1587.

Hillenkamp, F, Karas, M, Beavis, RC, and Chait, BT (1991). Matrix-assisted laser desorption ionization mass-spectrometry of biopolymers. *Anal Chem* **63**:A1193–A1202.

Hofstadler, SA, Swanek, FD, Gale, DC, Ewing, AG, and Smith, RD (1995). Capillary electrophoresis-electrospray ionization Fourier transform ion cyclotron resonance mass spectrometry for direct analysis of cellular proteins. *Anal Chem* **67**:1477–1480.

Hofstadler, SA, Severs, JC, Smith, RD, Swanek, FD, and Ewing, AG (1996). Analysis of single cells with capillary electrophoresis electrospray ionization Fourier transform ion cyclotron resonance mass spectrometry. *Rapid Commun Mass Spectrom* **10**:919–922.

Honig, RE (1985). The development of secondary ion mass spectrometry (SIMS). A retrospective. *Int J Mass Spectrom Ion Process* **66**:31–54.

Honour, JW (2003). Benchtop mass spectrometry in clinical biochemistry. *Ann Clin Biochem* **40**:628–638.

Hummon, AB, Amare, A, and Sweedler, JV (2005). Discovering new invertebrate neuropeptides using mass spectrometry. *Mass Spectrom Rev* **25**:77–89.

Ingram, JC, Bauer, WF, Lehman, RM, O'Connell, SP, and Shaw, AD (2003). Detection of fatty acids from intact microorganisms by molecular beam static secondary ion mass spectrometry. *J Microbiol Methods* **53**:295–307.

Jimenez, CR, van Veelan, PA, Li, KW, Wildering, WC, Gerearts, WP, Tjaden, UR, and van der Greef, J (1994). Neuropeptide expression and processing as revealed by direct matrix-assisted laser desorption ionization mass spectrometry of single neurons. *J Neurochem* **62**:404–407.

Jungnickel, H, Jones, EA, Lockyer, NP, Oliver, SG, Stephens, GM, and Vickerman, JC (2005). Application of TOF-SIMS with chemometrics to discriminate between four different yeast strains from the species *Candida glabrata* and *Saccharomyces cerevisiae*. *Anal Chem* **77**:1740–1745.

Jurchen, JC, Monroe, EB, Christie, MO, Losh, JL, Rubakhin, SS, and Sweedler, JV (2005). Massively parallel single-cell sized sample preparation for MALDI-MS profiling. 53rd ASMS Conference on Mass Spectrometry, San Antonio, TX.

Karas, M, and Hillenkamp, F (1988). Laser desorption ionization of proteins with molecular masses exceeding 10,000 daltons. *Anal Chem* **60**:2299–2301.

Kruse, R, and Sweedler, JV (2003). Spatial profiling invertebrate ganglia using MALDI MS. *J Am Soc Mass Spectrom* **14**:752–759.

Li, L, Garden, RW, Romanova, EV, and Sweedler, JV (1999). *In situ* sequencing of peptides from biological tissues and single cells using MALDI-PSD/CID analysis. *Anal Chem* **71**:5451–5458.

Li, L, Golding, RE, and Whittal, RM (1996). Analysis of single mammalian cell lysates by mass spectrometry. *J Am Chem Soc* **118**:11662–11663.

Li, L, Moroz, TP, Garden, RW, Floyd, PD, Weiss, KR, and Sweedler, JV (1998). Mass spectrometric survey of interganglionically transported peptides in *Aplysia Peptides* **19**:1425–1433.

Lindner, B, and Seydel, U (1983). Mass spectrometric analysis of drug-induced changes in Na^+ and K^+ contents of single bacterial cells. *J Gen Microbiol* **129**:51–55.

Ma, PWK, Garden, RW, Niermann, JT, O'Connor, M, Sweedler, JV, and Roelofs, WL (2000). Characterizing the Hez-PBAN gene products in neuronal clusters with immunocytochemistry and MALDI MS. *J Insect Physiol* **46**:221–230.

Marvin, LF, Roberts, MA, and Fay, LB (2003). Matrix-assisted laser desorption/ionization time-of-flight mass spectrometry in clinical chemistry. *Clin Chim Acta* **337**:11–21.

McArthur, SL, Vendettuoli, MC, Ratner, BD, and Castner, DG (2004). Methods for generating protein molecular ions in ToF-SIMS. *Langmuir* **20**:3704–3709.

McDonnell, LA, Piersma, SR, Maarten Altelaar, AF, Mize, TH, Luxembourg, SL, Verhaert, PD, van Minnen, J, and Heeren, RM (2005). Subcellular imaging mass spectrometry of brain tissue. *J Mass Spectrom* **40**:160–168.

Mirgorodskaya, E, O'Connor, PB, and Costello, CE (2002). A general method for precalculation of parameters for sustained off resonance irradiation/collision-induced dissociation. *J Am Soc Mass Spectrom* **13**:318–324.

Mogensen, KB, Klank, H, and Kutter, JP (2004). Recent developments in detection for microfluidic systems. *Electrophoresis* **25**:3498–3512.

Monroe, EB, Jurchen, JC, Rubakhin, SS, and Sweedler, JV (2005). Vitamin E imaging and localization in the neuronal membrane. *J Am Chem Soc* **127**:12152–12153.

Monroe, EB, Koszczuk, BA, Losh, JL, Sweedler, JV (2006a) Measuring salty samples without adducts with MALDI MS. *Int J Mass Spectrom*, in press and online: doi:10.1016/j.ijms.2006.08.019.

Monroe, EB, Jurchen, JC, Koszczuk, BA, Losh, JL, Rubakhin, SS, and Sweedler, JV (2006b) Massively parallel sample preparation for the MALDI MS analyses of tissues. *Anal Chem* **78**:6826–6832.

Nagy, G, Gelb, LD, and Walker, AV (2005). An investigation of enhanced secondary ion emission under Au$(n)^+$ ($n = 1$–7) bombardment. *J Am Soc Mass Spectrom* **16**:733–742.

Neupert, S, and Predel, R (2005). Mass spectrometric analysis of single identified neurons of an insect. *Biochem Biophys Res Commun* **327**:640–645.

Novikov, A, Caroff, M, Della-Negra, S, Depauw, J, Fallavier, M, Le Beyec, Y, Pautrat, M, Schultz, JA, Tempez, A, and Woods, AS (2005). The Au(n) cluster probe in secondary ion mass spectrometry: Influence of the projectile size and energy on the desorption/ionization rate from biomolecular solids. *Rapid Commun Mass Spectrom* **19**:1851–1857.

Nygren, H, Johansson, BR, and Malmberg, P (2004a). Bioimaging TOF-SIMS of tissues by gold ion bombardment of a silver-coated thin section. *Microsc Res Tech* **65**:282–286.

Nygren, H, Malmberg, P, Kriegeskotte, C, and Arlinghaus, HF (2004b). Bioimaging TOF-SIMS: Localization of cholesterol in rat kidney sections. *FEBS Lett* **566**:291–293.

Ostrowski, SG, Van Bell, CT, Winograd, N, and Ewing, AG (2004). Mass spectrometric imaging of highly curved membranes during tetrahymena mating. *Science* **305**:71–73.

Oyedepo, AC, Brooke, SL, Heard, PJ, Day, JC, Allen, GC, and Patel, H (2004). Analysis of boron-10 in soft tissue by dynamic secondary ion mass spectrometry. *J Microsc* **213**:39–45.

Palmer-Toy, DE, Sarracino, DA, Sgroi, D, LeVangie, R, and Leopold, PE (2000). Direct acquisition of matrix-assisted laser desorption/ionization time-of-flight mass spectra from laser capture microdissected tissues. *Clin Chem* **46**:1513–1516.

Pierson, J, Norris, JL, Aerni, HR, Svenningsson, P, Caprioli, RM, and Andren, PE (2004). Molecular profiling of experimental Parkinson's disease: Direct analysis of peptides and proteins on brain tissue sections by MALDI mass spectrometry. *J Proteome Res* **3**:289–295.

Quong, JN, Knize, MG, Kulp, KS, and Wu, KJ (2004). Molecule-specific imaging analysis of carcinogens in breast cancer cells using time-of-flight secondary ion mass spectrometry. *Appl Surf Sci* **231–232**:424–427.

Redeker, V, Toullec, JY, Vinh, J, Rossier, J, and Soyez, D (1998). Combination of peptide profiling by matrix-assisted laser desorption/ionization time-of-flight mass spectrometry and immunodetection on single glands or cells. *Anal Chem* **70**:1805–1811.

Reyzer, ML, Hsieh, Y, Ng, K, Korfmacher, WA, and Caprioli, RM (2003). Direct analysis of drug candidates in tissue by matrix-assisted laser desorption/ionization mass spectrometry. *J Mass Spectrom* **38**:1081–1092.

Roddy, TP, Cannon, DM, Meserole, CA, Winograd, N, and Ewing, AG (2002). Imaging of freeze-fractured cells with *in situ* fluorescence and time-of-flight secondary ion mass spectrometry. *Anal Chem* **74**:4011–4019.

Rubakhin, SS, Garden, RW, Fuller, RR, and Sweedler, JV (2000). Measuring the peptides in individual organelles with mass spectrometry. *Nat Biotechnol* **18**:172–175.

Rubakhin, SS, Greenough, WT, and Sweedler, JV (2003). Spatial profiling with MALDI MS: Distribution of neuropeptides within single neurons. *Anal Chem* **75**:5374–5380.

Rubakhin, SS, Jurchen, JC, Monroe, EB, and Sweedler, JV (2005). Imaging mass spectrometry: Fundamentals and applications to drug discovery. *Drug Discovery Today* **10**:823–837.

Rubakhin, SS, Churchill, JD, Greenough, WT, Sweedler, JV (2006) Profiling signaling peptides in single mammalian cells using mass spectrometry. *Anal Chem* **78**:7267–7272.

Schmidt, PF, Fromme, HG, and Pfefferkorn, G (1980). LAMMA-investigations of biological and medical specimens. *Scan Electron Microsc* 623–634.

Schwartz, SA, Weil, RJ, Johnson, MD, Toms, SA, and Caprioli, RM (2004). Protein profiling in brain tumors using mass spectrometry: Feasibility of a new technique for the analysis of protein expression. *Clin Cancer Res* **10**:981–987.

Seydel, U, and Linder, B (1981). Application of the laser microprobe mass analyzer (LAMMA) to qualitative and quantitative single cell analysis. *Int J Quant Chem* **20**:505–512.

Seydel, U, and Lindner, B (1988). Monitoring of bacterial drug response by mass spectrometry of single cells. *Biomed Environ Mass Spectrom* **16**:457–459.

Shimizu, M, Ojima, N, Ohnishi, H, Shingaki, T, Hirakawa, Y, and Masujima, T (2003). Development of the single-cell MALDI-TOF (matrix-assisted laser desorption/ionization time-of-flight) mass-spectroscopic assay. *Anal Sci* **19**:49–53.

Sickmann, A, Mreyen, M, and Meyer, HE (2003). Mass spectrometry—A key technology in proteome research. *Adv Biochem Eng Biotechnol* **83**:141–176.

Siuzdak, G (1998). Probing viruses with mass spectrometry. *J Mass Spectrom* **33**:203–211.

Sjovall, P, Lausmaa, J, Nygren, H, Carlsson, L, and Malmberg, P (2003). Imaging of membrane lipids in single cells by imprint-imaging time-of-flight secondary ion mass spectrometry. *Anal Chem* **75**:3429–3434.

Skoog, DA, Holler, FJ, and Nieman, TA (1997). *Principles of Instrumental Analysis*. Brooks/Cole, New York.

Sleno, L, and Volmer, DA (2004). Ion activation methods for tandem mass spectrometry. *J Mass Spectrom* **39**:1091–1112.

Stoeckli, M, Farmer, TB, and Caprioli, RM (1999). Automated mass spectrometry imaging with a matrix-assisted laser desorption ionization time-of-flight instrument. *J Am Soc Mass Spectrom* **10**:67–71.

Stoeckli, M, Chaurand, P, Hallahan, DE, and Caprioli, RM (2001). Imaging mass spectrometry: A new technology for the analysis of protein expression in mammalian tissues. *Nat Med* **7**:493–496.

Stoeckli, M, Staab, D, Staufenbiel, M, Wiederhold, KH, and Signor, L (2002). Molecular imaging of amyloid beta peptides in mouse brain sections using mass spectrometry. *Anal Biochem* **311**:33–39.

Sung, WC, Makamba, H, and Chen, SH (2005). Chip-based microfluidic devices coupled with electrospray ionization–mass spectrometry. *Electrophoresis* **26**:1783–1791.

Tanaka, K, Waki, H, Ido, Y, Akita, S, Yoshida, Y, and Yoshida, T (1988). Protein and polymer analysis up to m/z 100.000 by laser ionisation time-of-flight mass spectrometry. *Rapid Commun Mass Spectrom* **2**:151–153.

Thompson, CE, Jungnickel, H, Lockyer, NP, Stephens, GM, and Vickerman, JC (2004). ToF-SIMS studies as a tool to discriminate between spores and vegetative cells of bacteria. *Appl Surf Sci* **231–232**:420–423.

Todd, PJ, Schaaff, TG, Chaurand, P, and Caprioli, RM (2001). Organic ion imaging of biological tissue with secondary ion mass spectrometry and matrix-assisted laser desorption/ionization. *J Mass Spectrom* **36**:355–369.

Touboul, D, Halgand, F, Brunelle, A, Kersting, R, Tallarek, E, Hagenhoff, B, and Laprevote, O (2004). Tissue molecular ion imaging by gold cluster ion bombardment. *Anal Chem* **76**:1550–1559.

Touboul, D, Kollmer, F, Niehuis, E, Brunelle, A, and Laprevote, O (2005). Improvement of biological time-of-flight-secondary ion mass spectrometry imaging with a bismuth cluster ion source. *J Am Soc Mass Spectrom*.

Valaskovic, GA, Kelleher, NL, and McLafferty, FW (1996). Attomole protein characterization by capillary electrophoresis–mass spectrometry. *Science* **273**:1199–1202.

van Strien, FJ, Jespersen, S, van der Greef, J, Jenks, BG, and Roubos, EW (1996). Identification of POMC processing products in single melanotrope cells by matrix-assisted laser desorption/ionization mass spectrometry. *FEBS Lett* **379**:165–170.

van Veelan, PA, Jimenez, CR, Li, KW, Wildering, WC, Gerearts, WP, Tjaden, UR, and van der Greef, J (1993). Direct peptide profiling of single neurons by matrix-assisted laser desorption-ionization mass spectrometry. *Organic Mass Spectrom* **28**:1542–1546.

Winograd, N (2005). The magic of cluster SIMS. *Anal Chem* **77**:143A–149A.

Worster, BM, Yeoman, MS, and Benjamin, PR (1998). Matrix-assisted laser desorption/ionization time-of-flight mass spectrometric analysis of the pattern of peptide expression in single neurons resulting from alternative splicing of the FMRFamide gene. *Eur J Biochem* **10**:3498–3507.

Wucher, A, Sun, S, Szakal, C, and Winograd, N (2004). Molecular depth profiling of histamine in ice using a buckminsterfullerene probe. *Anal Chem* **76**:7234–7242.

Wysocki, VH, Resing, KA, Zhang, Q, and Cheng, G (2005). Mass spectrometry of peptides and proteins. *Methods* **35**:211–222.

Xie, CX, Mattson, MP, Lovell, MA, and Yokel, RA (1996). Intraneuronal aluminum potentiates iron-induced oxidative stress in cultured rat hippocampal neurons. *Brain Res* **743**:271–277.

Xu, BJ, Caprioli, RM, Sanders, ME, and Jensen, RA (2002). Direct analysis of laser capture microdissected cells by MALDI mass spectrometry. *J Am Soc Mass Spectrom* **13**:1292–1297.

Yanagisawa, K, Xu, BJ, Carbone, DP, and Caprioli, RM (2003). Molecular fingerprinting in human lung cancer. *Clin Lung Cancer* **5**:113–118.

CHAPTER

13

OUTLOOKS OF ULTRASENSITIVE DETECTION IN BIOANALYSIS

XIAO-HONG NANCY XU

In summary, ultrasensitive detection represents one of most exciting research fields and plays a vital role in advancing an array of scientific disciplines, including biological, biomedical, chemical, environmental, and materials sciences and engineering. The representative research topics that are exceedingly challenging and most likely to create the profound impacts in ultrasensitive bioanalysis are outlined below:

13.1. SINGLE MOLECULE DETECTION AND ITS APPLICATION

Detection of some types of single molecules (e.g., high-quantum-yield fluorophor, fluorescence protein molecules) in aqueous solution and room temperature, and using these fluorophors to tag the molecules of interest for SMD has become nearly routine in a wide variety of research laboratories (Zander et al., 2002). As apparatus and instrumentation required for SMD becomes less sophisticate, SMD has become a powerful tool that is more accessible to a wide variety of researchers. However, characterization of individual unknown single molecules and qualitative analysis of individual molecules remain a forbidden challenge. Thus, the development of highly sensitive qualitative characterization means, such as mass spectrometer and NMR, that are able to characterize the chemical properties (e.g., mass and structure) of single molecules remains crucial to advance the application of SMD to a wide array of research fields. The research effort of this kind will create the opportunity to address numerous fundamental questions and revolutionize the way that we conduct sample analysis.

Furthermore, it is essential to develop nonintrusive means for effective manipulation of SM to better understand molecular interactions and molecular assembly. The research effort gearing toward the application of micro- and nano-fluidic device for trapping and guiding the motion of individual molecules will continue thriving and will become a primary driving force for probing the molecular interactions, which will open up the new possibility of better understanding of biological function of molecular machinery and assembly of new materials one-molecule-at-a-time.

Study of interactions of single molecules in real time is essential to understand an array of chemical and biochemical mechanisms that are crucial to address a wide variety of fundamental questions and advance our understanding about the function of living cell.

New Frontiers in Ultrasensitive Bioanalysis. Edited by Xiao-Hong Nancy Xu
Copyright © 2007 John Wiley & Sons, Inc.

Such research will also aid (a) the design of highly selective molecule sensors and (b) assembly of new materials at the molecule level. The interactions of individual molecules have been studied primarily using fluorescence resonance energy transfer (FRET) (Tinnefeld and Sauer, 2005; Weiss, 2000). The demanding experimental conditions that are required by FRET greatly limit its applications to studying the interactions of wide variety of individual molecules. New tools and probes will be needed to explore the possibility of study of individual molecular interactions, especially specific interactions, such as ligand–receptor interactions on living cell surface.

Tracking the dynamics of individual molecules is vital to understand a wide variety of biochemical pathways, cellular function, and signaling. Currently, the most popular fluorescent probes, such as fluorophor, fluorescence protein molecule, and quantum dot, suffer photodecomposition and hence provide brief lifetime for tracing the dynamics of individual molecules. New probes, such as non-photobleaching noble metal nanoparticles, have emerged and have been used to study the ligand–receptor interactions on single living cells (Xu and Patel, 2005) and probe membrane transport of single living cells (Xu et al., 2004; Xu and Patel, 2004). Development of exceedingly bright, non-photobleaching, and biocompatible probes for studying intracellular function will no doubt become an important thrust of ultrasensitive bioanalysis.

SMD has been widely used for analysis of biomolecules in buffer solution and probing the function and mechanism of individual biomolecules, such as activity of single enzyme molecules, catalysis properties of single RNA molecules, and binding affinity of protein molecules, as outlined in Chapters 1–3. To probe the biological activities of individual molecules and determine their roles in cellular function, it is essential to study biological activities of individual molecules in living cells. Despite the numerous efforts, detection of single molecules in single living cells remains a great challenge, in part due to the complex components of single living cells. Nevertheless, in recent years, SMD of single living cells has been successfully demonstrated and has become a powerful tool to unravel critical biological pathways, such as membrane transport and gene expression, as illustrated in Chapters 2 and 3.

In the coming years, research effort of depicting cellular pathways in single living cells at the single-molecule level (Xu et al., 2003, 2004; Yu et al., 2006) will continue to flourish and become a dominative driving force for the further development of new means of SMD and ultrasenstive bioanalysis. It is quite certain that the research of this kind will be a leading topic in the coming decades, which will generate new knowledge in biological, biomedical, and chemical sciences.

13.2. ROLES OF NANOSCIENCE AND NANOTECHNOLOGY

The unification of micro- and nano-fluidic devices with ultrasensitive detection will play a significant role in miniaturization of conventional sampling and detection systems, leading to the reduction of detection volume and the amount of samples that is required for the detection. This new research approach has showed a great promise of profiling the content of single living cells and organelles, which is essential to better understand the role of individual molecules in cellular function and identification of molecular markers for disease diagnosis and treatment.

The development of micro- and nano-fabrication devices coupled with ultrasensitive detection also offers the possibility of multiplexing and high-throughput analysis of biomolecules in individual living cells and organelles, as illustrated in Chapter 6. The technology of this kind, such as DNA and protein microarray, has demonstrated its great importance of advancing genomic and proteomic research. The research effort, aiming to develop high-throughput detection array with high sensitivity and selectivity, will continue thriving and will certainly play a vital role in biomedical research, such as identification of biomarkers for disease diagnosis and treatment.

Furthermore, the recent development of nanoparticle technologies, such as single-nanoparticle optics and multiple functional nanoparticles, provides the unique possibility of real-time *in vivo* and *in vitro* imaging and sensing of individual biomolecules with sub-100-nm spatial resolution and millisecond-to-nanosecond time resolution (Gao et al., 2004; Xu et al., 2004). It also creates the new platform of multiplexing analysis of complex samples using single nanoparticles, offering the new possibility of advancing miniaturization from "lab-on-a-chip" to "lab-on-a-dot". This new approach will definitely lead to the development of new analytical techniques for ultrasensitive bioanalysis and open up new opportunities for better understanding of biological mechanisms and cellular pathways in single living cells. Numerous challenges await to be overcome prior to the realization of such major advanced in ultrasensitive detection.

On the other hand, the development of ultrasensitive detection will undoubtedly invigorate the development of nanotechnology and nanomaterials for molecular bioanalysis. Thus, research programs using interdisciplinary approaches will maximize the possibility of developing new nano tools for molecular detection and characterization, as well as using the new tools to study biological function of individual biomolecules in single living cells.

13.3. ANALYSIS OF SINGLE LIVING CELLS

In general, sizes of individual cells range from 0.5 µm in diameter with 2 µm in length of bacterial cells to about 10 µm in diameter of mammal cells. Such a compact space is further divided into numerous tangled tiny compartments by highly selective and sophisticate membranes. Thus, the structure of living cells is highly complex. A single living cell contains numerous molecules, and these molecules move purposely among compartments of tiny cellular space to meet the demand of cellular function. The amount and types of molecules in a living cell changes constantly over time. Therefore, it is essential to qualitative and quantitative analysis of entire content of individual living cells with sufficient spatial and temporal resolution, offering molecular dynamic profile of single living cells in real time. To meet such exceedingly demanding challenge, the detection means must be nonintrusive, highly selective, and sensitive, as well as offer sufficient temporal and spatial resolutions to map the contents of a single living cell in real time. To this end, new tools to analyze the contents of individual living cells in real time will be continuously developed.

As illustrated in previous chapters, a wide variety of tools and approaches, such as single-molecule detection (Chapters 1–3), single nanoparticle assays (Chapters 3–5), micro- and nano- technologies (Chapter 6), multiplexing strategy (Chapters 7 and 8), metal ion sensors (Chapter 9), electrochemical detection approaches (Chapters 10 and 11), and mass spectrometry imagining tools (Chapter 12) have been developed for molecular analysis of individual cells, aiming to depict the role of individual molecules in specific cellular pathways and function.

To completely understand the cellular function and depict the roles of individual molecules in cellular function, it is inevitable to profile the entire content of individual cells with spatial and temporal resolution and hence project the trajectories of molecular profiles of individual cells in real time. The unification of a wide variety of techniques and approaches will be needed to accomplish this unprecedented challenging task. Apparently, this research field is still in its infancy, and numerous challenges and opportunities lay ahead. Significant breakthrough in experimental technology and theoretical simulation will be essential for making progress in this field.

Nevertheless, the question about the possibility and necessity of profiling the entire content of individual cells with molecular sensitivity and spatial resolution in real time has been raised. Is there any fundamental limitation that may prevent us from achieving this ultimate goal? Will this approach be the most effective way to depict the cellular function?

A new engineering approach has recently been proposed by treating a living cell as a black box. One may be able to input the specific information into this "black box" (e.g., altering gene code, signaling molecules) and then look at the output of the "black box," such as which function of cells may be altered or which diseases may be caused, and then finally work one's way backward to figure out the role of individual molecules in cellular function (Ho, 2005). One may argue that the similar approach can be used to treat the disease by altering the different types of drugs. Consequently, such an approach may somehow effectively pin down the role of primary molecules in cellular function with no need of profiling the entire content of living cells.

Though the approach appears quite refreshing and certainly worth the efforts of pursuing, similar approaches have been practiced by biological and biomedical community for decades. For example, by creating mutants, one hopes to find out which gene and protein may be responsible for a particular cellular pathway. By creating the animal model and studying the efficacy of certain drugs, one can identify the effective treatment of diseases without knowing the mechanisms of cellular function and the mode of action of drugs. This seems to generate effective short-term benefit and meet the need of urgency.

However, for the long term, it will be far more benefit if one finally develops the effective tools that can profile the contents of individual living cells in real time and fully understand the role of individual molecules and their interaction in cellular function. Such new knowledge will be essential to the design of bioinspired self-functioning devices and materials, and assembly of living cells one-molecule-at-a-time, which will lead to vital breakthrough in a wide variety of research fields.

REFERENCES

Gao, X, Cui, Y, Levenson, RM, Chung, LW, and Nie, S (2004): In vivo cancer targeting and imaging with semiconductor quantum dots. *Nat Biotechnol* **22**:969–976.

Ho, CM (2005). Center for Cell Mimetic Space Exploration, Nanoscale Science and Engineering 2005 NSF Grantees Conference.

Tinnefeld, P, and Sauer, M (2005). Branching out of single-molecule fluorescence spectroscopy: Challenges for chemistry and influence on biology. *Angew Chem Int Ed* **44**:2642–2671 and references therein.

Weiss, S (2000). Measuring conformational dynamics of biomolecules by single molecule fluorescence spectroscopy. *Nat Struct Biol* **7**:724–729 and references therein.

Xu, XH, and Patel, R (2004). Nanoparticles for live cell dynamics. In: Nalwa, HS (ed). *Encyclopedia of Nanoscience and Nanotechnology*, Vol. 7. American Scientific Publishers, pp. 189–192 and references therein.

Xu, XH, and Patel, R (2005). Imaging and assemble of nanoparticles in biological systems. In: Nalwa, HS (ed.)., *Handbook of Nanostructured Biomaterials and Their Applications in Nanobiotechnology*, Vol. 1. American Scientific Publishers, Stevenson Ranch, CA, pp. 435–456 and references therein.

Xu, XH, Brownlow, W, Huang, S, and Chen, J (2003). Single-molecule detection of efflux pump machinery in *Pseudomonas aeruginosa*. *Biochem Biophys Res Communi* **305**:79–86.

Xu, XH, Brownlow, W, Kyriacou S, Wan Q, and Viola J (2004). Real-time probing of membrane transport in living microbial cells using single nanoparticle optics and living cell imaging. *Biochemistry* **43**(32):10400–10413.

Yu, J, Xiao, J, Ren, XJ, Lao, K, and Xie, XS (2006). Probing gene expression in live cells, one protein molecule at a time. *Science* **311**:1600–1603.

Zander, C, Enderlein, J, and Keller, RA (2002). *Single Molecule Detection in Solution: Methods and Application*, 1st edition. Wiley-VCH, Berlin, and references therein.

INDEX

A

ABI PRISM® 3700 DNA analyzer, 162
β-actin gene, 32
γ-actin gene, 32
active extrusion systems, in living organisms, 42
adrenal chromaffin cells
    utilized widely for amperometric studies, 217
Ag nanoparticles, 54–56
aging effect
    impact on Ca^{2+} efflux and epinephrine release, 218
    on fight-or-flight response, 218
AlexaFluor series of dyes, 148, 157
amperometric detection
    of exocytosis at single cells, 216
amperometry, 215. *See also* electrochemistry
    along with patch clamp techniques, 226
    for measuring secretion from cells, 216
    for monitoring vesicle fusion in cells, 217
analog-to-digital converter (ADC), 151
analytes probes, 198
annihilation ECL mechanism, 235, 236
Aplysia californica, 136, 277
array fabrication, 173–175
ASH1 mRNA localization, 33
Auramine O, 199
autofluorescence emission, 82
avalanche photodiodes (APDs), 45
aztreonam (AZT), 56

B

benzyl ether, 72
binding assays with ECL detection
    schematics of design of, 239
bingo card display, 17
bioconjugation methods, 74
biological systems, amperometry in, 216, 217
biomolecular interactions, 91
biomolecules
    ionization, by MALDI MS, 271
    ultrasensitive analysis by ECL, 235
biomolecules, 41
biosensor approach, advantages of, 205

blackbody infrared radiative dissociation (BIRD), 275
blinking phenomena, of fluorescent proteins, 46
boro-float gel plates, 162
broad excitation spectrum, 46

C

caffeine, impact on exocytosis, 221
Calmodulin, 200
cancer research, 71
Candida glabrata, 281
capillary array electrophoresis (CAE), 162
capillary electrophoresis with laser-induced fluorescence detection (CE-LIF), 120
capillary electrophoresis (CE), 120
capillary gel electrophoresis (CGE), 144, 153
carbocyanines, 146
carboxy-4',5'-dichloro-2',7'-dimetoxyfluorescien (JOE), 144
carboxyfluorescein (FAM), 144
carboxytetramethylrhodamine (TAMRA), 144
carboxy-X-rhodamine (ROX), 144
cascade steps, 200
CCD cameras, 163, 171
CD34+ progenitor cells, 84
CD4+ lymphocytes, 81, 84
CdSc nanoparticle scaffolds, 104–105
β-cells, measuring quantal release of serotonin by amperometry, 223
cells
    and interacellular analysis of, 196, 197
    quantitation of analytes in, 198
    signaling for exocytosis, role of Ca^{2+}, 222
cell–cell communication, 42
cellular functions, 41, 42
CEQ™ Series Genetic Analysis sys, 148
c-fos gene, 32
charge-coupled device (CCD) camera, 240
chelator, characterstictics of, 205
chemical analysis in single cells
    mass spectroscopy for, 195
4-chloro-7-nitrobenzo-2-1-diazole (NBD), 144
chloramphenicol, 57–60, 62, 63
cholesterol homeostasis, 197

New Frontiers in Ultrasensitive Bioanalysis. Edited by Xiao-Hong Nancy Xu
Copyright © 2007 John Wiley & Sons, Inc.

chromaffin cells
  study of fight-or-flight response in by amperometry, 218
chromatography, 19
chromogranin A, 217
chromosome Y, 29
chymotrypsin (ChT), 102
c-jun gene, 32
cluster ion sources, 285
collisionally induced dissociation (CID), 275
competition assay, 239
confocal and multiphoton microscopies
  for detection of trace metal ions, 196
confocal fluorescence microscopy (CFM), 43, 45
conjugation ratio determination
  of PSA with Tag, Ru(bpy)$_3^{2+}$–NHS, 246
constant fraction discriminator (CFD), 151
coreactant ECL mechanism, 236, 237
  used for analytical applications, 237
core-shelled nanoparticles, 93
cross-linked with epichlorohydrin (CLIO), 81
Cryofixation, by flash freezing, 280
cumulative distribution method, 5
Cy3 molecule, 44
Cy3-cAMP molecules, 44
Cy5.5 dye, 154
cycle time, defined, 4
cyclic voltammetry, fast-scan, 225
Cy-dye series, 148
cytolysins, 74

D

Dansylamide, 199
data analysis techniques
  for bulk systems and synchronous starts, 6, 11
  for prebulk systems, 6–8, 10, 11
  in single-molecule case, 8, 9
  noise considerations, 9, 10
dideoxynucleotides (ddNTPs), 143
direct immunoassay, 239
DNA bio-bar codes, 79
DNA biosensors, 239
DNA fingerprint, 13
DNA sequencing, 141, 142
  fluorescence detection for, 142–148
DNA:polymerase complex, 98
DNA–nanoparticle complex, 98
Drosophila oocytes and embryos, 36
dye primer /dye-terminator DNA sequencing, 142, 143
dynamic light scattering (DLS), 98, 103

E

E. coli, 108
ECL analyzer, basic components of, 241
ECL annihilation reactions, 236
ECL based bioassays, 258
  for measuring enzyme activity schematics of, 240
  for ultrasensitive bioanalysis
    by using enzyme-based ECL biosensors, 254
    tumor markers analysis using solution-phase ECL assay, 246
ECL biosensors
  for detection of DNA and protein, 250–254
ECL detection system, 241, 242
  future major advances of, 258
ECL DNA biosensors, 252, 253
ECL immunoassay
  for detecting molecular forms of PSA, 248
ECL instrumentation, 240, 241
ECL labels, 240
ECL ORIGEN analyzer, 242
ECL probes, 240
ECL reaction mechanism
  of PSA–Ru(bpy)$_3^{2+}$, 247
ECL sensors, enzyme-based, 254, 255
ECL studies and research historic overveiw of, 242–245
ECL surface-phase immunoassays, 238
ECL systems
  fundamental mechanisms of ECL reactions, 245
  Ru(bpy)$_3^{2+}$/TPrA system, 255, 256
ECL technique in bioanalysis
  applications and unique features of, 237, 238
  comparison to other detection processes, 239
  fundamentals of, 235
ECL technique, 235
  for sensitive quantification of enzyme β-lactamase, 245
  utilization in various fields, 244
electrochemical detection of exocytosis, types of, 215
  and interactions of cellular compartments, 215
  challenges in, 229
  in cells and possibilities in, 217
Electrochemiluminescence (ECL), 235
  active emitters for ultrasensitive bioanalysis, 237
  ultrasensitive analytical tool in bioanalysis, 258
electrochemistry
  inside and outside single nerve cells, 215, 216
  role in identification and quantification of neurotransmitters, 215
electron capture dissociation (ECD), 275
electron transfer dissociation (ETD), 275
electron-beam lithography, 136
electrophoresis, 17, 144, 153, 162
electrophysiological techniques
  for analyzing individual cells, 275
electroporation technique, 76
Electrospray ionization (ESI), 274
  solution-based ionization technique, 274
electrospray ionization MS (ESI MS), 270, 283
  and advantages over other techniques, 274
  instrumentation and sample handling, 286

electrowetting phenomenon, 133
ELISA-based amplification scheme, 177
ELISA-based enzymatic method, for detecting species, 180
engineered transferrin receptor (ETR), 84
enzyme-based ECL assay
 illustration of, 254
epidermal growth factor (EGF), 82
erbB/HER receptor-mediated signal transduction, 82
ES-D3 stem cells, 124, 126
EtBr molecule, 51–54, 100
 validation of single molecules using, 62, 63
EtBr molecules, 43, 48, 49
ethidium bromide (Eb), 100
eukaryotes, 42, 49, 52
Euler integration methods, 181
exocytosis
 analysis of Munc18 protein role, 219, 220
 detection of role of F-actin by amperometry, 217, 218
 kiss and run aided by high Ca^{2+}, 228
 role in cell-to-cell communication, 215
 technique for measuring, 215
exocytosis in single cells
 study by patch amperometry, 226
exocytosis, stages of, 215
expressed probes, 200
 utilizes energy transfer assays with an *expressible* indicator, 200, 201

F

F-actin, 217
F-actin breakage
 increase in catecholamine release, 218
Fast-scan cyclic voltammetry (FSCV), 215. *See also* electrochemistry
 used to identify electroactive species release during exocytosis, 225, 226
femtomolar oligonucleotide detection, 183–185
FePt magnetic nanoparticles, 72
FePt nanoparticles, 111
fiber optics
 and metal ion biosensors, 208, 209
fight-or-flight response, 218
FITC fluorophore, 84
flow cytometry, 13, 14, 198
fluorescein, 144
fluorescence correlation spectroscopy (FCS), 5, 43, 45
fluorescence detection principle, 207
fluorescence detection, for DNA sequencing
 dye-primer/dye-terminator chemistry, 142, 143
 fluorescent dyes, 143–145
 instrumental formats, 161–163
 near-IR fluorescent dyes, 145–148
 strategies, 149–161
fluorescence intensity autocorrelation function, 8, 10

fluorescence intensity, of a fluorophor, 42
fluorescence microscopy and spectroscopy, 202
 detection configurations of, 43–45
 detection probes, 45–47
fluorescence polarization immunoassay, 199, 206
fluorescence probes, limitations, 45, 46
fluorescence quantum dots, 43
fluorescence ratio methods, 199
fluorescence resonance energy transfer (FRET), 296
fluorescence sensing of Cu(II)
 schematic of, 207
fluorescence-activated cell sorting, 198
fluorescent compounds
 target proteins for, 199
fluorescent *in situ* hybridization (FISH) technique, 31
fluorescent *in vivo* hybridization (FIVH) technique, 32
fluorescent indicator, 199
fluorescent nucleotide analogs, 29
fluorescent probes, 197, 201, 296
 analytes for, 197
fluorescent proteins, 29
 blinking phenomena of, 46
fluorescent resonance energy transfer (FRET), 111, 112
fluorescent thymidine analogs, 29
fluorescent-labeled inhibitors or agonists, 199
fluorophores
 nanomolar concentration of, 201
four-lifetime/one-lane DNA sequencing, 153, 154
fragmentation method, 276
freeze-fracture techniques, 274, 286
FRET donor/acceptor pair, 19

G

GaAlAs diode laser, 151
β-galactosidase (β-Gal), 105, 106
gadolinium compounds, 71
Gd-based paramagnetic liposomes, 84
gel electrophoresis, 16
gene expression, 41
gephyrin, 81
GFP-exuperantia, 35
GFP-MS2 proteins, 33, 36
GFP-polyA-binding protein II (PABP2), 35
GFP-U1A protein, 35
glutathione (GSH), 105, 106
glycine, 81
gold nanoparticles, 72, 79, 98, 99, 102, 107, 108
 action with bivalent lectins, 111
 in HIV recognition, 110, 111
 with PSA antibodies and DNA, 100, 101
 in study on carbohydrate–protein interactions, 108
 with thrombin, 112
 synthesis of MPCs of, 93, 94
Green fluorescence protein (GFP), 43, 46, 200
GWC technologies, 171

H

heavy-atom-modified tricarbocyanine dyes, 146
*Hind*III digest of λ phage DNA, 13, 14
HIV-TAT peptide, 76, 81
HTS Biosystems, 171
Human genome project (HGP), 141, 162
human respiratory syncytial virus (RSV), 77
hydrogen-bonding efficiency, 96, 97

I

immunoassays
  by using antibody, 198
  ELISA and RIA, 198
immunofluorescent stains, 198
in situ detection, of singular RNA molecules, 31–36
in vitro detection and imaging techniques, for multiplexed biological detection, 77–81
individual cells biology, 202
infrared fluorophores, 209
infrared multiphoton dissociation (IRMPD), 275
in-source decay (ISD), 275. *See also* single-cell analyses
intact organisms, transdermal measurement in, 209
intensified CCD detector, 43, 44, 51, 54
intermediary metabolites, 198
intracellular analysis
  approaches involved in, 200–202, 204–208
  PEBBLE, 203
intracellular localization, prospect of, 198
IP$_3$ probes, 198
IRD 700 dye, 148, 154, 159
IRD 800 dye, 148, 157

K

kinase activity assay, 240
kinesin, 43
kiss and run phenomenon, 219, 228

L

lab on a chip device, 21
labeled antibodies, 71
labeling process, 206
lacI GFP fusion, 30
lacO sequence, 30, 31
β-lactam, 57
lactose operon repressor protein *(lacI)*, 30
Langmuir adsorption isotherm, 175, 176, 181, 182
laser manipulation techniques, 122
laser microprobe mass spectrometry (LMMS)
  application of, 276
laser photolysis, 130
laser radiation, 162
laser surgery, 122, 123
laser tweezers. *See* optical trapping technique
latex beads and dye molecules, 81
lifetime discrimination methods, 150–153
  combination of color-discrimination and time-resolved methods, 157–160
  four-lifetime/one-lane method, 153, 154
  two-lifetime/two-lane DNA sequencing, 154–157
lipid imaging, for very small cells, 281
liposome cell model
  advantages of, 225
  for exocytosis, 223
  fusion pore formation in, 223, 224
  studied data as compared to PC12 cells, 225
live cell, intracellular analysis, 198
  and detection of chromosomes, 29–31
LMMS analysis, 276
localized surface plasmon resonance (LSPR) spectra, of nanoparticles, 47
localized surface plasmon resonance spectra (LSPRS), of Ag nanoparticles, 55, 56
Lymnaea stagnalis, 277

M

M13 vectors, 143
macromolecular scaffolds, 91
magnetic nanoparticles, 80, 81, 83, 84
MALDI MS imaging (MSI), 271, 279
malignant cancers studies, by using single-cell SIMS, 281
Marcus electrontransfer (ET) theory, 236. *See also* ECL annihilation reactions
Mass spectrometry (MS), 269
  analyte identification with, 275, 276
  for single cell analysis, sample simplicity benefits for, 270
  fundamentals of, 270, 271
  initial analysis of, 276
  quantitative analysis with, 275
mass-to-charge ratio (m/z), 270
Mast cell vesicles, 217
Mast cells
  amperometric studies in, 220
    correlation of ca2+ levels and vesicle secretion, 221
  Ca^{2+} and ryanodine receptors role in, 220
  calcium imaging with amperometry for studying exocytosis, 222
  serotonin release
    impact of caffeine, 220
matrix assisted laser desorption ionization MS (MALDI MS), 270, 271
  time-of-flight mass spectrometry (MALDI-TOF), 121
maximum likelihood theorem, 13
11-mercaptoundecylamine (MUAM), 173
MegaBACE, 162
membrane permeability, in living organisms, 42

mercaptoacetic acid, 74
metal ion binding
  impact of mutagenegsis, 206
metal ion biosensors, 205
  adapted to use with fiber optics, 208
metal ions
  abundant in cells, 205
  in traces in living organisms, 195
  present as trace or ultra trace levels, 205
  quantitative imaging of, 195
  trafficking in cells, 195
metal nanoparticles, 72, 79
metal-chelate affinity chromatography (MCAC), 107
MexAB-OprM membrane pump, 48, 49, 61, 62
MexB protein, 48
Michaelis–Menten model, 229
micro-and nano-fluidic device, application of, 295
microvials, along with fast-scan cyclic voltammetry, 229
Mie theory, 55
mixed monolayer protected clusters (MMPCs), 92
  fabrication and properties, 92–96
  in modulating DNA activity, 97–102
  surface recognition of protein and peptides, 102–112
molecular beacon technology, 201
molecular bioanalysis
  and roles of nanoscience and nanotechnology, 296, 297
molecular imprint-imaging, 281
molecule fluorescent probes, 199
monolayer protected clusters (MPCs), 92
  fabrication and properties, 93–96
moving average method, 9
MS2 protein, 33
multi channel analyzer (MCA), 151
multicellular organisms
  and cellular structure, 197
multicellular organisms biology
  in vivo measurements of, 208, 209
multidrug resistance (MDR), 42
multiplexed optical encoding, 71
multisubstrate extrusion systems, 42
Munc18, role in exocytosis, 220
Murray place-exchange method, 93
myosin, 43

N

nalB-1 mutant, 49
nanocrystals (NCs), 257
nanoparticle probes, for biological detection
  delivering and targeting, 74–76
  in cancer research, 71
  in vitro detection and imaging, 77–81
  intracellular and *in vivo* applications, 81–84
  probe preparation, 72–74

nanoparticle technologies, 297
nanos mRNA, 36
nanosurgery, 121, 123
  cause of cell death in, 125
  cell viability after, 124, 125
naphthalocyanine (NPcs) family of compounds, 147
Nd:YAG laser, 123, 124, 148
N-dodecyl-PEI2 conjugates, 99, 100
near-IR fluorescent dyes, 145–148
near-IR tricarbocyanines, 154
near-IR-Br dye, 157
nerve growth factor (NGF), 279
NG108 stem cells, 124, 126
N-hydroxysuccinimide ester of 9-fluorenylmethoxycarbonyl (Fmoc-NHS), 173
N-hydroxysulfosuccinimide (NHS), 172
nitrilotriacetic acid (NTA), 107
nitrilotriacetic acid (NTA)-modified magnetic nanoparticles, 111
noble metal nanoparticles, 47
nonexcitable cells
  effect of Ca^{2+} on, 221
non-photobleaching noble metal nanoparticles, 296
N-succinimidyl 3-(2-pyridyldithio)-propionamido (SPDP), 172
nucleotide bases, 142

O

octylamine-modified low-molecular-weight polyacrylic acid, 72
oligo(ethylene glycol) (OEG) ligands, 104
oligonucleotide primers, 154
OprM protein, 48
optical dark-field microscopy, 49
optical mapping, 14–17
optical methods
  and other approaches for quantitative imaging of metal ions, 195, 196
  for trafficking of metal ions, 195
optical trapping technique, 122
Orconectes limosus, 278
oxygen tension
  inside cells, 199

P

pancreatic β-cell
  differentiation of fast and slow exocytosis in, 222
  to study exocytosis with amperometric detection in, 222
PAS-complex marker, 45
patch amperometry, 226
  added advantage in, 227
  electrical configuration of, 227
  initial use of, 228

patch clamp technique, 201
  for recording capacitance of cell during exocytosis, 226
PDMS microchannels, 174
PEBBLE. *See* probes encapsulated by biologically localized embedding
PEG derivatized phospholipids, 72
PEGylated ABC triblock copolymer, 82
peptide analysis
  identification of post-translational modifications (PTMs), 277
phallotoxins, for F-actin, 199
pharmacophores, 199
phenyl ether, 72
pheochromocytoma (PC)12 cells
  dopamine levels increase
    impact on vesicles size, 219
  used for amperometric studies, 219
    change in volume of vesicles, 219
phospholipid environment, 41
phospholipids-PEG block copolymer coatings, 82
phosphor plates, 163
photoablation, 122, 123
photobleaching
  loss of fluorescence emission, 202
photobleaching propensity
  and uses of fluorophores, 201
photobleaching technique, 7
photomultiplier tube (PMTs), 144, 240
phthalocyanines (Pc) family of compounds, 147
picoliter microvials, 230
platinum acetylacetonate Pt(acac)$_2$, 72
*Pme*I digest of a mixture of λ phage, 14–16
2-kDa polyethylenimine (PEI2), 98
60-kDa polyethylenimine (PEI), 98
poly(dimethylsiloxane) (PDMS) channels, 127, 134
polyanhydrides, 72
polydimethylsiloxane (PDMS) microfluidic networks, 173
polymerase chain reaction (PCR), 79
  product detection, 186
polymer-coated nanoparticle, 74
post-source decay (PSD), 275. *See also* single-cell analyses
post-translational modifications (PTMs), 277
principal component analysis (PCA)
  utilized in differentiation of gram positive and negative bacteria, 281
probability theory, 6
probe targets, 197
probe volumes, 11–13
probes encapsulated by biologically localized embedding
  advantages of, 204
  and introductory routes inside cell, 204
  different ways to make that, 203
  fluorescent indicator into and advantages of, 202

  for intracellular measurements, 202
  issues involved in, 205
  PEBBLE-based biosensors, 203
prokaryotes, 42, 49
prostate-specific antigen (PSA), 79, 246
protease activity assay, 240
proteins and polysaccharides
  localization and identification by, 198
PSA-free marker, 45
PSA-free monoclonal antibody (MAB)
  impact on ECL emission intensity, 249
Pseudomonas aeruginosa, 42, 47–49, 56, 57, 61

Q

quantum dots (QDs), 46, 47, 71, 74, 81–83, 100, 111, 257
  NIR, 82, 83
  semiconductor type, 72

R

R6G fluorescence dye molecule, 56
R6G fluorophor, 43, 45
R6G molecules, 79
Raman scattering, 145
random-walk theory, 45
rapid lifetime determination method (RLD), 152
rapid prototyping techniques, 135
Rayleigh scattering, 58, 145
redox probes, 199
Rhodamine 6G, 202
rhodamine-based dyes, 144
ribosomal peptidyl transferase, 57
RNA polymerase II catalytic subunits, 30
RNA–DNA heteroduplexes, 177, 178, 181, 183
RNA–protein complexes (mRNPs), 33
RNase H hydrolysis, of RNA microarrays, 183
RNase H surface enzyme kinetics, 177–182
Ru(bpy)$_3^{2+}$/TPrA system
  study mechanism of, 255, 256, 257

S

S. aureus, 14
Saccharomyces cerevisiae, 281
Sandwich immunoassay, 239
Sanger chain termination reactions, 153
Sanger sequencing reactions, 144
secondary ion mass spectrometry (SIMS), 270, 273
  cholesterol ions and phospholipid ions detection by, 283
  for mapping amino acids, 281
  SIMS imaging, 273
  single-cell research, 279
self-assembled alkanethiol monolayers (SAMs), 172, 238
sensor, properties of, 205
signal transduction, 41
signaling cascades, 200

signal-to-noise ratio, 10, 137
silver (Ag) nanoparticles, 54
   identification of, 60
   preparation, 57
   Rayleigh scattering of, 58
silver enhancement technique, 80
single molecule detection (SMD), and its applications, 295, 296
single nanoparticle optics, 297
single nerve cells
   electrochemistry inside and outside of, 215
   methods involved in, 276
single-cell analyses, 270
   analysis of, 297, 298
   by MALDI MS, 271
   challenges for ESI MS, 274
   and neural communication, 215
   MS combined with other detection techniques, 274–276
   study of interactions of, 295, 296
   tracking the dynamics of, 296
single-cell analysis via mass spectrometry, 270
single-cell discoveries
   made using electrochemistry, 231
single-cell MALDI MS
   Aplysia as neurobiological model, 277
   Periplaneta americana as experimental model, 278
single-cell model systems
   analyzed by using electrochemical methods, 232
single-cell molecular analysis via SIMS, 285
single-cell MS, 270
   advances in sample preparation for, 286
   and alternative sampling approaches, 283, 284
   application of, 286
   ionization methods in, 272
single-cell nanosurgery, 121, 123, 133
single-cell scale, chemical imaging on, 286
single-cell SIMS analysis, benefits of, 273
single-cell systems, 269
single-molecule and ensemble measurements
   application of DNA fingerprinting, 13–17
   detectors, 22
   for identification of species, 21
   model system of
      data analysis, 4–11
      equilibrium and kinetics, 3, 4
      generation of synthetic data, 4
      probe volumes, 11–13
   in medical diagnostics, 20
   for sample preparation and detection, 21, 22
   statistical chemistry, 19, 20
   virtual sorting, 17–19
single-molecule detection (SMD), 41
   detection configurations, 43–45
   detection probes, 45–47
   probing efflux pump machinery of single living bacterial cells, 48–54
   requirements, 42
   sizing the membrane transport of single living cells in real time, 54–63
single-molecule experiments, 1–3
single-molecule fluorescence detection technique, 201
single-photon avalanche diode (SPAD), 145, 152, 157
site-directed mutagenesis, 206
SNARE, 220
sol–gel formulations, 203. *See also* PEBBLE
sol–gel PEBBLES, 203
solid-phase extraction (SPE), 283
solution-phase ECL immunoassay, 238, 249
   for detection of molecular forms of serum PSA, 249
   for detection of tumor markers, 246, 247, 248, 249, 250
spectrally resolved fluorescence lifetime imaging microscopy (SFLIM), 43, 45
SPION-Tat particles, 84
statistical chemistry, 19
streptavidin, 112
subcellular compartments centers, 119, 120
   analysis of contents, 136, 137
   extraction of, 122–125
   nanoscale manipulations of, 125–136
   research past/present, 120, 121
   technology platform for profiling of, 121
sulfosuccinimidyl 4-(*N*-maleimidomethyl) cyclohexane-1-carboxylate (SSMCC), 172
superparamagnetic iron oxide nanoparticles (SPION), 83
surface plasmon fluorescence spectroscopy (SPFS), 181
surface plasmon resonance imaging (SPRI) methodology
   and direct detection of DNA hybridization adsorption, 175–177
   for RNase H activity, 177–182
   overview, 170–175
   ultrasensitive DNA detection with, 182–187
surface-enhanced Raman scattering (SERS), 79
surface-phase ECL assay, 238
   used to detect activity of enzyme, 238
sustained off-resonance irradiation (SORI), 275

T

T4 GT7 genomes, 15, 16
T-branch channels, 127–129
tetracycline responsive element (TRE), 31
tetramethylrhodamine, 144
Texas Red, 144
TGF-β receptors, 84

thin-layer total internal reflection fluorescence microscopy (TL-TIR-FM), 43, 44, 49, 51
thiol-modified biomolecules, 172
thiols, 93
 on the self-assembled monolayer (SAM) surface, 94
time-correlated single-photon counting (TCSPC) technique, 151
time-course fluorescence spectroscopy, 48, 49
time-dependent fluorescence intensity fluctuations, 2
time-resolved fluorescence scanning detectors, 162, 163
time-to-amplitude converter (TAC), 151
total fluorescent intensity (TFI), 32
total internal reflection (TIR), 12
trifluoroacetic acid (TFA), 273
trimethylammonium-functionalized surfactants, 103
tri-n-octylphosphine oxide (TOPO), 72
Tsien group, 200
 redox probe of, 199
TSPY gene, 186, 187
tumor markers
 analysis using solution-phase ECL assay, 246
tunable narrow emission spectrum, 46
two-lifetime/two lane DNA sequencing, 154–157

U

U1A protein, 34, 35

V

vesicles volume change
 with increase and decrease in dopamine, 219
virtual sorting, 17–19

W

Watson–Crick hydrogen bonding, 97
WellRED dye set, 148
wireless ECL detector, 241
 schematics of, 243

X

Xenopus embryos, 82
x-ray crystallography, 42, 54

Y

YOYO-1, 14

CHEMICAL ANALYSIS

A SERIES OF MONOGRAPHS ON ANALYTICAL CHEMISTRY
AND ITS APPLICATIONS

Series Editor
J. D. WINEFORDNER

Vol. 1 **The Analytical Chemistry of Industrial Poisons, Hazards, and Solvents.** *Second Edition.* By the late Morris B. Jacobs
Vol. 2 **Chromatographic Adsorption Analysis.** By Harold H. Strain (*out of print*)
Vol. 3 **Photometric Determination of Traces of Metals.** *Fourth Edition*
Part I: General Aspects. By E. B. Sandell and Hiroshi Onishi
Part IIA: Individual Metals, Aluminum to Lithium. By Hiroshi Onishi
Part IIB: Individual Metals, Magnesium to Zirconium. By Hiroshi Onishi
Vol. 4 **Organic Reagents Used in Gravimetric and Volumetric Analysis.** By John F. Flagg (*out of print*)
Vol. 5 **Aquametry: A Treatise on Methods for the Determination of Water.** *Second Edition* (*in three parts*). By John Mitchell, Jr. and Donald Milton Smith
Vol. 6 **Analysis of Insecticides and Acaricides.** By Francis A. Gunther and Roger C. Blinn (*out of print*)
Vol. 7 **Chemical Analysis of Industrial Solvents.** By the late Morris B. Jacobs and Leopold Schetlan
Vol. 8 **Colorimetric Determination of Nonmetals.** *Second Edition.* Edited by the late David F. Boltz and James A. Howell
Vol. 9 **Analytical Chemistry of Titanium Metals and Compounds.** By Maurice Codell
Vol. 10 **The Chemical Analysis of Air Pollutants.** By the late Morris B. Jacobs
Vol. 11 **X-Ray Spectrochemical Analysis.** *Second Edition.* By L. S. Birks
Vol. 12 **Systematic Analysis of Surface-Active Agents.** *Second Edition.* By Milton J. Rosen and Henry A. Goldsmith
Vol. 13 **Alternating Current Polarography and Tensammetry.** By B. Breyer and H.H.Bauer
Vol. 14 **Flame Photometry.** By R. Herrmann and J. Alkemade
Vol. 15 **The Titration of Organic Compounds** (*in two parts*). By M. R. F. Ashworth
Vol. 16 **Complexation in Analytical Chemistry: A Guide for the Critical Selection of Analytical Methods Based on Complexation Reactions.** By the late Anders Ringbom
Vol. 17 **Electron Probe Microanalysis.** *Second Edition.* By L. S. Birks
Vol. 18 **Organic Complexing Reagents: Structure, Behavior, and Application to Inorganic Analysis.** By D. D. Perrin
Vol. 19 **Thermal Analysis.** *Third Edition.* By Wesley Wm.Wendlandt
Vol. 20 **Amperometric Titrations.** By John T. Stock
Vol. 21 **Reflctance Spectroscopy.** By Wesley Wm.Wendlandt and Harry G. Hecht
Vol. 22 **The Analytical Toxicology of Industrial Inorganic Poisons.** By the late Morris B. Jacobs
Vol. 23 **The Formation and Properties of Precipitates.** By Alan G.Walton
Vol. 24 **Kinetics in Analytical Chemistry.** By Harry B. Mark, Jr. and Garry A. Rechnitz
Vol. 25 **Atomic Absorption Spectroscopy.** *Second Edition.* By Morris Slavin
Vol. 26 **Characterization of Organometallic Compounds** (*in two parts*). Edited by Minoru Tsutsui
Vol. 27 **Rock and Mineral Analysis.** *Second Edition.* By Wesley M. Johnson and John A. Maxwell
Vol. 28 **The Analytical Chemistry of Nitrogen and Its Compounds** (*in two parts*). Edited by C. A. Streuli and Philip R.Averell
Vol. 29 **The Analytical Chemistry of Sulfur and Its Compounds** (*in three parts*). By J. H. Karchmer
Vol. 30 **Ultramicro Elemental Analysis.** By Güther Toölg
Vol. 31 **Photometric Organic Analysis** (*in two parts*). By Eugene Sawicki
Vol. 32 **Determination of Organic Compounds: Methods and Procedures.** By Frederick T. Weiss
Vol. 33 **Masking and Demasking of Chemical Reactions.** By D. D. Perrin

Vol. 34. **Neutron Activation Analysis**. By D. De Soete, R. Gijbels, and J. Hoste
Vol. 35 **Laser Raman Spectroscopy**. By Marvin C. Tobin
Vol. 36 **Emission Spectrochemical Analysis**. By Morris Slavin
Vol. 37 **Analytical Chemistry of Phosphorus Compounds**. Edited by M. Halmann
Vol. 38 **Luminescence Spectrometry in Analytical Chemistry**. By J. D.Winefordner, S. G. Schulman, and T. C. O'Haver
Vol. 39. **Activation Analysis with Neutron Generators**. By Sam S. Nargolwalla and Edwin P. Przybylowicz
Vol. 40 **Determination of Gaseous Elements in Metals**. Edited by Lynn L. Lewis, Laben M. Melnick, and Ben D. Holt
Vol. 41 **Analysis of Silicones**. Edited by A. Lee Smith
Vol. 42 **Foundations of Ultracentrifugal Analysis**. By H. Fujita
Vol. 43 **Chemical Infrared Fourier Transform Spectroscopy**. By Peter R. Griffiths
Vol. 44 **Microscale Manipulations in Chemistry**. By T. S. Ma and V. Horak
Vol. 45 **Thermometric Titrations**. By J. Barthel
Vol. 46 **Trace Analysis: Spectroscopic Methods for Elements**. Edited by J. D.Winefordner
Vol. 47 **Contamination Control in Trace Element Analysis**. By Morris Zief and James W. Mitchell
Vol. 48 **Analytical Applications of NMR**. By D. E. Leyden and R. H. Cox
Vol. 49 **Measurement of Dissolved Oxygen**. By Michael L. Hitchman
Vol. 50 **Analytical Laser Spectroscopy**. Edited by Nicolo Omenetto
Vol. 51 **Trace Element Analysis of Geological Materials**. By Roger D. Reeves and Robert R. Brooks
Vol. 52 **Chemical Analysis by Microwave Rotational Spectroscopy**. By Ravi Varma and Lawrence W. Hrubesh
Vol. 53 **Information Theory as Applied to Chemical Analysis**. By Karl Eckschlager and Vladimir Stepanek
Vol. 54 **Applied Infrared Spectroscopy: Fundamentals, Techniques, and Analytical Problemsolving**. By A. Lee Smith
Vol. 55 **Archaeological Chemistry**. By Zvi Goffer
Vol. 56 **Immobilized Enzymes in Analytical and Clinical Chemistry**. By P. W. Carr and L. D. Bowers
Vol. 57 **Photoacoustics and Photoacoustic Spectroscopy**. By Allan Rosencwaig
Vol. 58 **Analysis of Pesticide Residues**. Edited by H. Anson Moye
Vol. 59 **Affity Chromatography**. By William H. Scouten
Vol. 60 **Quality Control in Analytical Chemistry**. *Second Edition*. By G. Kateman and L. Buydens
Vol. 61 **Direct Characterization of Fineparticles**. By Brian H. Kaye
Vol. 62 **Flow Injection Analysis**. By J. Ruzicka and E. H. Hansen
Vol. 63 **Applied Electron Spectroscopy for Chemical Analysis**. Edited by Hassan Windawi and Floyd Ho
Vol. 64 **Analytical Aspects of Environmental Chemistry**. Edited by David F. S. Natusch and Philip K. Hopke
Vol. 65 **The Interpretation of Analytical Chemical Data by the Use of Cluster Analysis**. By D. Luc Massart and Leonard Kaufman
Vol. 66 **Solid Phase Biochemistry: Analytical and Synthetic Aspects**. Edited by William H. Scouten
Vol. 67 **An Introduction to Photoelectron Spectroscopy**. By Pradip K. Ghosh
Vol. 68 **Room Temperature Phosphorimetry for Chemical Analysis**. By Tuan Vo-Dinh
Vol. 69 **Potentiometry and Potentiometric Titrations**. By E. P. Serjeant
Vol. 70 **Design and Application of Process Analyzer Systems**. By Paul E. Mix
Vol. 71 **Analysis of Organic and Biological Surfaces**. Edited by Patrick Echlin
Vol. 72 **Small Bore Liquid Chromatography Columns: Their Properties and Uses**. Edited by Raymond P.W. Scott
Vol. 73 **Modern Methods of Particle Size Analysis**. Edited by Howard G. Barth
Vol. 74 **Auger Electron Spectroscopy**. By Michael Thompson, M. D. Baker, Alec Christie, and J. F. Tyson
Vol. 75 **Spot Test Analysis: Clinical, Environmental, Forensic and Geochemical Applications**. By Ervin Jungreis
Vol. 76 **Receptor Modeling in Environmental Chemistry**. By Philip K. Hopke

Vol. 77	**Molecular Luminescence Spectroscopy: Methods and Applications** (*in three parts*). Edited by Stephen G. Schulman
Vol. 78	**Inorganic Chromatographic Analysis**. Edited by John C. MacDonald
Vol. 79	**Analytical Solution Calorimetry**. Edited by J. K. Grime
Vol. 80	**Selected Methods of Trace Metal Analysis: Biological and Environmental Samples**. By Jon C.VanLoon
Vol. 81	**The Analysis of Extraterrestrial Materials**. By Isidore Adler
Vol. 82	**Chemometrics**. By Muhammad A. Sharaf, Deborah L. Illman, and Bruce R. Kowalski
Vol. 83	**Fourier Transform Infrared Spectrometry**. By Peter R. Griffiths and James A. de Haseth
Vol. 84	**Trace Analysis: Spectroscopic Methods for Molecules**. Edited by Gary Christian and James B. Callis
Vol. 85	**Ultratrace Analysis of Pharmaceuticals and Other Compounds of Interest**. Edited by S. Ahuja
Vol. 86	**Secondary Ion Mass Spectrometry: Basic Concepts, Instrumental Aspects, Applications and Trends**. By A. Benninghoven, F. G. Rüenauer, and H.W.Werner
Vol. 87	**Analytical Applications of Lasers**. Edited by Edward H. Piepmeier
Vol. 88	**Applied Geochemical Analysis**. By C. O. Ingamells and F. F. Pitard
Vol. 89	**Detectors for Liquid Chromatography**. Edited by Edward S.Yeung
Vol. 90	**Inductively Coupled Plasma Emission Spectroscopy: Part 1: Methodology, Instrumentation, and Performance; Part II: Applications and Fundamentals**. Edited by J. M. Boumans
Vol. 91	**Applications of New Mass Spectrometry Techniques in Pesticide Chemistry**. Edited by Joseph Rosen
Vol. 92	**X-Ray Absorption: Principles,Applications,Techniques of EXAFS, SEXAFS, and XANES**. Edited by D. C. Konnigsberger
Vol. 93	**Quantitative Structure-Chromatographic Retention Relationships**. By Roman Kaliszan
Vol. 94	**Laser Remote Chemical Analysis**. Edited by Raymond M. Measures
Vol. 95	**Inorganic Mass Spectrometry**. Edited by F.Adams,R.Gijbels, and R.Van Grieken
Vol. 96	**Kinetic Aspects of Analytical Chemistry**. By Horacio A. Mottola
Vol. 97	**Two-Dimensional NMR Spectroscopy**. By Jan Schraml and Jon M. Bellama
Vol. 98	**High Performance Liquid Chromatography**. Edited by Phyllis R. Brown and Richard A. Hartwick
Vol. 99	**X-Ray Fluorescence Spectrometry**. By Ron Jenkins
Vol. 100	**Analytical Aspects of Drug Testing**. Edited by Dale G. Deustch
Vol. 101	**Chemical Analysis of Polycyclic Aromatic Compounds**. Edited by Tuan Vo-Dinh
Vol. 102	**Quadrupole Storage Mass Spectrometry**. By Raymond E. March and Richard J. Hughes (*out of print: see Vol. 165*)
Vol. 103	**Determination of Molecular Weight**. Edited by Anthony R. Cooper
Vol. 104	**Selectivity and Detectability Optimization in HPLC**. By Satinder Ahuja
Vol. 105	**Laser Microanalysis**. By Lieselotte Moenke-Blankenburg
Vol. 106	**Clinical Chemistry**. Edited by E. Howard Taylor
Vol. 107	**Multielement Detection Systems for Spectrochemical Analysis**. By Kenneth W. Busch and Marianna A. Busch
Vol. 108	**Planar Chromatography in the Life Sciences**. Edited by Joseph C. Touchstone
Vol. 109	**Fluorometric Analysis in Biomedical Chemistry: Trends and Techniques Including HPLC Applications**. By Norio Ichinose, George Schwedt, Frank Michael Schnepel, and Kyoko Adochi
Vol. 110	**An Introduction to Laboratory Automation**. By Victor Cerdá and Guillermo Ramis
Vol. 111	**Gas Chromatography: Biochemical, Biomedical, and Clinical Applications**. Edited by Ray E. Clement
Vol. 112	**The Analytical Chemistry of Silicones**. Edited by A. Lee Smith
Vol. 113	**Modern Methods of Polymer Characterization**. Edited by Howard G. Barth and Jimmy W. Mays
Vol. 114	**Analytical Raman Spectroscopy**. Edited by Jeanette Graselli and Bernard J. Bulkin
Vol. 115	**Trace and Ultratrace Analysis by HPLC**. By Satinder Ahuja
Vol. 116	**Radiochemistry and Nuclear Methods of Analysis**. By William D. Ehmann and Diane E.Vance

Vol. 117 **Applications of Fluorescence in Immunoassays**. By Ilkka Hemmila
Vol. 118 **Principles and Practice of Spectroscopic Calibration**. By Howard Mark
Vol. 119 **Activation Spectrometry in Chemical Analysis**. By S. J. Parry
Vol. 120 **Remote Sensing by Fourier Transform Spectrometry**. By Reinhard Beer
Vol. 121 **Detectors for Capillary Chromatography**. Edited by Herbert H. Hill and Dennis McMinn
Vol. 122 **Photochemical Vapor Deposition**. By J. G. Eden
Vol. 123 **Statistical Methods in Analytical Chemistry**. By Peter C. Meier and Richard Züd
Vol. 124 **Laser Ionization Mass Analysis**. Edited by Akos Vertes, Renaat Gijbels, and Fred Adams
Vol. 125 **Physics and Chemistry of Solid State Sensor Devices**. By Andreas Mandelis and Constantinos Christofides
Vol. 126 **Electroanalytical Stripping Methods**. By Khjena Z. Brainina and E. Neyman
Vol. 127 **Air Monitoring by Spectroscopic Techniques**. Edited by Markus W. Sigrist
Vol. 128 **Information Theory in Analytical Chemistry**. By Karel Eckschlager and Klaus Danzer
Vol. 129 **Flame Chemiluminescence Analysis by Molecular Emission Cavity Detection**. Edited by David Stiles, Anthony Calokerinos, and Alan Townshend
Vol. 130 **Hydride Generation Atomic Absorption Spectrometry**. Edited by Jiri Dedina and Dimiter L. Tsalev
Vol. 131 **Selective Detectors: Environmental, Industrial, and Biomedical Applications**. Edited by Robert E. Sievers
Vol. 132 **High-Speed Countercurrent Chromatography**. Edited by Yoichiro Ito and Walter D. Conway
Vol. 133 **Particle-Induced X-Ray Emission Spectrometry**. By Sven A. E. Johansson, John L. Campbell, and Klas G. Malmqvist
Vol. 134 **Photothermal Spectroscopy Methods for Chemical Analysis**. By Stephen E. Bialkowski
Vol. 135 **Element Speciation in Bioinorganic Chemistry**. Edited by Sergio Caroli
Vol. 136 **Laser-Enhanced Ionization Spectrometry**. Edited by John C. Travis and Gregory C. Turk
Vol. 137 **Fluorescence Imaging Spectroscopy and Microscopy**. Edited by Xue Feng Wang and Brian Herman
Vol. 138 **Introduction to X-Ray Powder Diffractometry**. By Ron Jenkins and Robert L. Snyder
Vol. 139 **Modern Techniques in Electroanalysis**. Edited by Petr Vanýek
Vol. 140 **Total-Reflction X-Ray Fluorescence Analysis**. By Reinhold Klockenkamper
Vol. 141 **Spot Test Analysis: Clinical, Environmental, Forensic, and Geochemical Applications**. *Second Edition*. By Ervin Jungreis
Vol. 142 **The Impact of Stereochemistry on Drug Development and Use**. Edited by Hassan Y. Aboul-Enein and Irving W.Wainer
Vol. 143 **Macrocyclic Compounds in Analytical Chemistry**. Edited by Yury A. Zolotov
Vol. 144 **Surface-Launched Acoustic Wave Sensors: Chemical Sensing and Thin-Film Characterization**. By Michael Thompson and David Stone
Vol. 145 **Modern Isotope Ratio Mass Spectrometry**. Edited by T. J. Platzner
Vol. 146 **High Performance Capillary Electrophoresis: Theory, Techniques, and Applications**. Edited by Morteza G. Khaledi
Vol. 147 **Solid Phase Extraction: Principles and Practice**. By E. M. Thurman
Vol. 148 **Commercial Biosensors: Applications to Clinical, Bioprocess and Environmental Samples**. Edited by Graham Ramsay
Vol. 149 **A Practical Guide to Graphite Furnace Atomic Absorption Spectrometry**. By David J. Butcher and Joseph Sneddon
Vol. 150 **Principles of Chemical and Biological Sensors**. Edited by Dermot Diamond
Vol. 151 **Pesticide Residue in Foods: Methods, Technologies, and Regulations**. By W. George Fong, H. Anson Moye, James N. Seiber, and John P. Toth
Vol. 152 **X-Ray Fluorescence Spectrometry**. *Second Edition*. By Ron Jenkins
Vol. 153 **Statistical Methods in Analytical Chemistry**. *Second Edition*. By Peter C. Meier and Richard E. Züd
Vol. 154 **Modern Analytical Methodologies in Fat- and Water-Soluble Vitamins**. Edited by Won O. Song, Gary R. Beecher, and Ronald R. Eitenmiller
Vol. 155 **Modern Analytical Methods in Art and Archaeology**. Edited by Enrico Ciliberto and Guiseppe Spoto

Vol. 156 **Shpol'skii Spectroscopy and Other Site Selection Methods: Applications in Environmental Analysis, Bioanalytical Chemistry and Chemical Physics.** Edited by C. Gooijer, F. Ariese and J.W. Hofstraat

Vol. 157 **Raman Spectroscopy for Chemical Analysis.** By Richard L. McCreery

Vol. 158 **Large (C> = 24) Polycyclic Aromatic Hydrocarbons: Chemistry and Analysis.** By John C. Fetzer

Vol. 159 **Handbook of Petroleum Analysis.** By James G. Speight

Vol. 160 **Handbook of Petroleum Product Analysis.** By James G. Speight

Vol. 161 **Photoacoustic Infrared Spectroscopy.** By Kirk H. Michaelian

Vol. 162 **Sample Preparation Techniques in Analytical Chemistry.** Edited by Somenath Mitra

Vol. 163 **Analysis and Purification Methods in Combination Chemistry.** Edited by Bing Yan

Vol. 164 **Chemometrics: From Basics to Wavelet Transform.** By Foo-tim Chau, Yi-Zeng Liang, Junbin Gao, and Xue-guang Shao

Vol. 165 **Quadrupole Ion Trap Mass Spectrometry.** *Second Edition.* By Raymond E. March and John F. J. Todd

Vol. 166 **Handbook of Coal Analysis.** By James G. Speight

Vol. 167 **Introduction to Soil Chemistry: Analysis and Instrumentation.** By Alfred R. Conklin, Jr.

Vol. 168 **Environmental Analysis and Technology for the Refining Industry.** By James G. Speight

Vol. 169 **Identification of Microorganisms by Mass Spectrometry.** Edited by Charles L. Wilkins and Jackson O. Lay, Jr.

Vol. 170 **Archaeological Chemistry.** *Second Edition.* By Zvi Goffer

Vol. 171 **Fourier Transform Infrared Spectrometry.** *Second Edition.* By Peter R. Griffiths and James A. de Haseth

Vol. 172 **New Frontiers in Ultrasensitive Bioanalysis: Advanced Analytical Chemistry Applications in Nanobiotechnology, Single Molecule Detection, and Single Cell Analysis.** Edited by Xiao-Hong Nancy Xu